The Regulation of Membrane Lipid Metabolism

2nd Edition

Guy A. Thompson, Jr.

CRC Press
Boca Raton Ann Arbor London Tokyo

Library of Congress Cataloging-in-Publication Data

Thompson, Guy A., 1931–
The regulation of membrane lipid metabolism/author, Guy A. Thompson, Jr. — 2nd ed.
p. cm.
Includes bibilographical references and index.
ISBN 0-8493-4561-8
1. Membrane lipids—Metabolism—Regulation. I. Title.
QP752.M45T46 1992
574.87′5—dc20
91-46562
CIP

Direct all inquiries to CRC Press, Inc., 2000 Corporate Blvd., N. W., Boca Raton, Florida, 33431.

International Standard Book Number 0-8493-4561-8

Library of Congress Card Number 91-46562
Printed in the United States 1 2 3 4 5 6 7 8 9 0
Printed on acid-free paper

PREFACE

Not so many years ago there appeared to be relatively few major challenges remaining in the scientific discipline of membrane lipid metabolism. Only the seasoned observer could then recognize what we all see clearly today — that lipid metabolism, through its many effects upon the physical properties of membranes, exerts a crucial, if indirect, regulatory role in numerous important cellular activities.

About ten years ago, as the concept of "membrane fluidity" rose to its current well-deserved popularity, the study of lipid metabolism assumed a new significance, for we could at least hope to assess, via physical chemical measurements, the consequences of even small changes in lipid composition. It became evident that each lipid type makes a unique contribution to the membrane's overall effectiveness, and minor alterations in lipid content may well have physiological repercussions more subtle than can be appreciated even with our present day knowledge.

More recently the focus of many laboratories has shifted from effects caused by large assemblages of lipids to more direct action by individual lipid types. The activity of inositol lipids and their hydrolysis products in transmembrane signaling is a much-studied case in point, and the previously unrecognized ability of certain lipids to promote the intracellular translocation of proteins is another phenomenon of current interest.

An ever increasing information flow regarding the metabolism of diverse types of membrane lipids renders it more and more difficult to keep track of the progress being made. This book represents an attempt to summarize the current status of the field with special emphasis on regulatory interactions. It is designed to bridge the gap between general textbook and comprehensive review article. I write primarily for those students who, having survived a basic course in biochemistry, now wish to apply their hard-earned knowledge to problems involving membranes. Hopefully, some of the growing number of more senior investigators who find themselves drawn almost unwillingly into the study of lipids from a peripheral field will also benefit from this condensed presentation. Last and not least, I confidently intend this volume to serve me personally as a readily available pocket guide to all the facts I should remember from day to day but can't.

My goal being thus limited to assessing the current status of the field, I have chosen to quote references that are illustrative but not necessarily comprehensive. I have not uniformly chronicled the worthy efforts of pioneers in lipid research, nor have I included more than an occasional sortie into comparative biochemistry. To do so would increase the length (and the price) of this work well beyond tolerable limits. Instead, I have endeavored to reference, at the beginning of each section, one or more in depth reviews that should compensate for my conscious omissions.

The general organization of the book is designed to introduce, for the novice, a few basic details of the physical, chemical, and physiological breakthroughs that provide the lipid biochemist with this exciting new challenge. I have tried to present, in some logical order, those reactions whereby the biosynthesis and catabolism of the major classes of membrane structural lipids can be regulated. (Regrettably, but perhaps not surprisingly, firm evidence for control mechanisms in several cases is rather sparse as yet.) The two following chapters provide clear-cut, albeit still incompletely understood, examples of how regulation of lipid metabolism does participate in maintaining the well-being of organisms under various types of stress.

The present volume is an extensively updated second edition of the 1980 monograph by the same name. Some parts of the book, such as Chapter 2 on fatty acid synthesis, have undergone only slight revision, incorporating recent advances in hormonal and other types of regulation. On the other extreme, Chapter 8 on protein-bound lipids is entirely new since that research area was essentially nonexistent in 1980.

In this effort as in my first, questions predominate over answers in the final product. The difference is that questions being asked today concern not whether lipids have a discrete function but, rather, how many functions do they have.

As I sat day after day in the University library, pondering the interpretations of one or another finding, my eyes seemed drawn increasingly to one of the many literary quotations inscribed on the ornate ceiling beams. From Lewis Carroll's *Alice in Wonderland,* it reads: " 'If there's no meaning in it,' said the King, 'that saves a world of trouble, you know, as we needn't try to find any.' " This thought, I admit, did seem more and more attractive to me near the end, and probably was responsible for the brevity of certain sections of the book. But I do wish to emphasize, before you, too, entertain creeping doubts, that in general I find the King's admonition quite inappropriate to this topic. There is without question important meaning behind these sometimes inscrutable experimental findings. Now that I have survived my exercise in writing with this belief still intact, I look forward, as I hope you do, to learning more about what that meaning is.

GAT

THE AUTHOR

Guy A. Thompson, Jr., Ph.D. is a member of the Department of Botany at the University of Texas at Austin. He received his B.S. degree in chemistry from Mississippi State University in 1953 and his Ph.D. degree in biochemistry from the California Institute of Technology in 1959. After receiving post-doctoral training at the University of Manchester, England, and the University of Washington, he joined the faculty at the Department of Biochemistry, University of Washington. In 1967, he became affiliated with the University of Texas, where he holds the rank of Professor.

Dr. Thompson is a member of the American Society of Biological Chemists, the American Chemical Society, and other professional societies fostering biological research. He has published extensively in the field of lipid and membrane biochemistry.

ACKNOWLEDGMENTS

I should like to thank Eileen Thompson for converting my handwritten draft into camera-ready form.

DEDICATION

To Eileen, Sally, Jill, and Jeremy

TABLE OF CONTENTS

Chapter 1

A RATIONALE GOVERNING THE REGULATION OF LIPID METABOLISM

I. INTRODUCTION

My aim in writing this introductory chapter is to reinforce and hopefully even enhance your interest in the regulation of membrane lipid metabolism. While you aren't by any means the first to experience curiosity about this subject, you are likely to be in the first generation of observers who will truly understand why membrane lipid metabolism must be regulated. Without that understanding, there has until now obviously been no rational way to approach the problem.

Thus we have, in my opinion, at last crossed the threshold into an era of knowing what to look for. There is suddenly a new appreciation of how such a seemingly random arrangement of countless lipid species can bring order to a membrane's activities simply by controlling its physical state. As we shall see, "simply" may really not be such a good adverb to use in this context, considering the previously unimagined complexities now being discovered. But the basic concept is a simple and satisfying one that permits the formulation of many testable hypotheses. Therefore, at this auspicious moment it seems worth-while to bring existing evidence into sharp focus and take stock of how well equipped we are to define the role of lipids in regulating membrane activities.

II. TISSUE-SPECIFIC DISTRIBUTION OF LIPIDS

Let me begin with an incontrovertible statement of fact. Membrane lipid composition is, beyond all doubt, regulated. This long-standing realization has in itself been impetus enough for continued investigation through the years, but a full appreciation of just how sensitively organisms do control their lipid composition came only with the advent of modern analytical methods. It had been known for years that each animal and plant species possesses specific lipids in characteristic proportions. Then, in the 1960s, improving methodology permitted sufficiently quantitative analyses to establish the concept of a tissue specificity for lipids, as opposed to the once more plausible belief in a species specificity.[1] Quite adequate data now exist,[1,2] to confirm that tissues having identical functions also have remarkably similar lipid compositions (Figure 1A). And conversely, tissues within the same organism that have different functions are distinctly different in their lipid contents (Figure 1B).

The logical next step was the demonstration that even within a particular cell, where individual metabolic functions are distributed among the various organelles, lipid compositions differ from one functional membrane type to another. These differences are also well catalogued now (Figure 2).

As indicated in Figure 2, the sterol content, like the phospholipid content, varies from one membrane to another. Plasma membranes are characteristically enriched in sterols. The phospholipid molar ratios of selected rat membranes (Table 1) are typical for those membrane types not only of mammals, but of many other organisms.

Further refinements have recently become routine thanks to the development of gas chromatographic columns stable at high temperatures[7] and even more important, reverse phase high performance liquid chromatography.[8] Using such methods it is now straightforward to determine the quantitative distribution of individual molecular species present within

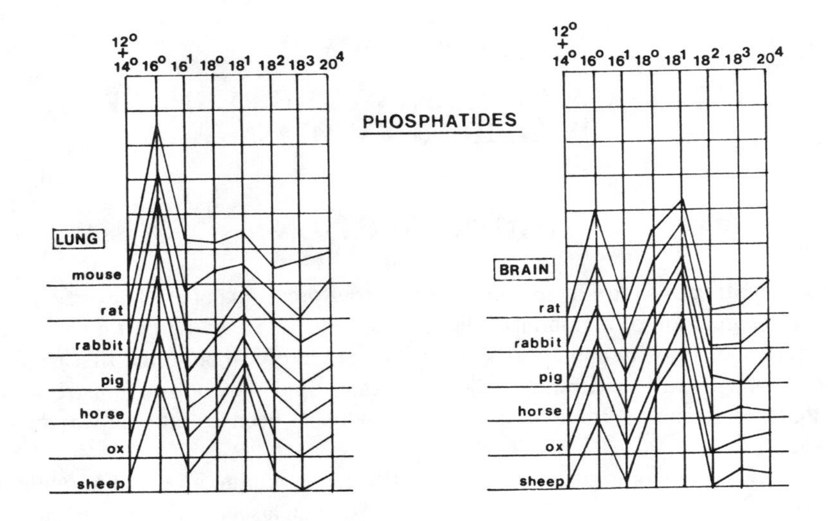

A

FIGURE 1. (A) Fatty acid patterns of the phospholipid fractions from lung and brain tissues of several mammalian species. On the abscissa, the fatty acids are indicated by means of the number of carbon atoms and the number of double bonds. On the ordinate, weight percentages of fatty acid methyl esters are plotted. The distance between two horizontal lines corresponds to 10% of the total fatty acids. (B) Comparison of the fatty acid patterns of phospholipids from several tissues of three animal species. Included are ascites, benzopyrene, Brown-Pearce tumors and hepatoma. (From van Deenen, L. L. M., *Prog. Chem. Fats Other Lipids,* 8 (Part 1), 1, 1965. With permission.)

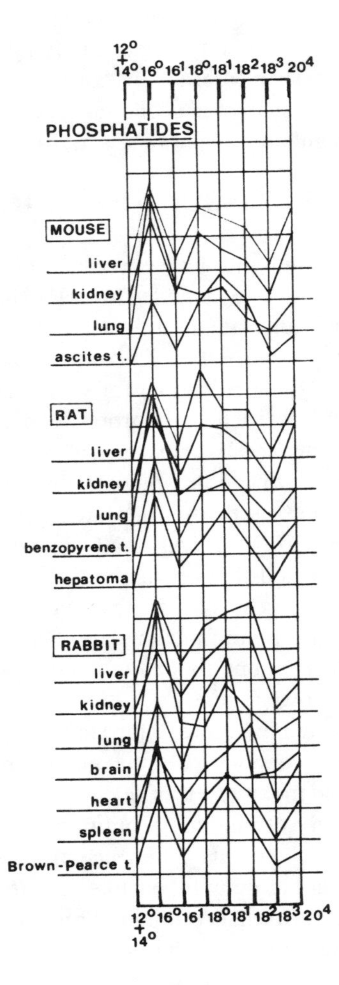

B

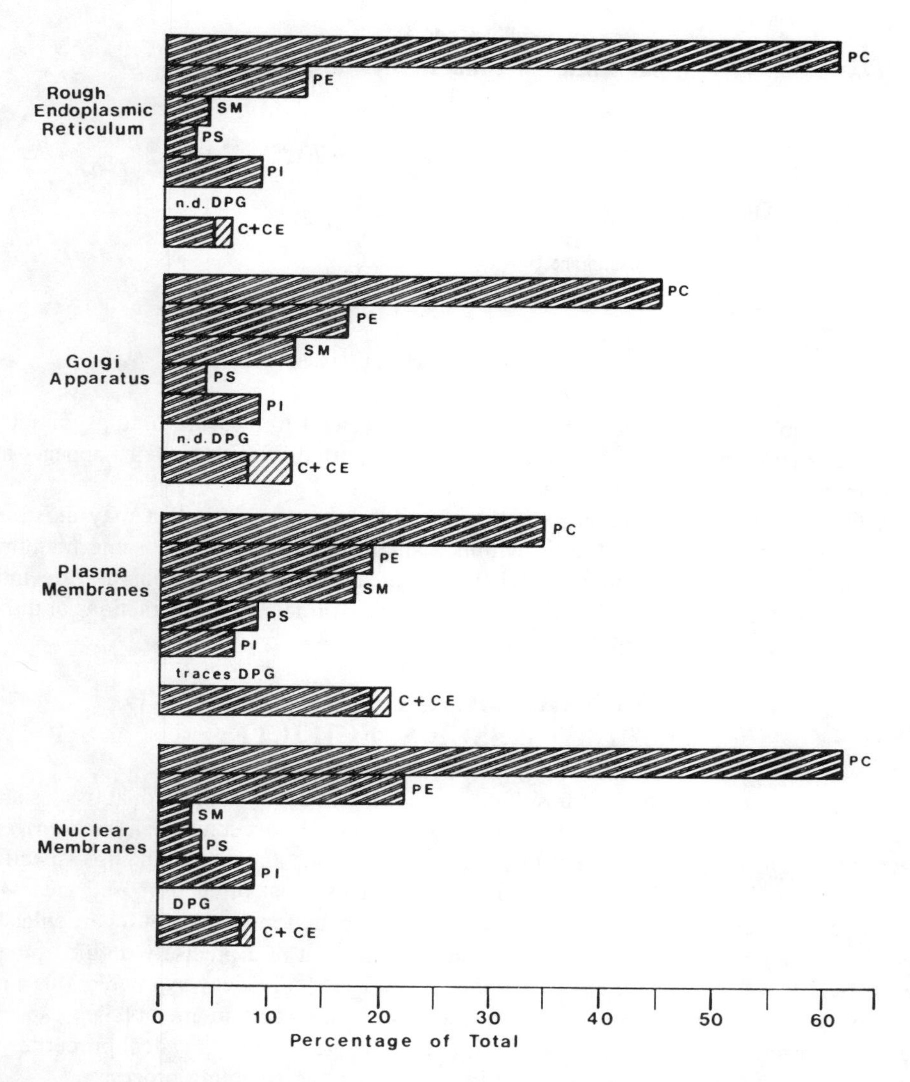

FIGURE 2. Phospholipid and sterol composition of rough endoplasmic reticulum, Golgi apparatus, plasma membranes, and nuclear membranes from rat liver. Individual phospholipid values are given as percentages of the total lipid phosphorus. PC = phosphatidylcholine, PE = phosphatidylethanolamine, SM = sphingomyelin, PI = phosphatidylinositol, DPG = diphosphatidylglycerol, C = cholesterol, CE = cholesterol esters (light cross hatching). Sterols are given as percentages of total lipids by weight. (From McMurray, W. C., *Form and Function of Phospholipids,* Ansell, B. G., Dawson, R. M. C., and Hawthorne, J. N., Eds., Elsevier, Amsterdam, 1973, 205. With permission.)

each lipid class. Data of this type are useful not only to assess the subtle but meaningful changes in molecular species composition triggered by environmental factors (see Chapter 11), but also to recognize differential rates of metabolism among different molecular species of a particular lipid class, as illustrated in Figure 3.[9]

The current goal of the analyst is to establish what assortment of lipid molecules is present in the particular membrane domain under study. It has been clear for some time that inhomogenieties exist between different portions of the same membrane. The best characterized of these differences are between the inner and the outer lipid monolayer of plasma membranes. In the relatively easily analyzed erythrocyte of certain mammals, virtually all of certain lipids, such as sphingomyelin and phosphatidylserine, are confined to one side of the

TABLE 1
Molar Ratios of Cholesterol/Phospholipid in Rat Tissues

Tissue	Choles/phospholipid	Ref.
Brain myelin	1.32[a]	6
Erythrocyte	0.89	6
Liver plasma membrane	0.83	4
Liver mitochondria	0.11	6
Liver ER	0.09	5

[a] Including cerebrosides, the sterol/phospholipid ratio becomes 0.83.

membrane (Table 2).[10] It is technically much more difficult to measure lipid asymmetry in intracellular membranes, but in at least some cases reported such asymmetry appears to be significant (reviewed by Zachowski and Devaux).[10]

The last, still elusive challenge is to identify any heterogeneities that may exist, even momentarily, in the lipid distribution within a single monolayer of the same membrane. Recent physical chemical findings (see below) indicate that many membranes are indeed heterogeneous in this respect, either as a result of specific protein-lipid interactions or through an environmentally-induced lipid phase separation.

III. THE PHYSIOLOGICAL SIGNIFICANCE OF LIPID TISSUE SPECIFICITY

A specificity governing the synthesis and positioning of lipids in individual tissues and in functionally different membranes within a cell may therefore be accepted as a noncontroversial matter of fact. The benefit to the cell of this pronounced tissue and membrane lipid specificity is much less well established. The potential advantages most often proposed are (1) the absolute requirement of some enzyme for one or more bound molecules of a particular lipid in order to maintain functional activity, (2) the requirement of a precisely defined physical state, or fluidity, of each membrane determined in a very sensitive way by interactions among the individual lipids and proteins, (3) the need to render certain membranes especially susceptible or resistant to environmentally-induced changes in their physical properties, and (4) the direct participation of select lipids in transmembrane signaling processes.

A. SPECIFIC LIPIDS REQUIRED FOR ENZYME ACTIVITY

The notion that certain enzymes require specific lipids for full activity gained highest favor some 20 years ago. At that time, investigators succeeded in purifying a number of membrane-bound proteins through solubilization with detergents. Not infrequently, enzymatic activity could be restored only by adding back a characteristic lipid.[11,12] In other cases, a variety of natural lipids and, sometimes, synthetic detergents would reactivate the enzyme.

It has usually not been possible to decide whether the lipids needed for enzyme functioning act on the protein molecules in a fashion directly and specifically altering its tertiary structure in some prescribed fashion, or in a less-specific envelopment of the protein within a tightly bound, hydrophobic microenvironment, sometimes referred to as a lipid annulus, so that the overall effect is optimally conducive to the required substrate-induced perturbations in the protein's three-dimensional structure.

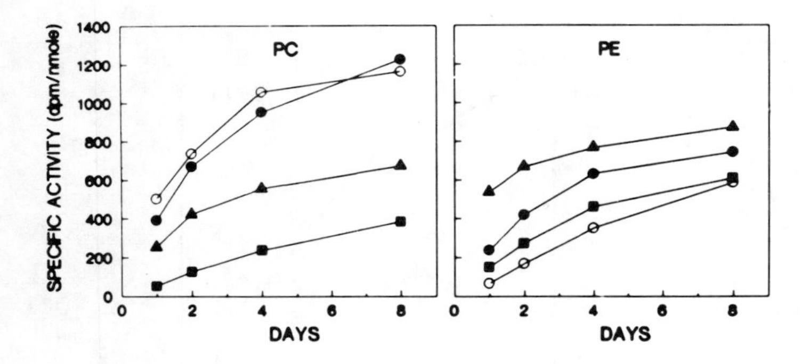

FIGURE 3. Time course of [2-^{3}H]glycerol incorporation into 22:6/22:6 (○), 18:1/22:6 (●), 16:0/22:6 (▲), and 18:0/22:6 (■) molecular species of frog retinal rod outer segment phosphatidylcholine (PC) and phosphatidylethanolamine (PE). (From Louie, K., Wiegand, R. D., and Anderson, R. E., *Biochemistry,* 27, 9014, 1988. With permission.)

TABLE 2
Asymmetry in Red Blood Cells

	% on the outer layer			
Species	**PC**	**SM**	**PE**	**PS**
Man	55	85	5	0
	70	n.d.[a]	n.d.	19
	78	79	21	8
	76	82	20	0
	56	n.d.	18	n.d.
Rat	63–75	100	0	n.d.
	62	100	20	6
Mouse	57	85	20	0
Monkey	67	82	13	n.d.
Pig		75		
Ox, Sheep		50–60		

[a] Not determined.

(From Zachowski, A. and Devaux, P. F., *Comments Mol. Cell. Biophys.*, 6, 63–90, 1989. With permission.)

An example of the former type of interaction is the specific requirement of 3-hydroxybutyrate dehydrogenase for phosphatidylcholine in order to attain optimal activity. Recent studies have suggested that the lipid enhances enzyme activity by facilitating binding of the cofactor NAD by at least one order of magnitude.[13] In this respect phosphatidylcholine seems to function as an allosteric activator. Another well known example of specific lipid binding involves the association of cardiolipin with mitochondrial cytochrome oxidase.[14] Although not absolutely required for activity,[15] cardiolipin is particularly effective in reconstituting enzyme activity and is thought to maintain the protein in an active conformation. Chemical modification of cardiolipin reduced its selective binding to cytochrome oxidase slightly, but the derivatives were still strongly preferred over phosphatidylcholine.[16] Clearly, cardiolipin's interaction with cytochrome oxidase is dictated by structural features additional to those of purely electrostatic origin.

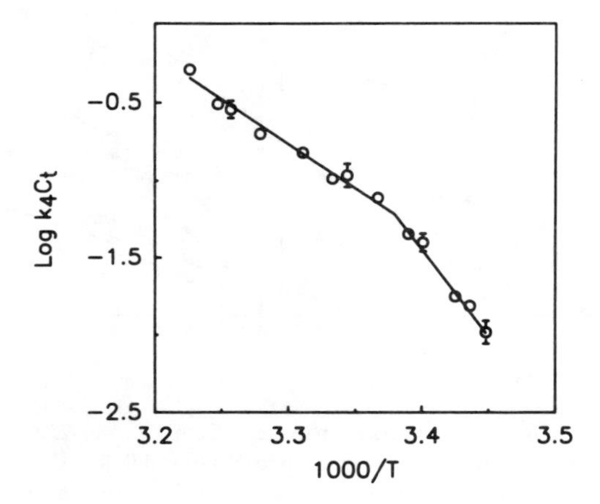

FIGURE 4. Temperature dependence of k_4C_t, a parameter related to the rat erythrocyte membrane glucose carrier's ability to change from an inward-facing conformation to an outward-facing conformation, expressed in an Arrhenius plot. (From Whitesell, R. R., Regen, D. M., Beth, A. H., Pelletier, D. K., and Abumrad, N. A., *Biochemistry,* 28, 5618–5625, 1989. With permission.)

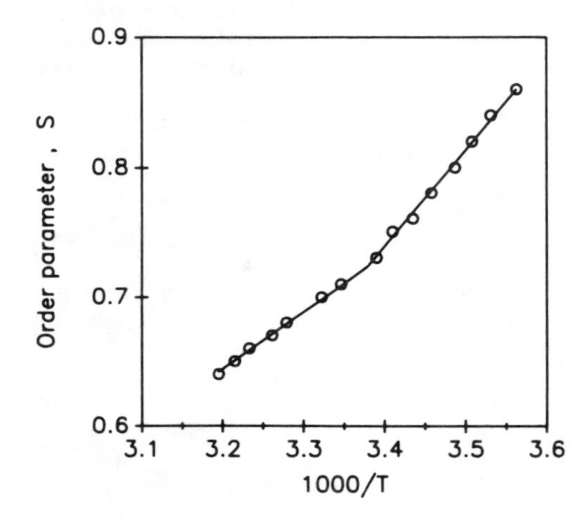

FIGURE 5. Temperature dependence of rat erythrocyte membrane lipid order, expressed as the order parameter of the spin probe 5-nitroxylstearate. (From Whitesell, R. R., Regen, D. M., Beth, A. H., Pelletier, D. K., and Abumrad, N. A., *Biochemistry,* 28, 5618–5625, 1989. With permission.)

B. REGULATION OF THE GENERAL PHYSICAL PROPERTIES OF MEMBRANES

Despite the apparent requirement of some enzymes for specific lipids, it is generally held that the more typical role of membrane lipids is to provide an environment of proper viscosity and surface ionic millieu for optimal enzyme function. The operation of the erythrocyte membrane glucose carrier exemplifies the role played by membrane fluidity. In a thorough analysis, it was discovered that the rate-limiting step (change of the carrier from an in-facing to an out-facing position) required a much greater activation energy below 23°C than above (Figure 4).[17] Membrane fluidity, as inferred using the electron spin resonance probe 5-nitroxylstearate, also exhibited a transition near 23°C. (Figure 5). The close correlation indicates that membrane fluidity may affect the ease of carrier conformational change.

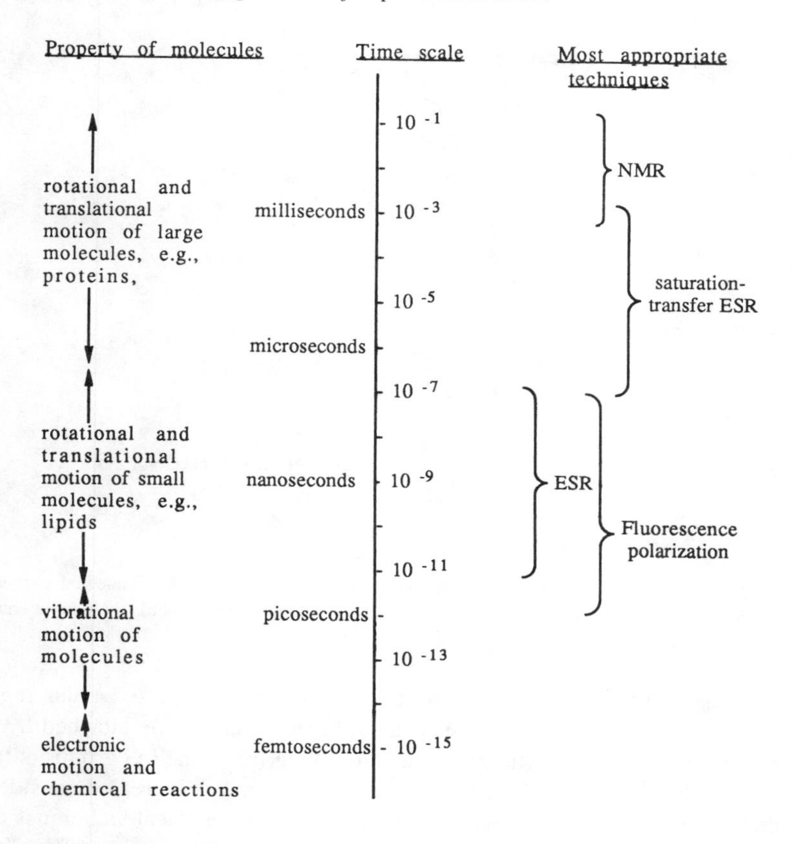

FIGURE 6. Common physical techniques and the types of molecular motion they detect.

Effects of this type depend upon the properties of lipid molecules not in direct contact with an enzyme as well as those that might be bound tightly to it. It is this concept, formulated as the fluid mosaic model for membrane structure,[18] that sparked the greatly amplified membrane lipid research in the 1970s and 1980s. Thanks to recent technological advances, it is now possible to measure by a variety of sensitive physical chemical techniques the fluidity of living membranes as well as isolated lipids.

The methodology for probing membrane fluidity has been well reviewed in the cases of differential scanning calorimetry,[19] X-ray diffraction,[20] nuclear magnetic resonance,[21] electron spin resonance,[22] fluorescence depolarization,[23] infrared and laser-Raman spectrometry,[24] and freeze-fracture electron microscopy,[25] as well as other tools. For a particular application, one or two of the above techniques often have decided advantages over the others. For example, calorimetry provides vital information on the rate and extent of lipid phase changes but gives no data on other properties. The techniques for measuring molecular motion and order are not all appropriate for the same time scale (Figure 6). Because each of the above techniques has its limitations, it is highly desirable in any serious study to seek confirmation of basic findings through the application of two or more complementary approaches. Closer correlations between the findings recorded using distinctively different techniques have been found frequently enough to encourage confidence in the interpretations.[26,27]

One of the more valuable contributions of the physical chemical studies has been the realization of how lipid molecules move within the membrane. In some respects, the most satisfactory evidence regarding relative motion of various atoms within a single lipid molecule came from ^{13}C-nuclear magnetic resonance spectrometry. It was possible to establish the

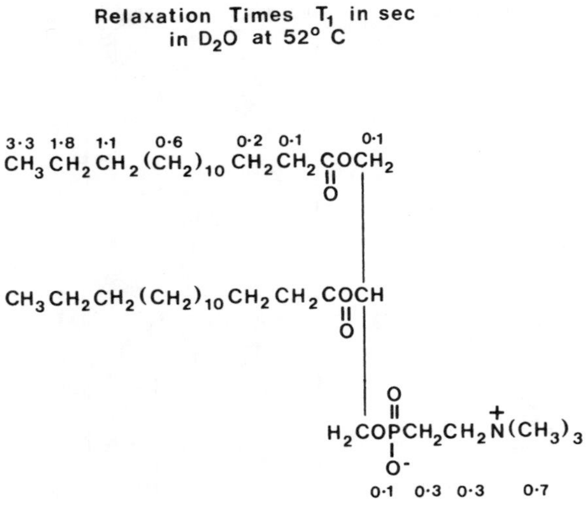

Relaxation Times T_1 in sec
in D_2O at 52° C

3·3 1·8 1·1 0·6 0·2 0·1 0·1
$CH_3CH_2CH_2(CH_2)_{10}CH_2CH_2COCH_2$ (C=O)

$CH_3CH_2CH_2(CH_2)_{10}CH_2CH_2COCH$ (C=O)

$H_2COP(=O)(O^-)CH_2CH_2\overset{+}{N}(CH_3)_3$
0·1 0·3 0·3 0·7

FIGURE 7. ^{13}C-Nuclear magnetic resonance relaxation times T_1 (in seconds) of selected carbon atoms of dipalmitoylphosphatidylcholine in D_2O at 52°C. The increased relaxation times towards the methyl end of the acyl chains indicate an increased motion of the carbon atoms.

existence of a flexibility gradient, beginning with the relatively fixed polar regions of a phospholipid lying near the membrane surface and extending out the attached hydrocarbon side chains, indicating increasing flexibility toward the terminal methyl groups at the middle of the lipid bilayer[28] (Figure 7). The fluidity of a membrane is therefore considered to be generally greater at the center than near the periphery. Other physical techniques confirmed this view. Unfortunately, many of the molecular probes used to report on fluidity are not localized at a fixed depth in the membrane, and the signals emanating from them necessarily constitute an average over various depths. Physical chemists are still laboring to resolve the experimental difficulties created by the inherent microheterogeneity in fluidity. One common strategy is to use a variety of spin-labeled probes that become localized at different depths within the lipid bilayer. Thus Subczynski et al.[41] gauged the effects of cholesterol on the physical properties of distearoyl-phosphatidylcholine by separate measurements utilizing Tempone for the aqueous interface, Tempocholine phosphatidic acid ester (T-PC) and 5-doxylstearic acid spin label (5-SASL) for the region in and near the polar head groups, and 16-doxylstearic acid (16-SASL) for the central region of the lipid bilayer (Figure 8).

In addition to information regarding the movement in place of the lipid components, a considerable stock of data is available on the translational movement of lipids laterally in the plane of the membrane. Although rates vary somewhat due to the particular circumstances involved, diffusion coefficients in phospholipid bilayers are in the range of 10^{-9} to 10^{-8} cm^2/sec.[29] Lateral movements of lipids in native membranes occur at about the same rate. At these rates, each molecule will exchange with its neighbor approximately 10^7 times per second.

Another dynamic motion of membrane lipids that interests biologists is the sudden translation of a molecule from one side of the lipid bilayer to the other. Despite the need to penetrate a seemingly unfavorable energy barrier, active transbilayer exchange of membrane lipids occurs in most natural membranes. This exchange of intact molecules, sometimes referred to as flip-flop, is never as rapid as lateral movement of the same molecules. Nevertheless, the transbilayer exchange has been observed to be as much as four orders of magnitude faster in biological membranes[30] than it is in artificial lipid bilayers.[31] It is thought that interaction of the lipid polar head group with an integral membrane protein might facilitate transbilayer exchange.

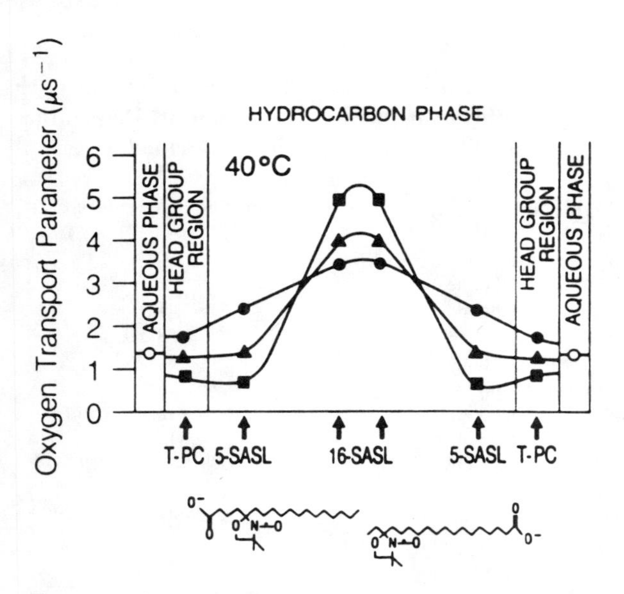

FIGURE 8. The effects of cholesterol on the transport of O_2 across egg yolk phosphatidylcholine at 40°C as evaluated using three different ESR spin probes (see text). The oxygen transport parameter was used as a monitor of membrane fluidity, reporting on the translational diffusion of small molecules in the membrane. Oxygen transport was measured in different regions of the membrane by monitoring the bimolecular collision rate between molecular oxygen and the spin probes, thereby altering the probes' spin-lattice relaxation time. The findings showed that oxygen diffusion increased (as reflected by higher O_2 transport parameter values) in the central region of the membrane as the cholesterol content rose from 0 (●) to 27.5 (▲) to 50 (■) mol%. However, near the membrane surface the cholesterol effect was reversed. (Modified from Subczynski, W. K., Hyde, J. S., and Kusumi, A., *Biochemistry*, 30, 8578–8590, 1991. With permission.)

Assuming that all membrane lipids participate in transbilayer exchange, the rate of flip-flop, as measured in functional membranes by physical chemical or labeling techniques, seems too rapid to account for the existence of any large lipid compositional differences between the opposite sides of the bilayer. Yet, as described above, compelling evidence is available to demonstrate that many biological membranes are asymmetric with respect to lipid distribution. The dynamics of maintaining this asymmetry will be discussed in Chapter 9.

One of the most important concepts to develop through consideration of the fluid mosaic model is that of coexisting regions of different fluidity in the same membrane. It has been known for many years that artificial bilayers made from pure lipids of the same type found in membranes will undergo a transition from a relatively fluid state, generally termed the liquid-crystalline phase, to a more rigid state, called the gel phase, when the temperature of the preparation is lowered to a characteristic point. This change is referred to as a phase transition. In a biological membrane, lipids are invariably present as complex mixtures, with each component possessing its own characteristic phase transition temperature. Chilling such a membrane usually leads, not to a simultaneous cocrystallization of all lipids, but rather to a lateral migration of the more readily gelled lipid classes into rigid assemblages capable of growing by accretion of similar, but slightly lower, freezing lipid species upon further chilling. This process is termed a phase separation. Thus, a membrane may under certain conditions contain coexisting domains of liquid-crystalline and gel phases in equilibrium.

By their very nature, the hydrophobic amino acids making up the surfaces of bilayer-embedded integral membrane proteins are constrained in an ordered but noncrystalline arrangement. Most of these proteins do not fit well into the crystal-like gel phase of lipids and are treated as impurities. When a phase separation commences, integral proteins are excluded from the gel phase by lateral movement in the plane of the membrane. This can lead ultimately

TABLE 3
The Phase Transition Temperatures of Several Pure Phospholipids in Aqueous Suspension, as Determined by Differential Scanning Calorimetry

Lipid species		Transition temperature (°C)	Ref.
12:0/12:0	phosphatidylcholine	0	38
14:0/14:0	phosphatidylcholine	23	38
16:0/16:0	phosphatidylcholine	41	38
18:0/18:0	phosphatidylcholine	å 58	38
20:0/20:0	phosphatidylcholine	66	34
18:0/18:1	phosphatidylcholine	6	34
18:0/18:2	phosphatidylcholine	–16	34
18:0/18:3	phosphatidylcholine	–13	34
18:0/20:4	phosphatidylcholine	–13	34
18:1/18:1	phosphatidylcholine	–22	38
14:0/14:0	phosphatidylethanolamine	48	38
16:0/16:0	phosphatidylethanolamine	60	38
14:0/14:0	phosphatidylglycerol	23[a]	38
16:0/16:0	phosphatidylglycerol	41[a]	38

[a] Transition temperature very dependent upon the salt present. These values are for the Na^+-free form.

to high densities of proteins congregated in bilayer domains containing the least crystalline lipid species.

C. THE CONTRIBUTION OF DIFFERENT LIPID TYPES TO MEMBRANE PHYSICAL PROPERTIES

The quantitative interactions of different lipid species coexisting in a bilayer have been most extensively studied using phospholipids. Even here we have rarely progressed beyond the examination of binary mixtures. Almost every aspect of a phospholipid's structure can exert a profound effect on its interaction with other lipid molecules. The phase transition temperature (melting point) of pure phospholipid vesicles suspended in an aqueous medium is highly dependent upon the length and the degree of unsaturation of the phospholipid's component fatty acids (Table 3). It also depends upon the nature of the polar head group, as shown by the 25° difference in phase transition temperature between phosphatidylcholine and phosphatidylethanolamine preparations having identical fatty acid complements. Many additional phase transition temperatures are listed in reference 33.

If two very similar phospholipid species, such as 16:0/16:0 phosphatidylcholine and 18:0/18:0 phosphatidylcholine, are mixed at a high temperature and chilled, they will cocrystallize at a temperature intermediate between the phase-transition temperature of either component. However, if the fatty acids of one species are as many as four carbon atoms longer than those of the other, or if one species of the binary mixture contains unsaturated fatty acids, then cocrystallization will not occur, and each species will behave at least semi-independently.[35]

Obviously the natural mixtures of phospholipids existing in cellular membranes must interact with each other through a variety of modes, including not only associations formed within the same lipid monolayer but also, through acyl chain interdigitation, with the opposite lipid monolayer of the membrane.[36] It is clear that relatively small changes in the lipid composition of a particular membrane could alter its physical properties significantly. Basic physical chemical studies of pure phospholipids and relatively simple mixtures are steadily moving ahead.[37,38]

A considerable store of information has also been accumulated on the association of phospholipids and sterols.[39] The presence of 50 mol% cholesterol in a membrane, as is usual in the plasma membrane of many cell types, leads to a pronounced condensing effect if the lipids are in the liquid crystalline state, or to a liquefying effect if they are in a gel state. Thus sterols exert a moderating influence, which in nature lessens the influence of interacting phospholipids on membrane physical properties. Differential scanning calorimetric studies showed that in mixed phospholipid bilayers cholesterol tends to associate with phosphatidylcholine species in preference to phosphatidylethanolamine and with the lower melting component of a two molecular species mixture of phosphatidylethanolamines.[40] The exact nature of the cholesterol interaction with phospholipid molecules is still uncertain, but recent ESR studies suggest that in fluid phase *cis*-unsaturated phosphatidylcholine bilayers cholesterol forms small oligomeric domains, which are less fluid and immiscible in the bulk phosphatidylcholine phase.[41] The extent of this domain formation can be altered by the presence of proteins.[42]

It is probably no accident that cells have evolved a pattern of lipid composition featuring a high cholesterol/phospholipid ratio in plasma membrane and a low ratio in intracellular membranes. The consequences of this are that environmental factors, e.g., changing temperature, exert a pronounced effect on the cholesterol-poor membranes, while fluidity change in the cholesterol-rich membranes is strongly resisted by the buffering effect of the sterol.[39]

D. THE EFFECTS OF A MEMBRANE'S PHYSICAL PROPERTIES ON ITS PHYSIOLOGICAL BEHAVIOR

Experimentally determined and theoretically predicted properties of phase-separated membranes[43] are of great interest in understanding biological activities of membranes. While most physiological functions proceed more expeditiously when the responsible membrane is largely in the liquid-crystalline phase, certain reactions are remarkably enhanced just at temperatures triggering extensive phase separation. This phenomenon is exemplified by the striking propensity for phospholipase A_2 to deacylate bilayers of pure and mixed phospholipids only near their phase transition temperatures (Figure 9) or under other circumstances giving rise to unstable lipid packing arrangements.[45] Functional properties, such as the transport of molecules across a bilayer structure, can also exhibit a maximal rate at the phase-transition temperature (Figure 10).[46] Even the association of lipids with hydrophobic proteins, e.g., the apolipoprotein A-1 from human plasma high density lipoprotein, can be enhanced by a factor of 500 to 1000 at the gel-liquid crystalline transition temperature.[47] And still more significant with respect to membrane biosynthesis is the observation that hydrophobic membrane proteins are more readily taken into lipid bilayers near the phase transition of the lipids (Figure 11).[48]

The opportunities for metabolic regulation through lipid compositional changes are unmistakably clear in these examples. It is also clear that membrane-bound enzymes respond differently to altered lipid patterns. Thus, various *Escherichia coli* oxidases were progressively inactivated by diet-induced changes in membrane fatty acid composition, but the associated dehydrogenase activities were not affected.[49] Recent findings tend to cast doubt on our earlier assumption that the preferred *in vivo* state of a membrane simply involves lipids entirely in a liquid crystalline phase. Increasing attention is being directed to the possible existence of small clusters of fairly strongly associating lipid molecules even at temperatures above any discernible phase separation.[41,46] It would seem possible that lipid molecules situated at the boundaries between the two physically distinct domains are especially mobile due to packing defects. This property may be responsible for heightened transport capabilities or enhanced enzymatic activity. If this is true, it will be essential to consider not merely the proportions of the different coexisting phases, but also the linear extent of the boundary separating them. This potentially important parameter depends upon the cooperativity be-

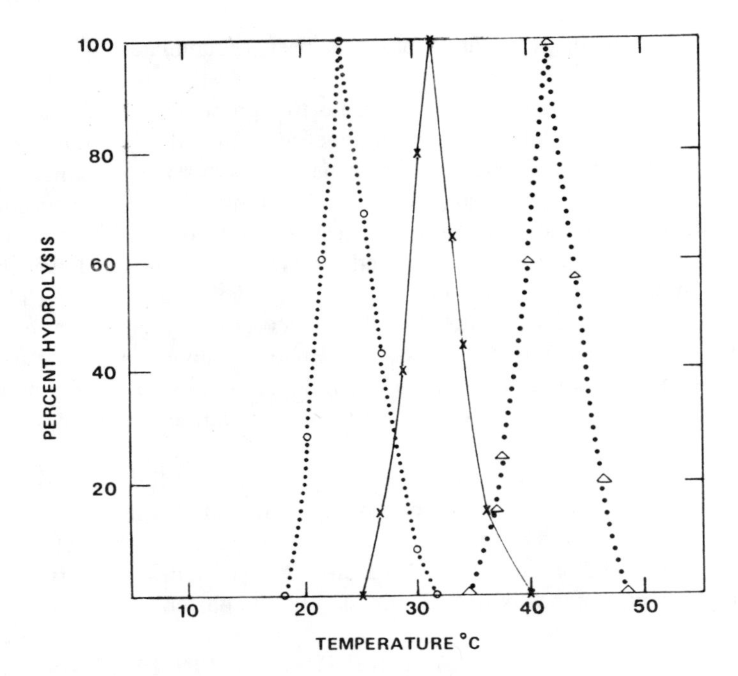

FIGURE 9. Temperature-dependent hydrolysis of dimyristoylphosphatidylcholine (○), dipalmitoylphosphatidylcholine (△), and an equimolar mixture of the two lipids (×). Vesicles of the lipids were incubated with phospholipase A_2 for 10 min. at various temperatures. The residual phosphatidylcholine was determined quantitatively, and the results are expressed as per cent of the initial amount of lipid that was hydrolyzed. (From Op den Kamp, J. A. F., Th. Kaverz, M., and van Deenen, L. L. M., *Biochim. Biophys. Acta,* 406, 196, 1975. With permission.)

tween elements of the microdomains (Figure 12), a property, like so many others, that is determined by the membrane's lipid and protein makeup.

Quite apart from the above-mentioned phase separations, in which interactions involve lipid molecules all present on one side of the lipid bilayer, one must consider the further possibility that the natural asymmetry of membrane lipids may in some situations lead to distinctively different fluidities in apposed membrane halves.[50] Aside from the more obvious effects upon enzymatic activity, the asymmetric membrane expansion accompanying perturbation of such a membrane might well induce cellular shape changes or other structural alterations.

E. THE DIRECT PARTICIPATION OF SPECIFIC LIPIDS IN CELLULAR SIGNALING PROCESSES

In the 1970s "membrane fluidity" and the cooperative behavior of vast lipid assemblages dominated the thinking of lipid biochemists. The 1980s saw the emergence of a new and contrasting focus emphasizing the direct involvement of very specific lipid metabolic pathways in transducing external signals into cellular responses. The importance of a quantitatively minor plasma membrane lipid, phosphatidylinositol 4,5-bisphosphate, as the source of inositol 1,4,5-trisphosphate and diacylglycerol, two second messengers leading to protein kinase activation, was fully recognized in the early 80s.[51] Thousands of laboratories are now exploiting these findings. The basic aspects of this signaling process are discussed in Chapter 4.

It now appears that another membrane lipid intermediate, sphingosine, can be a potent inhibitor of protein kinase action.[52] It is also known that the noncovalent binding of particular lipids can bring about a physiologically meaningful translocation of enzymes such as protein

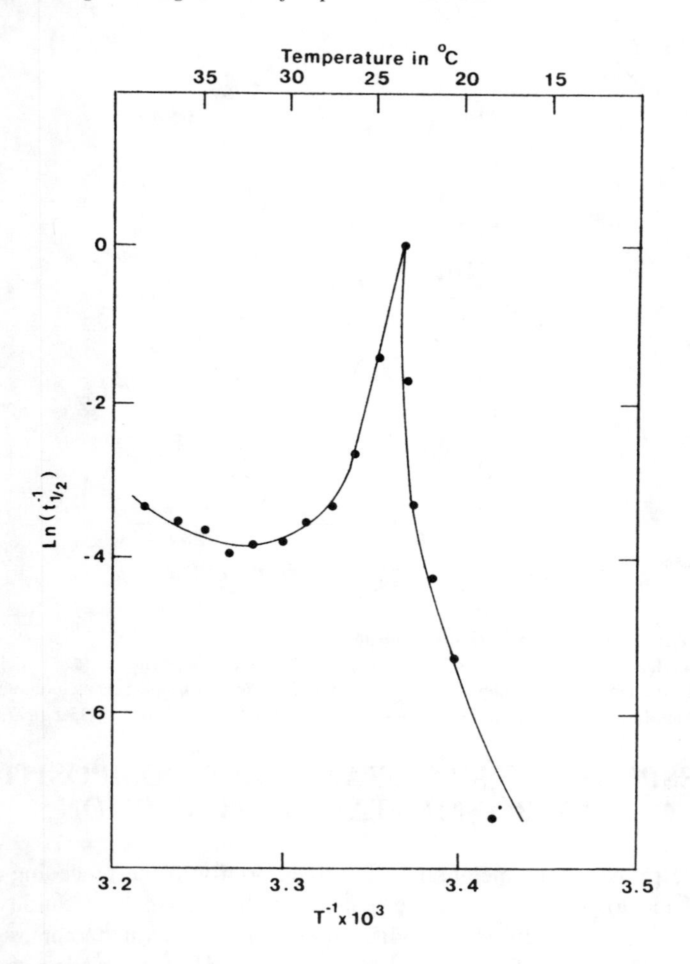

FIGURE 10. Temperature dependence of the rate of 8-anilino-1-naphthalene-sulfonate (ANS) permeation in the bilayers of dimyristoylphosphatidylcholine vesicles. The kinetics of ANS transport is illustrated by plotting the logarithm of the rate (the reciprocal of the half-time) vs. temperature. Sharply increased permeation occurs at the phase transition temperature of 24°C. (From Tsong, T. Y., Greenberg, M., and Kanehisa, M. I., *Biochemistry*, 16, 3115, 1977. With permission.)

kinase C,[53] diacylglycerol kinase,[54] choline phosphate cytidylyltransferase,[55] and phosphatidate phosphohydrolase[56] from a cytoplasmic environment into a more functional membrane association.

Several more recently detected cases of lipid participation in protein trafficking are under study. For example glycosphingolipids being packaged in the Golgi apparatus for transfer to the plasma membrane appear to sequester newly formed phosphatidylinositol-anchored proteins and chaperone them to the same destination.[57]

Perhaps the most unexpected lipid involvement unfolding during the 80s was the discovery that a previously unknown lipid, *sn*-1 alkyl, *sn*-2 acetyl phosphatidylcholine (platelet-activating factor), is responsible for activating not only platelets but also polymorphonuclear leukocytes, monocytes, and other cell types.[58] The action of platelet-activating factor is hormone-like in that it binds to specific cell surface receptors and triggers a G protein-mediated transmembrane signaling pathway.

It is likely that further examples of such highly specialized lipid functions will be forthcoming.

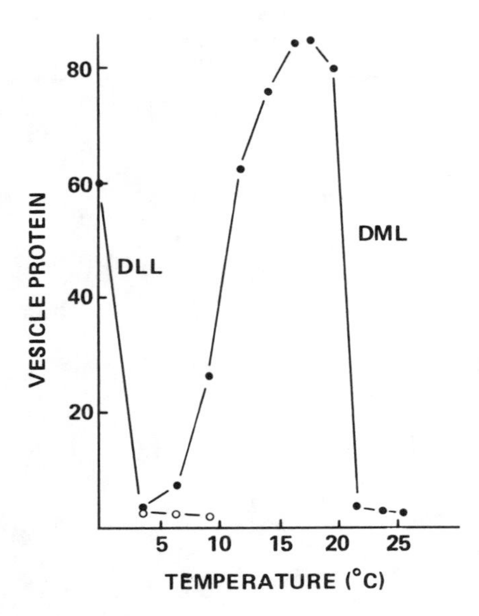

FIGURE 11. The effect of temperature on the incorporation of the M13 virus coat protein into dilaurylphosphatidylcholine (DLL) (○) or dimyristoylphosphatidylcholine (DML), (●) vesicles. Vesicles were mixed with [^{3}H]lysine-labeled coat protein at the indicated temperature. Following incubation, vesicle-bound radioactivity was measured by centrifugation. (From Wickner, W. T., *Biochemistry,* 16, 254–258, 1977. With permission.)

IV. RESPONSE OF MEMBRANE LIPID COMPOSITION TO ENVIRONMENTAL PERTURBATION

Most of the properties of membranes described briefly in the preceding paragraphs are intimately related to the fluidity of the constituent lipids. And the fluidity, it should be remembered, is subject to significant modification by many external factors, such as temperature, pH, cations, drugs, etc. There is evidence in Chapter 11 that most organisms are capable of at least partially reversing these fluidity modifications by altering selected reactions of lipid metabolism, thus shifting membrane fluidity back towards an optimal value for the new conditions. The authenticated role of lipid metabolism as a mechanism for regulating membrane fluidity in such instances provides a unifying theme of this book. Concisely stated, cellular regulation of lipid metabolism strives to maintain under all conditions a physical environment optimal for the proper physiological functioning of each membrane.

Recent quantitative lipid analyses of organisms acclimating to a variety of fluidity-perturbing factors (Chapter 11) furnish the first obvious manifestation of this fine tuning. Most commonly, the regulatory stratagem employed is an alteration of the lipid composition by increasing or decreasing the degree of fatty acid unsaturation. The effects of such alterations on membrane physical properties are relatively well known, but there is evidence that almost every other modifiable parameter of a membrane's lipid makeup also can be and is utilized by cells to provide a physical environment custom-tailored to their needs.

Unfortunately, at present we can only guess what membrane physical state is best suited to accommodate a given set of environmental conditions. It often appears, from our still rather insensitive measurements, that the fluidity of a membrane acclimated to a certain situation is significantly different from that attained by the same membrane under other circumstances. Thus, synaptosomal membranes of fishes acclimated to low temperatures appeared from fluorescence polarization measurements to be less fluid than equivalent membranes of high

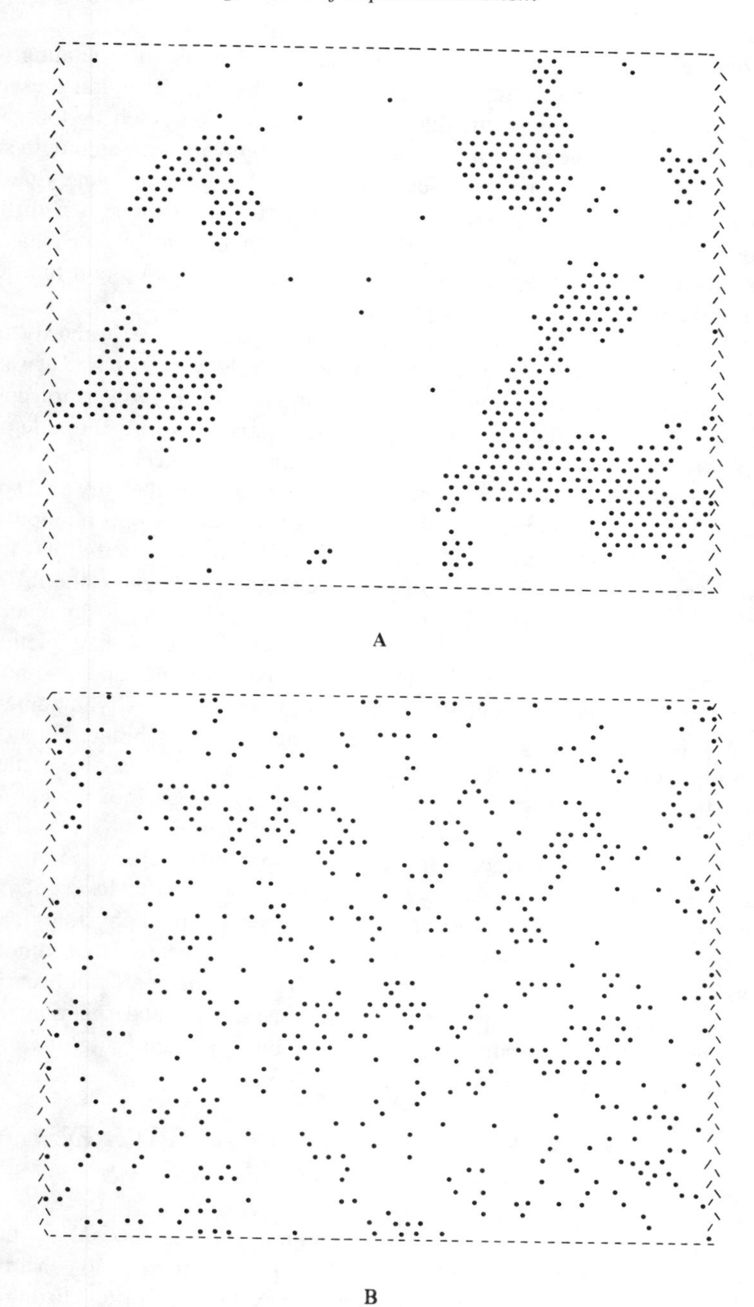

FIGURE 12. A model simulating high (A) and a low (B) cooperatively between lipid clusters (groups of disordered units linked together by nearest-neighbor bonds). In the less cooperative system, the total phase boundaries of the lattices increase, permitting faster permeation of transportable molecules. (From Tsong, T. Y., Greenberg, M., and Kanehisa, M. I., *Biochemistry,* 16, 3115–3121, 1977. With permission.)

temperature-acclimated individuals of the same species.[59] The same holds true for microsomal membranes of the protozoan *Tetrahymena*.[60] Of course, there is no reason to doubt that different temperatures demand different membrane properties in order to retain tight control of the metabolic reins, but before we can assess the role of lipids in regulating membrane fluidity we must be capable of accurately measuring fluidity. This is not at present an easy thing to do. In experiments involving fluorescence polarization, such as those described above,[59,60] the properties of the probe may themselves change with temperature in such a way as to render the extrapolation of experimentally observed data to absolute fluidity values unmanageably complex.[61] Other physical techniques suffer from their own difficulties of interpretation which, coupled with the fact that the membrane lipid bilayer is, as described earlier, not uniformly fluid throughout, still impose serious limitations on the quantitative assessment of a membrane's true physical state.

For the above reasons, accepting the postulate that lipids change primarily to regulate fluidity currently requires a certain amount of blind faith. However, I find it satisfying to believe that the complexities of cellular membrane structure have evolved for good reason. And it excites me to consider that the rapidly proliferating data favoring a regulatory role for membrane fluidity have been met by very little contradictory evidence.

I now wish to develop this concept somewhat further. Apart from the fatty acid composition of a membrane's lipids, there is much variability in the distribution of phospholipid polar head groups. We have seen earlier in this chapter examples of the tissue specificity (Figure 1A,B) and the subcellular specificity (Figure 2) of polar head groups. Because the different polar groups have different combinations of charge, e.g., a negative and a positive charge on phosphatidylcholine, sphingomyelin, and under some conditions, phosphatidylethanolamine, two negative charges and a positive charge on phosphatidylserine, and a single negative charge on phosphatidylglycerol and phosphatidylinositol, one may easily imagine specificity in their capacity for transporting charged molecules across a membrane. Hypotheses have been put forward to clarify this function of individual phospholipid classes, e.g., the proposed role of phosphatidylcholine in facilitating transmembrane H^+ and Cl^- movement.[62] The ability of lipid polar head groups to engage in such activities would certainly be affected by the ionic composition at the membrane surfaces. Short-term compositional changes in the lipid polar moieties may well be able to compensate for fluctuating environmental levels of various ions.

Understanding how changing membrane physical properties affect physiological responses also requires that we deal with certain short-lived nonbilayer structures that sometimes form, especially under conditions of rapid perturbation.[63] These ephemeral lipid rearrangements may be of critical importance in promoting membrane fusion and other physiologically significant membrane behavior, but the extreme difficulty in quantifying or even detecting their presence has retarded efforts to evaluate their role.

V. THE ROLE OF MEMBRANE LIPIDS IN ABNORMAL AND DISEASE STATES

The apparently tight metabolic control of membrane lipid composition, and therefore fluidity, seems to be upset in a number of diseases.[16] At present, there is no generally accepted pattern of change in lipid composition or membrane physical properties distinguishing diseased tissues from normal ones. Even in rather closely related defects, such as varieties of tumors, conflicting results have been obtained when comparing normal and transformed cells.[16] Some of the reported membrane fluidity differences between the two cell states may result from experimental artifacts,[64] but it appears definite that genuine fluidity alterations do exist in some instances.

The question of greatest pertinence in this context is whether membrane lipids are responsible for true fluidity differences that might be found in the abnormal state. An equivocal answer must be given at the present time. Most changes that have been reported are small, and often homogeneity of the membranes might be questioned. Since fluidity changes, as esti-

TABLE 4
Some Common Sphingolipidoses

Disease	Symptoms	Major lipid accumulation	Primary organ involvement	Enzyme deficiency
Gaucher's disease	Spleen and liver enlargement, erosion of long bones and pelvis, mental retardation (in infantile form)	Glucocerebroside	Liver, spleen, bone, brain (in infantile form)	Glucocerebroside β–glucosidase
Niemann-Pick disease	Liver and spleen enlargement, mental retardation	Sphingomyelin	Brain, liver, spleen	Sphingomyelinase
Krabbe's disease	Mental retardation, almost total absence of myelin	Galactocerebroside	Brain	Galactocerebroside β-galactosidase
Metachromatic leukodystrophy	Mental retardation, psychological disturbances (in adult form)	Sulfatide	Brain	Sulfatidase
Fabry's disease	Reddish-purple skin rash, kidney failure, pain in lower extremities	Ceramide trihexoside	Kidney	Ceramidetrihexoside β-galactosidase
Tay-Sachs disease	Mental retardation, blindness, muscular weakness	Ganglioside G_{M2}	Brain	Hexosaminidase A
Generalized gangliosidosis	Mental retardation, liver enlargement, skeletal deformities	Ganglioside G_{M1}	Brain, liver, bone	β-galactosidase

mated in certain ways, can be strongly influenced by the interaction of membrane proteins with the underlying cytoskeleton,[65] more extensive research will be required before lipid abnormalities can be unequivocally shown to constitute a primary defect in transformed cells.

Quite apart from derangement of lipid metabolism that may produce subtle effects on membrane fluidity, there are many gross abnormalities manifested by accumulations of one or more lipid. Examples of sphingolipid storage diseases are illustrated in Table 4.[66] As described in Chapter 7, many of these conditions appear to result from defects in a single enzymatic reaction. There is optimism that successful therapy and correction of these relatively straight-forward disorders may be forthcoming in the near future.

VI. RECAPITULATION

As we proceed to immerse ourselves in the intricate details of intermediary lipid metabolism and its regulation, we may easily lose sight of the benefits to be gained by such control. It seems likely that each membrane of every cell is poised in a physical state most conducive to satisfy in direct and indirect ways a wide variety of cellular needs. The molecular composition of each membrane lipid class is continuously retailored so as to contribute optimally to that end, even in the face of severe environmental stress.

If you find little evidence in the following chapters that such a wonderfully integrated state indeed exists, it is perhaps because of the approaches we have used to the problem. Almost all experimental enzymatic studies have been performed on *in vitro* preparations. While this was desirable and necessary as a beginning, now, having delineated most of the pathways reasonably well, we should make every effort to reconstitute the metabolic systems, ultimately returning to the intact tissue or organism. Only in that condition will the full gamut of physiological controls be in effect.

In this sense, we have a long way to go before membrane lipid regulation is understood, but the stirrings of comprehension are there. In some areas you will detect the necessary base of information to begin reconstructing the natural control systems. Progress will be rapid now.

REFERENCES

1. **van Deenen, L. L. M.,** Phospholipids and biomembranes, *Prog. Chem. Fats Other Lipids,* 8 (Part 1) 1–127, 1965.
2. **White, D. A.,** The phospholipid composition of mammalian tissues, in *Form and Function of Phospholipids,* Ansell, G. B., Dawson, R. M. C., and Hawthorne, J. N., Eds., Elsevier, Amsterdam, 1973, 441–482.
3. **McMurray, W. C.,** Phospholipids in subcellular organelles and membranes, in *Form and Function of Phospholipids,* Ansell, G. B., Dawson, R. M. C., and Hawthorne, J. N., Eds., Elsevier, Amsterdam, 1973, 205–251.
4. **Dorling, P. R. and LePage, R. N.,** A rapid high yield method for the preparation of rat liver cell plasma membranes, *Biochim. Biophys. Acta,* 318, 33–40, 1973.
5. **Keenan, T. W. and Morré, D. J.,** Phospholipid class and fatty acid composition of Golgi apparatus isolated from rat liver and comparison with other cell fractions, *Biochemistry,* 9, 19–25, 1970.
6. **Ashworth, L. A. E. and Green, C.,** Plasma membranes: phospholipid and sterol content, *Science,* 151, 210–211, 1966.
7. **Lynch, D. V. and Thompson, G. A., Jr.,** Analysis of phospholipid molecular species by gas chromatography and coupled gas chromatography-mass spectrometry, in *Modern Methods of Plant Analysis,* Vol. 3, Linskens, H. F., and Jackson, J. F., Eds. Springer-Verlag, Berlin, 1986, 100–120.
8. **Christie, W. W.,** *HPLC and Lipids,* Pergamon Press, Oxford, 1987, pp 272.
9. **Louie, K., Wiegand, R. D., and Anderson, R. E.,** Docosahexaenoate-containing molecular species of glycerophospholipids from frog retinal rod outer segments show different rates of biosynthesis and turnover, *Biochemistry,* 27, 9014–9020, 1988.
10. **Zachowski, A. and Devaux, P. F.,** Bilayer asymmetry and lipid transport across biomembranes, *Comments Mol. Cell. Biophys.,* 6, 63–90, 1989.
11. **Coleman, R.,** Membrane-bound enzymes and membrane ultrastructure, *Biochim. Biophys. Acta,* 300, 1–30, 1973.
12. **Fourcans, B. and Jain, M. K.,** Role of phospholipids in transport and enzymic reactions, *Adv. Lipid Res.,* 12, 147–226, 1974.
13. **Rudy, B., Dubois, H., Mink, R., Trommer, W. E., McIntyre, J. O., and Fleischer, S.,** Coenzyme binding by 3-hydroxybutyrate dehydrogenase, a lipid-requiring enzyme: lecithin acts as an allosteric modulator to enhance the affinity for coenzyme, *Biochemistry,* 28, 5354–5366, 1989.
14. **Knowles, P. F., Watts, A., and Marsh, D.,** Spin-label studies of head group specificity in the interaction of phospholipids with yeast cytochrome oxidase, *Biochemistry,* 20, 5888–5894, 1981.
15. **Watts, A., Marsh, D., and Knowles, P. F.,** Lipid-substituted cytochrome oxidase: no absolute requirement of cardiolipin for activity, *Biochem. Biophys. Res. Commun.,* 81, 403–409, 1978.
16. **Powell, G. L., Knowles, P. F., and Marsh, D.,** Spin-label studies on the specificity of interaction of cardiolipin with beef heart cytochrome oxidase, *Biochemistry,* 26, 8138–8145, 1987.
17. **Whitesell, R. R., Regen, D. M., Beth, A. H., Pelletier, D. K., and Abumrad, N. A.,** Activation energy of the slowest step in the glucose carrier cycle: break at 23°C and correlation with membrane lipid fluidity, *Biochemistry,* 28, 5618–5625, 1989.
18. **Singer, S. J. and Nicholson, G. L.,** The fluid mosaic model of the structure of cell membranes, *Science,* 175, 720–731, 1972.
19. **McElhaney, R. N.,** The uses of differential scanning calorimetry and differential thermal analysis in studies of model and biological membranes, *Chem. Phys. Membranes,* 30, 229–259, 1982.
20. **Blaurock, A. E.,** Evidence of bilayer structure and of membrane interactions from x-ray diffraction analysis, *Biochim. Biophys. Acta.,* 650, 167–207, 1982.
21. **Smith, I. C. P.,** Structure and dynamics of cell membranes as revealed by NMR techniques, in *Structure and Properties of Cell Membranes,* Vol. III, Benga, G., Ed., CRC Press, Boca Raton, 237–260, 1985.
22. **Gordon, L. M. and Curtain, C. C.,** Electron spin resonance analysis of model and biological membranes, *Advances in Membrane Fluidity,* Vol. I, Aloia, R. C., Curtain, C. C., and Gordon, L. M., Eds., Alan R. Liss, New York, 1988, 25–88.

23. **Lee, A. G.,** Membrane studies using fluorescence spectroscopy, in *Techniques in the Life Sciences,* Elsevier Biomedical, County Clare, Ireland, 1982, 1–49.
24. **Wong, P. T. T.,** Raman spectroscopy of thermotropic and high-pressure phases of aqueous phospholipid dispersions, *Annu. Rev. Biophys. Bioeng.,* 13, 1–24, 1984.
25. **Kitajima, Y. and Thompson, G. A., Jr.,** *Tetrahymena* strives to maintain the fluidity interrelationships of all its membranes constant, electron microscope evidence, *J. Cell Biol.,* 72, 744–755, 1977.
26. **Jacobson, K. and Papahadjopoulos, D.,** Phase transitions and phase separations in phospholipid membranes induced by changes in temperature, pH, and concentration of bivalent cations, *Biochemistry,* 14, 152–161, 1975.
27. **Martin, C. E. and Thompson, G. A., Jr.,** Use of fluorescence polarization to monitor intracellular membrane changes during temperature acclimation. Correlation with lipid compositional and ultrastructural changes, *Biochemistry,* 17, 3581–3586, 1978.
28. **Levine, Y. K., Birdsall, N. J. M., Lee, A. G., and Metcalfe, J. C.,** ^{13}C Nuclear magnetic resonance relaxation measurements of synthetic lecithins and the effects of spin-labeled lipids, *Biochemistry,* 11, 1416–1421, 1972.
29. **Houslay, M. D. and Stanley, K. K.,** Dynamics of Biological Membranes, Wiley, New York, 1982, p 41.
30. **Rothman, J. E. and Kennedy, E. P.,** Rapid transmembrane movement of newly synthesized phospholipids during membrane assembly, *Proc. Natl. Acad. Sci. U.S.A.,* 74, 1821–1825, 1977.
31. **Roseman, M., Litman, B. J., and Thompson, T. E.,** Transbilayer exchange of phosphatidylethanolamine for phosphatidylcholine and N-acetimidoylphosphatidylethanolamine in single-walled bilayer vesicles, *Biochemistry,* 14, 4826–4830, 1975.
32. **Grant, C. W. M.,** Lateral phase separations and the cell membrane, in *Membrane Fluidity in Biology,* Vol. 2, Aloia, R. C., Ed., Academic Press, New York, 1983, 131–150.
33. **Silvius, J. R.,** Thermotropic phase transitions of pure lipids in model membranes and their modification by membrane proteins, in *Lipid-Protein Interactions,* Vol. 2, Jost, P. C. and Griffith, O. H., Eds., Wiley, New York, 1982, 239–281.
34. **Coolbear, K. P., Berde, C. B., and Keough, K. M. W.,** Gel to liquid crystalline phase transitions of aqueous dispersions of polyunsaturated mixed-acid phosphatidylcholines, *Biochemistry,* 22, 1466–1473, 1983.
35. **de Kruyff, B., van Dijck, P. W. M., Demel, R. A., Schuijff, A., Brants, F., and van Deenen, L. L. M.,** Non-random distribution of cholesterol in phosphatidylcholine bilayers, *Biochim. Biophys. Acta,* 356, 1–7, 1974.
36. **Harwood, J. L.,** Trans-bilayer lipid interactions, *Trends Biochem. Sci.,* 14, 2–4, 1989.
37. **Keough, K. M. W.,** Modifications of lipid structure and their influence on mesomorphism in model membranes: the influence of hydrocarbon chains, *Biochem. Cell Biol.* 64, 44–49, 1986.
38. **Small, D. M.,** *The Physical Chemistry of Lipids,* Vol. 4 of Handbook of Lipid Research, Plenum, New York, 1986, pp 672.
39. **Presti, F. T.,** The role of cholesterol in regulating membrane fluidity, in *Membrane Fluidity in Biology,* Vol. 4, Aloia, R. C. and Boggs, J. M., Eds., Academic Press, New York, 1985, 97–146.
40. **van Dijck, P. W. M., de Kruyff, B., van Deenen, L. L. M., de Gier, J., and Demel, R. A.,** The preference of cholesterol for phosphatidylcholine in mixed phosphatidylcholine-phosphatidylethanolamine bilayers, *Biochim. Biophys. Acta,* 455, 576–587, 1976.
41. **Subczynski, W. K., Hyde, J. S., and Kusumi, A.,** Effect of alkyl chain unsaturation and cholesterol intercalation on oxygen transport in membranes: a pulse ESR spin labeling study, *Biochemistry,* 30, 8578–8590, 1991.
42. **Tampé, R., von Lukas, A., and Galla, H.-J.,** Glycophorin-induced cholesterol-phospholipid domains in dimyristoylphosphatidylcholine bilayer vesicles, *Biochemistry,* 30, 4909–4916, 1991.
43. **Houslay, M. D. and Stanley, K. K.,** Dynamics of Biological Membranes, Wiley, New York, 1982, 92–151.
44. **Op den Kamp, J. A. F., Th. Kaverz, M., and van Deenen, L. L. M.,** Action of pancreatic phospholipase A_2 on phosphatidylcholine bilayers in different physical states, *Biochim. Biophys. Acta,* 406, 169–177, 1975.
45. **Wilschut, J. C., Regts, J., Westenberg, H., and Scherphof, G.,** Action of phospholipases A_2 on phosphatidylcholine bilayers. Effects of the phase transition, bilayer curvature and structural defects, *Biochim. Biophys. Acta,* 508, 185–196, 1978.
46. **Tsong, T. Y., Greenberg, M., and Kanehisa, M. I.,** Anesthetic action on membrane lipids, *Biochemistry,* 16, 3115–3121, 1977.
47. **Pownall, H. J., Massey, J. B., Kusserow, S. K., and Gotto, A. M., Jr.,** Kinetics of lipid-protein interactions: interactions of apolipoprotein A-I from human plasma high density lipoproteins with phosphatidylcholines, *Biochemistry,* 17, 1183–1187, 1978.
48. **Wickner, W. T.,** Role of hydrophobic forces in membrane protein asymmetry, *Biochemistry,* 16, 254–258, 1977.
49. **Baldassare, J. J., Brenckle, G. M., Hoffman, M., and Silbert, D. F.,** Modification of membrane lipid. Functional properties of membrane in relation to fatty acid structure, *J. Biol. Chem.,* 252, 8797–8803, 1977.

50. **Singer, S. J.,** On the fluidity and asymmetry of biological membranes, in *Perspectives in Membrane Biology,* Estrada-O., S., and Gitler, C., Eds., Academic Press, New York, 1974, 131–147.
51. **Rana, R. S. and Hokin, L. E.,** Role of phosphoinositides in transmembrane signaling, *Physiol. Rev.*, 70: 115–164, 1990.
52. **Merrill, A. H., Jr. and Stevens, V. L.,** Modulation of protein kinase C and diverse cell functions by sphingosine — a pharmacologically interesting compound linking sphingolipids and signal transduction, *Biochim. Biophys. Acta,* 1010, 131–139, 1989.
53. **Lee, M.-H. and Bell, R. M.,** Phospholipid functional groups involved in protein kinase C activation, phorbol ester binding, and binding to mixed micelles, *J. Biol. Chem.*, 264, 14797–14805, 1989.
54. **Maroney, A. C. and Macara, I. G.,** Phorbol ester-induced translocation of diacylglycerol kinase from the cytosol to the membrane in Swiss 3T3 fibroblasts, *J. Biol. Chem.*, 264, 2537–2544, 1989.
55. **Kolesnick, R. N. and Hemer, M. R.,** Physiologic 1,2-diacylglycerol levels induce protein kinase C-independent translocation of a regulatory enzyme, *J. Biol. Chem.*, 265, 10900–10904, 1990.
56. **Cascales, C., Mangiapane, E. H., and Brindley, D. N.,** Oleic acid promotes the activation and translocation of phosphatidate phosphohydrolase from cytosol to particulate fractions of isolated rat hepatocytes, *Biochem. J.*, 219, 911–916, 1984.
57. **Rodriguez-Boulan, E. and Nelson, W. J.,** Morphogenesis of the polarized epithelial cell phenotype, *Science,* 245, 718–725, 1989.
58. **Prescott, S. M., Zimmerman, G. A., and McIntyre, T. M.,** Platelet-activating factor, *J. Biol. Chem.*, 265, 17381–17384, 1990.
59. **Cossins, A. R. and Prosser, C. L.,** Evolutionary adaptation of membranes to temperature, *Proc. Natl. Acad. Sci., U.S.A.* 75, 2040–2043, 1978.
60. **Martin, C. E. and Foyt, D. C.,** Rotational relaxation of 1,6-diphenylhexatriene in membrane lipids of cells acclimated to high and low growth temperatures, *Biochemistry,* 17, 3587–3591, 1978.
61. **Dale, R. E., Chen, L. A., and Brand, L.,** Rotational relaxation of the "microviscosity" probe diphenylhexatriene in paraffin oil and egg lecithin vesicles, *J. Biol. Chem.*, 252, 7500–7510, 1977.
62. **Robertson, R. N. and Thompson, T. E.,** The function of phospholipid polar groups in membranes, *FEBS Lett.*, 76, 16–19, 1977.
63. **Quinn, P. J.,** Principles of membrane stability and phase behavior under extreme conditions, *J. Bioenerg. Biomemb.*, 21, 3–19, 1989.
64. **Pessin, J. E., Salter, D. W., and Glaser, M.,** Use of a fluorescent probe to compare the plasma membrane properties in normal and transformed cells. Evaluation of the interference by triacylglycerols and alkyldiacylglycerols, *Biochemistry,* 17, 1997–2004, 1978.
65. **Edelman, G. M.,** Surface modulation in cell recognition and cell growth, *Science,* 192, 218–226, 1976.
66. **Svennerholm, L.,** Diagnosis of sphingolipidoses with labeled natural substrates, in *Enzymes of Lipid Metabolism,* Gatt, S., Freysz, L., and Mandel, P., Eds., Plenum Press, New York, 1978, 689–706.

Chapter 2

FATTY ACID BIOSYNTHESIS

I. INTRODUCTION

It will be obvious from reading Chapter 1 that fatty acids are in most cases the principal determinants of a membrane's internal physical state. The regulation of fatty acid synthesis is therefore a topic of great interest to the membrane biochemist. In the following discussion I will attempt to summarize the current state of our knowledge in this field.

The existence of a sensitive and multifaceted control system for fatty acid synthesis has been recognized for many years. Much of the most elegant work has been conducted using mammalian liver, adipose tissue, and mammary gland and avian liver — all tissues whose lipid metabolism is especially well suited for the large scale synthesis of storage triglycerides. Even under normal nutritional conditions, the fluctuations in triglyceride production and storage within such cells are enormous. In contrast, the need of these same cells for fatty acids to be used in membrane synthesis is relatively modest. It seems most probable that much of our detailed information regarding the control of fatty acid synthesis is more applicable to the regulation of storage fat metabolism than it is to membrane metabolism.

Having stated this disclaimer, I shall nevertheless proceed to incorporate findings made with the abovementioned tissues into my general discussion. Bearing in mind the universality of biochemical reactions, I shall assume that even the regulatory activities most conspicuous in liver and adipose tissue have some pertinence for establishing the membrane lipid composition of these and other cells.

Wherever possible I shall compare properties of the mammalian and avian systems with those of higher plants and bacterial systems that do not form significant amounts of neutral glycerides. The latter organisms currently furnish our clearest opportunities for examining fatty acid synthesis as a factor controlling membrane metabolism exclusively.

II. THE PATHWAY

The pathway for saturated fatty acid biosynthesis is basically the same in all organisms examined to date. The universal substrate is acetyl-CoA. In eukaryotic organisms, the level of acetyl-CoA available for fatty acid synthesis is often controlled by the cytosolic enzyme ATP citrate lyase, more commonly known as the citrate cleavage enzyme. Citrate transported from the mitochondria to the cytoplasm is cleaved to acetyl-CoA and oxalacetate by the reaction shown in Figure 1. This reaction, along with changes in the rate of citrate transport across the mitochondrial membranes, can be of immense importance in controlling the production of fatty acids, as indicated on p. 29.

The first step in the fatty acid biosynthetic pathway proper is the conversion of acetyl-CoA to malonyl-CoA. The reaction, catalyzed by the enzyme acetyl-CoA carboxylase, proceeds as shown in Figure 2.

As the enzyme that commits acetyl-CoA to the production of a fatty acid or a related product, acetyl-CoA carboxylase is well situated to be the key regulatory enzyme in the pathway.[1] Since the demonstration in 1958 by Wakil[2] that carboxylation of acetyl-CoA is required for its activation, much has been learned concerning this enzyme.[3] It is a complex molecule containing three protein subunits. Details of the subunit organization and of the reaction, which is rate limiting for the entire pathway of fatty acid synthesis,[4] will be considered during the discussion of control (Section IV).

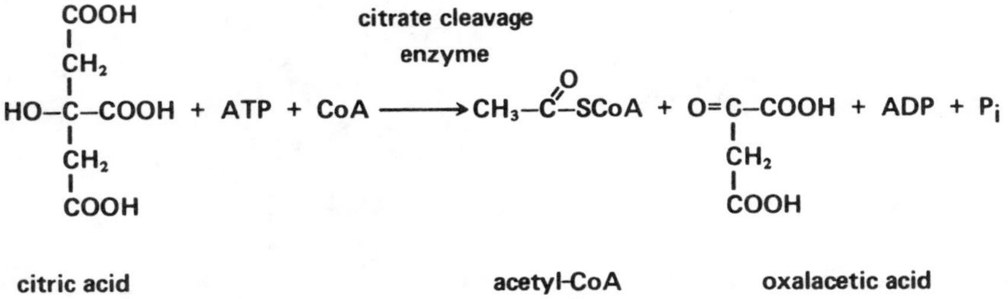

$$HOOC{-}CH_2{-}C(OH)(COOH){-}CH_2{-}COOH + ATP + CoA \xrightarrow{\text{citrate cleavage enzyme}} CH_3{-}C(=O){-}SCoA + O{=}C(COOH){-}CH_2{-}COOH + ADP + P_i$$

citric acid acetyl-CoA oxalacetic acid

FIGURE 1.

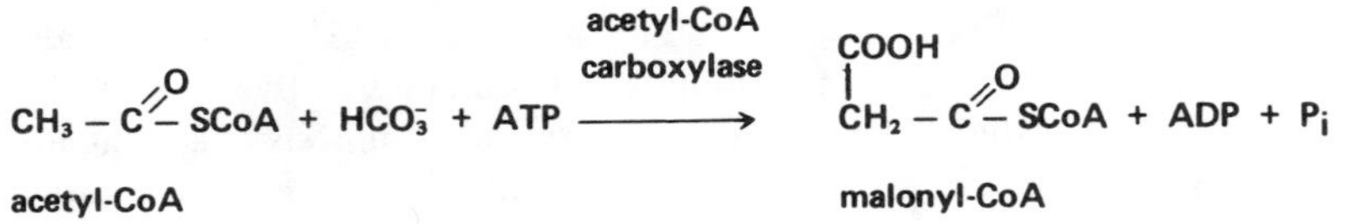

$$CH_3-C(=O)-SCoA + HCO_3^- + ATP \xrightarrow{\text{acetyl-CoA carboxylase}} HOOC{-}CH_2-C(=O)-SCoA + ADP + P_i$$

acetyl-CoA malonyl-CoA

FIGURE 2.

Malonyl-CoA is further utilized by a series of enzymes known collectively as the fatty acid synthase system. In most bacteria and in chloroplasts and other plastids of plants the fatty acid synthase components exist as a series of monofunctional polypeptides. However, in fungi the activities are grouped together on two nonidentical polypeptides, and in animals they are all integrated into a single multifunctional polypeptide. The individual reactions are shown in Figure 3.

From the initial step of the overall reaction, the substrate remains firmly bound to the acyl carrier protein (ACP). The unique properties of this small protein have been detailed by Prescott and Vagelos.[5] The acyl chain-binding end of the *E. coli* ACP is shown in Figure 4. The acyl group is bound to the terminal sulfhydral group of the 4′-phosphopantatheine residue.

Isolation of a discrete ACP from animal cells was attempted many times with rather equivocal results. Although small peptides containing 4′-phosphopantetheine were sometimes recovered from fatty acid synthase preparations, the existence of ACP as a discrete entity in animals was never demonstrated. We now know that ACP activity in animal cells is part of a large multifunctional protein possessing several partial activities in the overall fatty acid synthase reactions (see discussion below). After overcoming interpretation problems caused by proteolysis artifacts,[6] the yeast fatty acid synthase system was shown to be composed of only two polypeptides, each of which contains several activities.[7] Thus the α subunit, now known to have a mol wt of 207,863,[8] contains the 4′-phosphopantetheine-binding region, the ß-ketoacyl synthase and the ß-ketoacyl reductase functions, while the 220,007 mol wt subunit ß encompasses the acetyl transacylase, malonyl transacylase, ß-hydroxyacyl-ACP dehydratase, and enoyl reductase activities (Figure 5). Further structural details determined by Stoops et al.[11] indicated that the α and ß subunits must exist *in vivo* as an aggregate of $\alpha_6\beta_6$ (mol wt 2.4 $\times 10^6$) in order to have full enzymatic activity. Qureshi et al.[11] have likewise separated fatty acid synthase complexes from rat, human, and chicken liver as well as yeast into two nonidentical subunits of approximately equal molecular weight by affinity chromatography on Sepharose 6-aminocaproyl pantatheine. Subunits from the various sources were generally similar to each other and to those found by Schweizer et al.[9] except that the Qureshi et al. analysis placed both the enoyl reductase and ß-ketoacyl reductase on Schweizer's subunit α. Findings from pulse-labeling experiments indicate that the subunits of the intact synthase complex may be in equilibrium with a pool of free subunits.

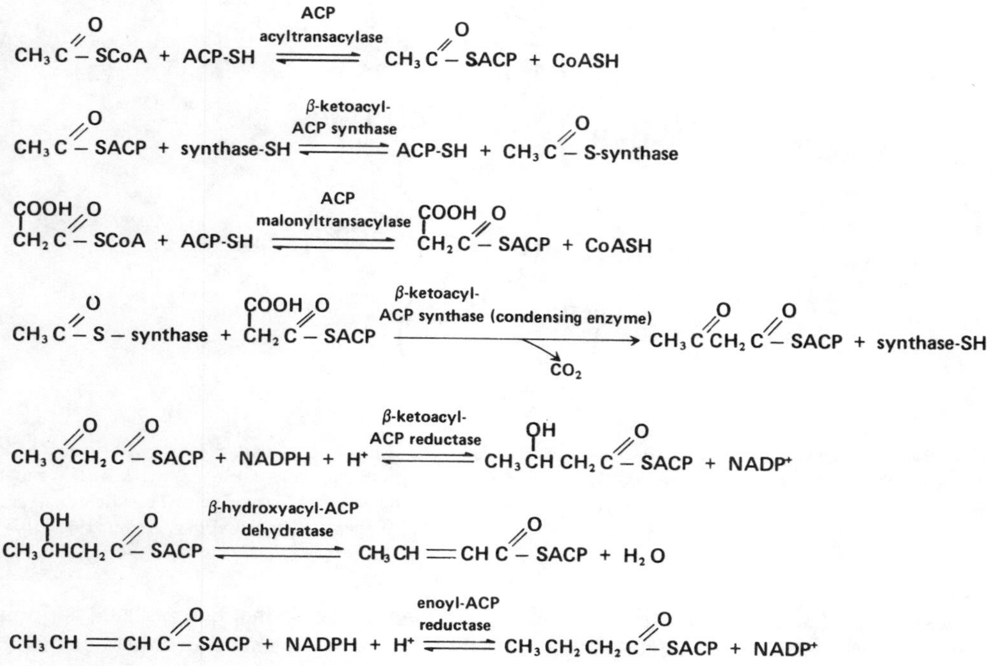

$$CH_3C(=O){-}SCoA + ACP{-}SH \xrightleftharpoons{\text{ACP acyltransacylase}} CH_3C(=O){-}SACP + CoASH$$

$$CH_3C(=O){-}SACP + \text{synthase-SH} \xrightleftharpoons{\beta\text{-ketoacyl-ACP synthase}} ACP{-}SH + CH_3C(=O){-}S\text{-synthase}$$

$$(COOH)CH_2C(=O){-}SCoA + ACP{-}SH \xrightleftharpoons{\text{ACP malonyltransacylase}} (COOH)CH_2C(=O){-}SACP + CoASH$$

$$CH_3C(=O){-}S{-}\text{synthase} + (COOH)CH_2C(=O){-}SACP \xrightarrow[\searrow CO_2]{\beta\text{-ketoacyl-ACP synthase (condensing enzyme)}} CH_3C(=O)CH_2C(=O){-}SACP + \text{synthase-SH}$$

$$CH_3C(=O)CH_2C(=O){-}SACP + NADPH + H^+ \xrightleftharpoons{\beta\text{-ketoacyl-ACP reductase}} CH_3CH(OH)CH_2C(=O){-}SACP + NADP^+$$

$$CH_3CH(OH)CH_2C(=O){-}SACP \xrightleftharpoons{\beta\text{-hydroxyacyl-ACP dehydratase}} CH_3CH{=}CHC(=O){-}SACP + H_2O$$

$$CH_3CH{=}CHC(=O){-}SACP + NADPH + H^+ \xrightleftharpoons{\text{enoyl-ACP reductase}} CH_3CH_2CH_2C(=O){-}SACP + NADP^+$$

Addition of C_2 units continues until C_{16} is reached; then acyl group is transferred to acceptor or cleaved to free fatty acid by palmitoyl-ACP thioesterase

Overall

$$\text{acetyl-CoA} + 7\ \text{malonyl-CoA} + 14\ NADPH + 14\ H^+ \longrightarrow CH_3(CH_2)_{14}COOH + 7\ CO_2 + 8\ CoA{-}SH + 14\ NADP + 6\ H_2O$$

FIGURE 3.

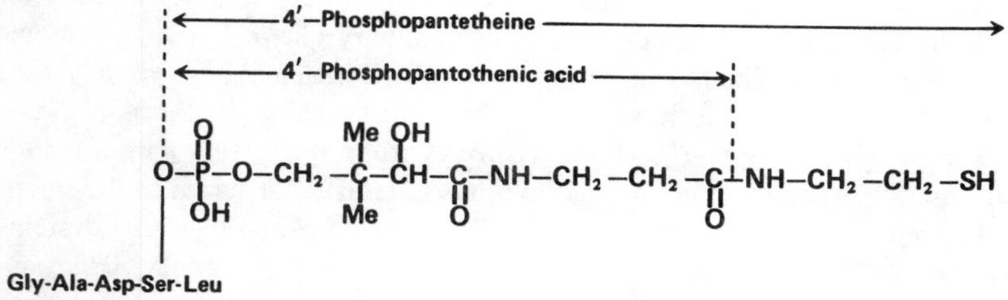

←— 4′–Phosphopantetheine —→

←— 4′–Phosphopantothenic acid —→

$$\text{Gly-Ala-Asp-Ser-Leu}{-}O{-}P(=O)(OH){-}O{-}CH_2{-}C(Me)_2{-}CH(OH){-}C(=O){-}NH{-}CH_2{-}CH_2{-}C(=O){-}NH{-}CH_2{-}CH_2{-}SH$$

FIGURE 4. Prosthetic group of the *Escherichia coli* acyl carrier protein.

Higher plants contain a fatty acid synthase of the bacterial type, with individual partial reactions catalyzed by separate proteins. These proteins appear to be localized entirely within the stroma compartment of plastids.[12] The product is almost exclusively palmitate, some of which is immediately elongated and desaturated to oleate.[13]

The existence of at least two different isoforms of plant ACP was recognized some years ago, but the significance of the observation was not readily apparent, especially since both isoforms are present in chloroplasts and are equally active in supporting fatty acid synthesis.

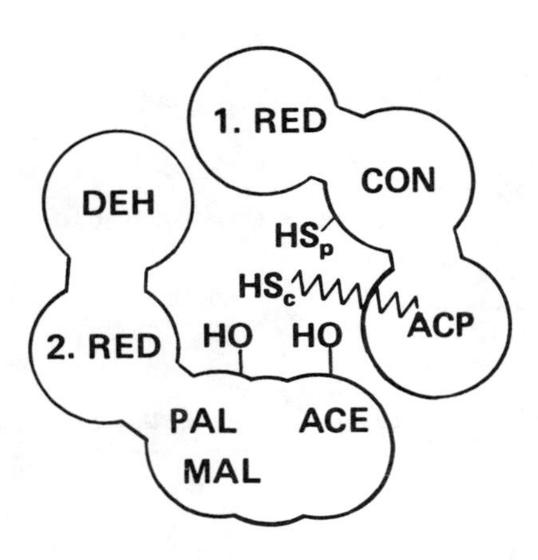

FIGURE 5. Active sites on the α and β subunits of yeast fatty acid synthase. α subunit: 1. RED = β-ketoacyl-ACP reductase; CON = condensing enzyme; ACP = acyl-carrier protein. β subunit: DEH = dehydratase; 2. RED = enoyl reductase, PAL, MAL, ACE = ACP-palmitoyl-, malonyl-, and acetyl-transacylase, respectively. (From Schweizer, E. and Lynen, F., *Transport by Proteins,* Sund, H. and Blaur, G., Eds., Walter de Gruyter-Verlag, Berlin, 1978, 103–118. With permission.)

A possible explanation comes from the findings of Guerra et al.[14] that one acylated isoform is the preferred substrate for acyl ACP:glycerol-l-phosphate acyltransferase while the other is the better substrate for acyl ACP thioesterase. If this *in vitro* specificity is operational *in vivo*, the ratio of the two isoforms could strongly influence the distribution of newly made fatty acids into glycerolipid synthesis within the plastid (via the acyltransferase) or the release of exportable free fatty acids (via the thioesterase).

Euglena gracilis contains three types of synthase complexes.[15] One is a multifunctional complex of the animal and yeast type and occurs in the cytoplasm. When *E. gracilis* is grown in light and contains chloroplast, a second type of synthase system is present in that organelle. It requires ACP and produces mainly palmitate and stearate, which it releases as the ACP derivatives. In most of its properties, it strongly resembles the *Escherichia coli* enzyme. A third synthase system, present in soluble form in both light and dark-grown cells, appears to function primarily as an elongation system lengthening C_{10}-C_{18} CoA derivatives.

The mechanism by which the fatty acid synthase system functions has been studied extensively in several systems, especially liver, mammary gland, *E. coli*, yeast, and several other microorganisms.[8,16] The mechanism proposed by Lynen and co-workers[17] for the yeast synthase is generally applicable to several of the other systems. Synthesis begins when the primer, an acetyl group, binds to a serine hydroxyl group in the first participating enzyme, ACP-acyl-transacylase, and a malonyl group binds to a different serine of the same enzyme. The acetyl group is transferred, first to the SH group of ACP and then to a cysteine residue on the ß-ketoacyl-ACP synthase. The malonyl moiety is transferred to the SH group of ACP, from whence it condenses with the acetyl primer, yielding acetoacetyl-ACP. This product then undergoes reduction, dehydration, and then another reduction as shown in Figure 3. After completing the initial cycle of reaction, the resulting C_4 acyl compound is transferred from ACP to the ß-ketoacyl-ACP synthase, thus freeing the ACP thiol group to accept the next malonyl-CoA. The cycle is repeated six or seven times until a fatty acyl chain of sixteen or eighteen carbon atoms is created.

Due to the close association of the two synthase homodimers, each acyl transfer from a pantetheine-SH to the cysteine-SH of ß-ketoacyl-ACP synthase involves moving the acyl chain from one subunit to the other (Figure 6). Thus two fatty acid synthesizing centers are present in one homodimer, with each requiring the participation of both subunits.

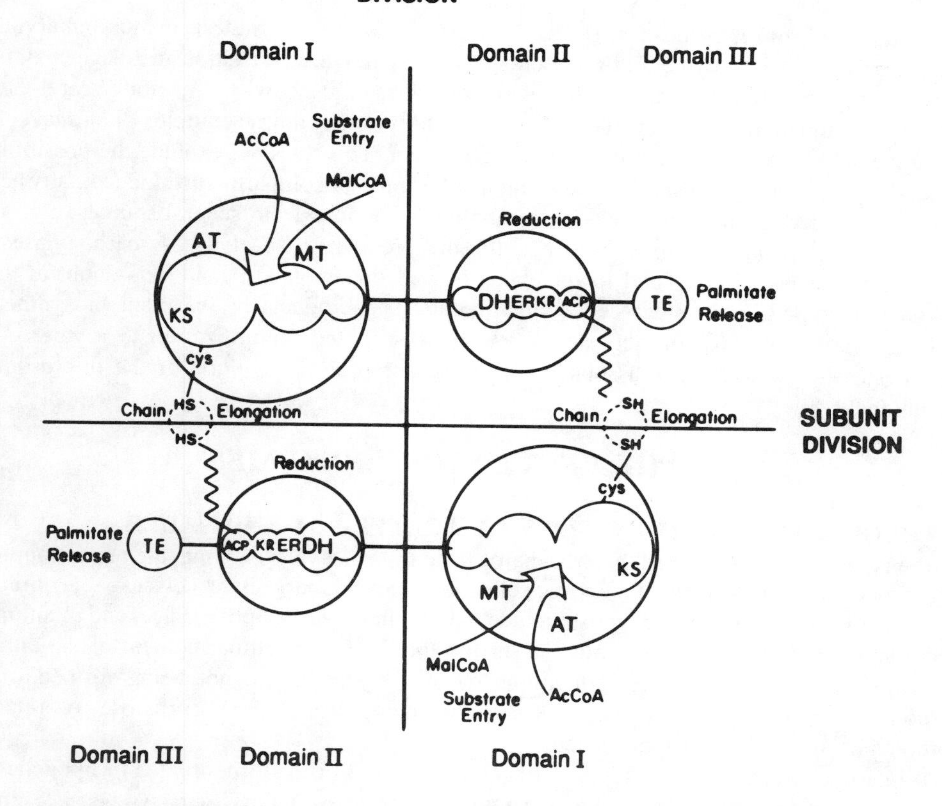

FIGURE 6. Functional map and organization of the subunit proteins of the chicken liver fatty acid synthase. The two subunits with their domains I, II, and III are drawn in an antiparallel arrangement (subunit division) so that two sites of palmitate synthesis are constructed (functional division). The abbreviations for the partial activities used are as follows: KS, β-ketoacyl synthase; AT, acetyl transacylase; MT, malonyl transacylase; DH, β-hydroxyacyl dehydratase; ER, enoyl reductase; KR, β-ketoacyl reductase; ACP, acyl carrier protein; TE, thioesterase. The wavy line represents the 4′-phosphopantetheine prosthetic group of ACP. (From Wakil, S. J., *Biochemistry,* 28, 4523–4530, 1989. With permission.)

Termination of the process is apparently controlled by the ß-ketoacyl-ACP synthase component. Palmitoyl-CoA was found to bind to the enzyme much less tightly than acetyl-CoA, and, in the *E. coli* systems, binding of acyl-ACP substrates decreased when the chain-length exceeded C_{10}. Palmitoyl-ACP, the principal product of *E. coli* fatty acid synthesis, showed no affinity for the enzyme.[18]

In yeast, the completed fatty acid is transferred from the fatty acid synthase complex to CoA. In *E. coli*, fatty acid synthesis is tightly coupled to the formation of phospholipids, and no acyl-CoA derivatives or free fatty acids exist under normal circumstances.[19,20] The animal fatty acid synthase releases a still different product, namely, free fatty acid. Release of the acyl chain from the ACP moiety of the synthase can be affected either by an integral thioesterase domain of the synthase complex, designated thioesterase I, or by separate monofunctional enzymes (thioesterase II), which in some tissues, such as mammary gland or uropygial gland, terminate fatty acid synthesis at medium chain length (C_8-C_{12}) rather than at the usual C_{16} product.

III. CELLULAR LOCATION

As indicated above, the primary source of carbon for fatty acid synthesis in most eukaryotes is the tricarboxylic acid cycle. Extramitochondrial enzymes, e.g., the citrate cleavage enzyme, convert products transported out of the mitochondria into acetyl-CoA. Both acetyl-CoA carboxylase and all the component enzymes of the fatty acid synthase complex of animal cells are universally found in the cytoplasmic compartment. This does not exclude the possibility of *de novo* fatty acid synthesis also occurring in other cellular compartments. Indeed, fatty acid synthase activity has been reported in membranous fractions from several sources, e.g., rat brain[21] and *Euglena gracilis*.[22] Evidence for the presence of acetyl-CoA carboxylase in chloroplasts of both higher and lower plants is well documented,[23] and it is doubtful that the enzyme functions in other cell compartments. The products of the plant biosynthetic pathway, fatty acyl-ACP or fatty acyl-CoA, are distributed for utilization to a variety of enzymes, most of which are associated with the chloroplast envelope or the microsomal membranes.

IV. THE REGULATORY ENZYMES

A. REGULATION OF THE SUPPLY OF ACETYL-CoA AND NADPH

It has long been recognized that activity of the citrate cleavage enzyme, the prime supplier of cytoplasmic acetyl-CoA in eukaryotic cells, fluctuates dramatically in cells capable of rapid lipogenesis. The tissues studied most include mammalian liver, adipose tissue, and mammary gland and certain lipid-accumulating yeasts and molds.[24] The cellular content of the citrate cleavage enzyme is low under conditions where fatty acids are not being formed from carbohydrate, but a rapid synthesis of the enzyme ensues if citrate, glucose, or related compounds are provided to the system.[24,25]

While maintaining a ready supply of acetyl-CoA is of critical importance in lipogenesis, it is not clear that citrate cleavage is a rate limiting step in fatty acid synthesis. Srere[24] interpreted a variety of data to indicate no direct cause and effect relationship between the activity of the citrate cleavage enzyme and the rate of fatty acid biogenesis. In fact, it seems possible that a high rate of fatty acid synthesis may instead cause the induction of more citrate cleavage enzyme by depleting the acetyl-CoA pool. The cases where fluctuations of the cleavage enzyme are greatest have involved cells that can be stimulated to form huge amounts of triglycerides. In any case, it is likely that the citrate cleavage enzyme is of greater significance in regulating the deposition of storage fats than it is in maintaining the more constant flow of fatty acids needed for membrane synthesis.

There have been scattered indications that other enzymatic steps, even prior to citrate cleavage, may play a regulatory role in fat synthesis. For example, citrate synthase is significantly inhibited by fatty acyl-CoA. While such inhibition has frequently been dismissed as being merely a nonspecific detergent effect, sensitive analyses employing spin-labeled fatty acyl-CoA analogues indicated site-specific binding of fatty acyl-CoA on the citrate synthase molecule.[26]

The supply of reducing power in the form of NADPH is also a potential rate limiting factor in fatty acid synthesis. The principal sources of cytoplasmic NADPH are the malic enzyme[27] and the dehydrogenases of the hexose monophosphate shunt. The cellular content of these enzymes also varies quite markedly in some animal tissues, depending upon the dietary or hormonal conditions.[25] However, as in the case of the citrate cleavage enzyme, the enhanced synthesis of NADPH-producing enzymes may well be needed only to permit the massive lipogenesis required for fat storage.

TABLE 1
Properties of the Acetyl-CoA Carboxylase Component Subunits

Component	Molecular weight	Partial reaction
Biotin carboxylase (BC)	98,000 (2 identical 51,000 subunits)	$HCO_3^- + ATP + CCP\text{-biotin} \overset{Me^{++}}{\rightleftharpoons} CCP\text{-biotin-}CO_2^- + ADP + P_i$
Carboxyl carrier protein (CCP-biotin)	45,000 (2 identical 22,500 subunits)	binding site for carboxyl group
Carboxyl transferase (CT) (transcarboxylase)	130,000 (2 identical 35,000 subunits and 2 identical 30,000 subunits)	$Acetyl\text{-CoA} + CCP\text{-biotin-}CO_2 \rightleftharpoons CCP\text{-biotin} + Malonyl\ CoA$

Acetyl-CoA, the direct substrate for fatty acid synthesis in the chloroplasts of plants, can be provided indirectly by pyruvate dehydrogenase in the cytoplasm, but the major supply is thought to be generated by glycolytic enzymes and pyruvate dehydrogenase located within the chloroplast.[28] The latter source is generally agreed to predominate in nongreen plant cells.[13]

B. REGULATION OF ACETYL-CoA CARBOXYLASE

There are two distinct types of regulations capable of affecting the formation of malonyl-CoA. One of these, known as short term or allosteric control, involves alterations in the activity of existing acetyl-CoA carboxylase molecules. A second type exerts its action over a longer term, usually by varying the absolute amount of the enzyme present in the cell. Details of acetyl-CoA carboxylase properties and regulation have been reviewed by Brownsey and Denton.[29,30] Since there are significant differences in prokaryotic and eukaryotic regulation of this enzyme, we shall consider the two classes of organisms separately.

1. Regulation of Acetyl-CoA Carboxylase in *Escherichia coli*

Appreciation of the allosteric type of regulation requires an understanding of the subunit structure of acetyl-CoA carboxylase. The nature of the subunits and the partial reactions that they catalyze was established principally by work carried out in the laboratories of Lane and Vagelos using the *E. coli* enzyme, which readily dissociates into three distinct protein subunits. Despite difficulties caused by proteolysis of the dissociated subunits during purification, it was possible to establish their molecular size and characterize the partial reactions catalyzed by each (Table 1).

It is envisioned that the biotinyl moiety of the CCP is capable of moving like a swinging arm from its binding site on BC to the corresponding site of CT (Figure 7). This movement would be possible due to the linkage of biotin to the CCP through the ε-amino group of lysine, thus providing a flexible side chain 14 Å in length separating the active C-2 position of biotin from the α-carbon of lysine.[32]

Regulation of bacterial acetyl-CoA carboxylase is quite different from that exerted in animal tissues (Section IV.B.2). Unlike the extensively studied animal acetyl-CoA carboxylases, the *E. coli* enzyme is neither activated by citrate nor inhibited by long chain fatty acyl-CoA derivatives. This difference can be rationalized by considering that the rate of bacterial fatty acid synthesis is closely coupled to cell growth only and does not function to facilitate fat production from carbon ingested by the cell in the form of carbohydrate, as is common in animal tissues.

A type of acetyl-CoA carboxylase regulation unique to bacteria was first detected in *E. coli*. This control involves the nucleotide derivatives guanosine 5′-diphosphate, 3′-diphosphate (ppGpp) and guanosine 5′-triphosphate, 3′-diphosphate (pppGpp). These two compounds are

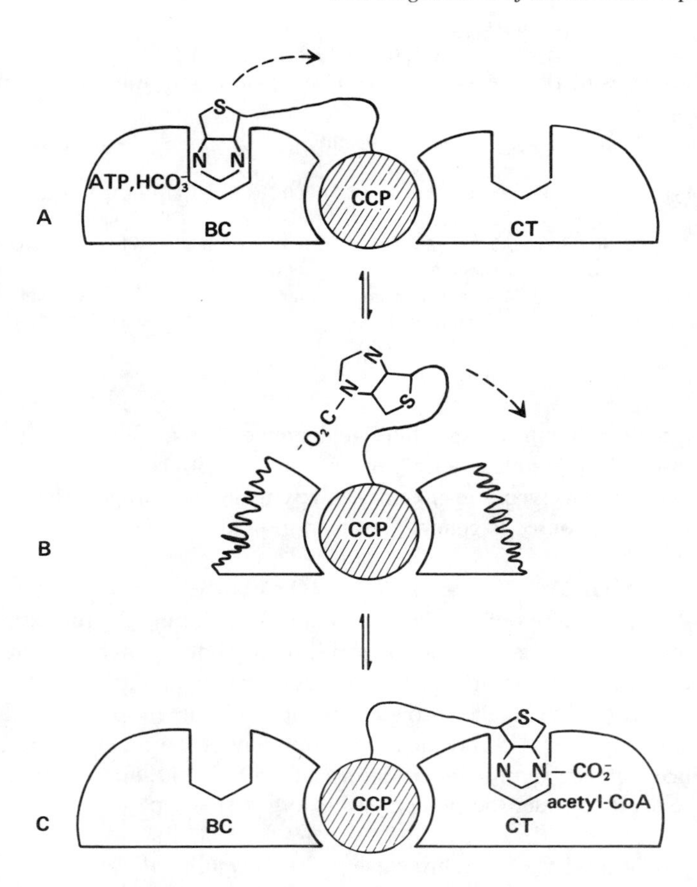

FIGURE 7. Model for intersubunit translocation of the carboxylated biotin prosthetic group of acetyl-CoA carboxylase. BC refers to biotin carboxylase, CCP to carboxyl carrier protein, and CT to carboxyl transferase. (From Guchhait, R. B., Moss, J., Sokolski, W., and Lane, M. D., *Proc. Natl. Acad. Sci. U.S.A.*, 68, 653-657, 1971. With permission.)

rapidly synthesized *in vivo* by bacterial ribosomes that have been prevented from forming proteins by the nonavailability of one or more required amino acids. A cessation of protein synthesis in some strains of *E. coli* can lead within 5 minutes to a rise in ppGpp concentration by 5- to 20-fold to a level as high as 4 m*M*.[33] It is now well established that ppGpp and pppGpp exert potent inhibitory effects on a variety of cellular activities, including RNA synthesis, carbohydrate synthesis, phospholipid synthesis, and fatty acid synthesis. In the latter case, stringent control of fatty acid synthesis is mediated by the action of ppGpp and pppGpp as negative effectors of acetyl-CoA carboxylase.[34] The inhibitory action is observed only on the carboxyl transferase partial reaction. The maximum extent of *in vitro* carboxylase activity inhibition is 50 to 60%. This occurs in the presence of a physiological level (2 m*M*) of ppGpp[35] and is quantitatively similar to the *in vivo* depression of fatty acid synthesis caused by amino acid deprivation.[34]

2. Regulation of Acetyl-CoA Carboxylase in Animals

a. Activation by Metabolites

It has been known since 1952 that fatty acid synthesis in animals is strongly activated by certain tricarboxylic acids.[36] The effect was subsequently shown in several laboratories to result from a marked increase in acetyl-CoA carboxylase activity. The most potent activators are citrate and isocitrate. Details of several years' intensive study of the activation phenomenon were drawn together by Lane et al.[3] Our current understanding of the process is summarized below, using the enzyme of chicken liver as an example.

In the absence of an activator, acetyl-CoA carboxylase exists in avian liver as a protomer having a molecular weight of about 500,000. Under denaturing conditions this protomer can be further dissociated into two apparently identical subunits. Chicken liver acetyl-CoA carboxylase has as its basic subunit a polypeptide of molecular weight 262,706, calculated from the nucleotide sequence of its cDNA.[37] Equivalent studies on rat liver acetyl-CoA carboxylase led to a deduced molecular weight of 265,200.[38]

Based on analyses of bound biotin, citrate, and acetyl-CoA, it appears that the chicken liver enzyme is roughly analogous in its organization to the *E. coli* carboxylase, having a single biotin prosthetic group, biotin carboxylating site, carboxyl transfer site, and (differing here from the bacterial enzyme) citrate activation site. In the absence of an activator such as citrate, the acetyl-CoA carboxylase exists entirely in its protomeric form, which is enzymatically inactive. The binding of citrate causes a conformational change leading to a shift in a protomer-polymer equilibrium toward the catalytically active polymer form. The average polymer consists of 20 protomeric units arranged as a helical, filamentous structure of high intrinsic viscosity (Figure 8). Considering estimates of the content of acetyl-CoA carboxylase in avian liver, a single cell has been calculated to contain 50,000 such filaments.

The isolated rat liver enzyme polymerizes in the presence of citrate to give a polymer of approximately 40 subunits. During isolation without rapid chilling it undergoes dephosphorylation and depolymerization to a less unstable octomer of molecular weight 2×10^6.[39] The dephosphorylated form of acetyl-CoA carboxylase does not require citrate for activity.

Much effort has been devoted to understanding the activating effect of citrate. Citrate appears to increase the reaction rate of bound substrates, i.e., it increases V_{max}, but has little effect on the enzyme's affinity for substrates (the K_m for neither ATP Mg^{+2}, HCO_3^-, nor acetyl-CoA are altered appreciably).[40] Citrate apparently induces a conformational change near the enzyme's active site, thus shielding the biotinyl prosthetic group. In fact, citrate and acetyl-CoA act synergistically to protect enzyme-bound biotin from inactivation by avidin, the protein from egg white known to bind free biotin almost irreversibly.[41] In the absence of citrate, the enzyme is sensitive to avidin.

It has generally been concluded that the binding of citrate triggers a conformational change in acetyl-CoA carboxylase that renders the biotin carboxylase and carboxyl transferase active sites more readily accessible to the biotinyl prosthetic group, and promotes the association of protomers into the characteristic filamentous polymer. While the metabolic consequences of the first action are obvious, those of the second are not. Lane and Moss[42] have speculated that the carboxylase filaments may act as an organizing matrix for other enzymes involved in lipogenesis.

b. Inhibition by Metabolites

The product of the acetyl-CoA carboxylase reaction, malonyl-CoA, inhibits the reaction competitively with respect to both the acetyl-CoA substrate and the citrate activator.[43] The K_i is approximately $10^{-5}M$. The inhibitory effect of malonyl-CoA toward citrate involves promoting a depolymerization of the citrate-stabilized active form of the enzyme.

Long-chain fatty acyl-CoA derivatives are also excellent inhibitors of the enzyme.[44] Salts of the free fatty acids are themselves not inhibitory at comparable levels. It is thought that a number of physiological states that depress liver fatty acid synthesis, e.g., fasting, ingestion of high fat diets, and diabetes, do so mainly through interaction of the resulting increased level of fatty acyl-CoAs with acetyl-CoA carboxylase.

Some investigators have questioned the *in vivo* significance of this interrelationship, since

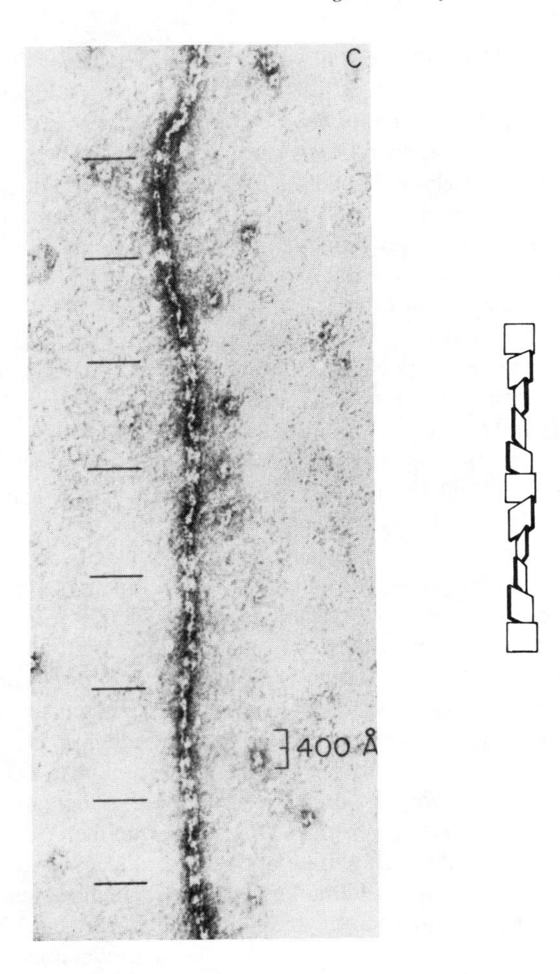

FIGURE 8. The filamentous form of avian liver acetyl-CoA carboxylase in the presence of citrate, as seen using the electron microscope. (From Lane, M. D., Moss, J., and Polakis, S. E., *Curr. Top. Cell Regul.*, 8, 139–195, 1974. With permission.)

the detergent properties of fatty acyl-CoAs can cause the inactivation of a variety of nonrelated enzyme preparations. However, there is good evidence supporting a physiological role for the CoA derivatives in regulating fatty acid biosynthesis. Goodridge[45] showed that palmitoyl-CoA is much more effective in inhibiting acetyl-CoA carboxylase than it is in affecting a number of other enzymes, including the lipogenic enzyme fatty acid synthetase, citrate cleavage enzyme, malic enzyme, and NADP-linked isocitrate dehydrogenase. Ogiwara et al.[46] were able to determine that complete inhibition of rat liver acetyl-CoA carboxylase followed the binding of one mole of palmitoyl-CoA per mole of enzyme. Citrate could displace the palmitoyl-CoA, restoring full activity.

c. *Regulation of Activity by Hormones*

A variety of experiments, reviewed by Volpe and Vagelos,[47] suggested that acetyl-CoA carboxylase activity can be controlled by hormone-induced changes in cyclic nucleotide concentrations. This phenomenon has been studied mainly with respect to the effects of glucagon, epinephrine, and insulin on lipogenesis in liver and adipose tissue. Evidence for an

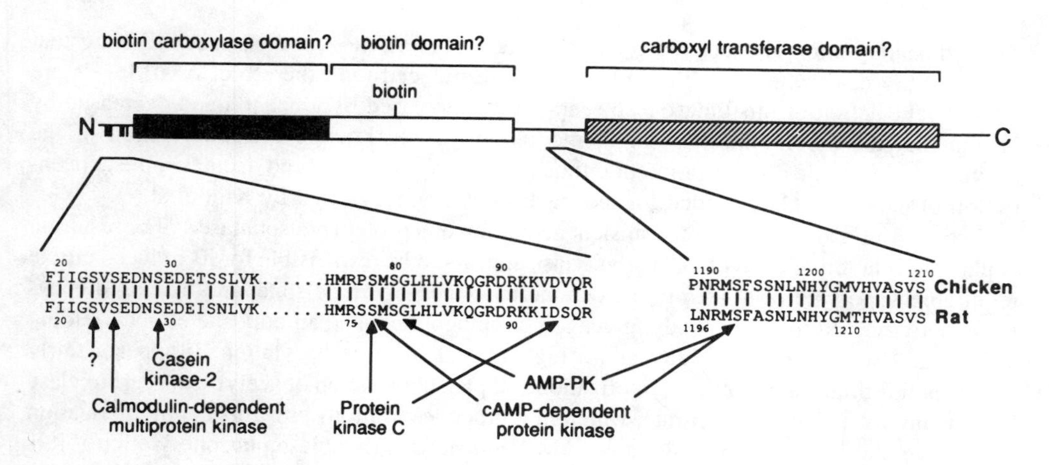

FIGURE 9. Proposed location of phosphorylation sites in acetyl-CoA carboxylase. Location of catalytic domains is tentative and is based on homologies with other proteins. (From Hardie, D. G., Carling, D., and Sim, A. T. R., *Trends Biochem. Sci.*, 14, 20–23, 1989. With permission.)

important, if indirect effect of glucocorticoids on lipogenesis has recently been reviewed[48] and will not be discussed here.

Glucagon causes a rapid drop in hepatic lipogenesis. Of the several potential regulatory enzymes tested, only acetyl-CoA carboxylase decreased in activity during this period. The same reduction in carboxylase activity could be produced by dibutyryl cAMP. These experiments implied, but did not prove, that a hormone-stimulated protein kinase can inactivate acetyl-CoA carboxylase by phosphorylating one or more subunits.

Further evidence for this mechanism of hormone inactivation of lipogenesis came from the laboratory of Kim. A protein kinase present in a partially purified rat liver acetyl-CoA carboxylase preparation catalyzed the phosphorylation of one subunit of the carboxylase, as determined by SDS-polyacrylamide gel electrophoresis.[49] Enzymatic activity was lost at a rate corresponding to the phosphorylation of this subunit, and activity could be restored by removing bound phosphate with hen oviduct phosphoprotein phosphatase.

Recent studies have implicated a variety of protein kinases and phosphatases in the regulation of acetyl-CoA carboxylase. Phosphorylation sites for some of the kinases are indicated in Figure 9. Identification of the physiologically significant sites has been rendered difficult by the action of kinases and phosphatases during carboxylase purification. These artifacts can be largely prevented by rapidly "freeze-clamping" the source tissue before enzyme purification begins.[39] Witters et al.[50] showed that glucagon causes a specific phosphorylation of rat hepatocyte acetyl-CoA carboxylase, as isolated by immunological techniques. However, inactivation of carboxylase through glucagon-induced phosphorylation of the enzyme was not indicated in a study by Watkins et al.[51] Instead, the hormone appeared to act at an undetermined site to depress the cellular citrate level, thus leading indirectly to carboxylase disaggregation. This apparent lack of effect may be explained by the finding that there are two different sites of phosphorylation on the enzyme, only one of which results in its inactivation (see below).

A slight loss of enzyme activity follows the phosphorylation of ser 77 plus ser 1200 (catalyzed by cAMP-dependent protein kinase), and a much more pronounced inactivation follows the phosphorylation of ser 79 and ser 1200. Dephosphorylation or proteolytic removal of the amino terminal domain containing ser 77/79 reverses the inhibition, implicating phosphorylation of ser 79 as the major reaction inhibiting acetyl-CoA carboxylase activity.[52]

This phosphorylation is catalyzed by a newly described AMP-activated protein kinase that reduces fatty acid synthesis when ATP levels are reduced (and therefore, AMP levels are high).[30] The action of this kinase also seems to be increased by glucagon and decreased by insulin, but the precise mechanism of modulation is not known. Phosphorylation of the enzyme increases the concentration of citrate needed for activation and decreases the concentration of palmitoyl-CoA needed for feedback inhibition (reviewed by Kim et al.[53]).

Type I insulin also triggers the transient activation of a protein phosphatase.[54] The resulting dephosphorylation of acetyl-CoA carboxylase appears to be responsible for its reduced citrate requirement, polymerization from the octameric to the polymeric state, and increased enzymatic activity.[52] In the same study, glucagon or epinephrene had an opposite effect, yielding carboxylase with a lower activity and a higher level of phosphorylation. It appears fairly certain at this time that both depolymerization and phosphorylation of acetyl-CoA carboxylase accompany its inactivation. Citrate, which in high concentrations prevents depolymerization (see above), also appears to inhibit any cAMP-enhanced carboxylase phosphorylation if it is present at a level as high as 10 to 20 m*M*.[56] Therefore, regulation of acetyl-CoA carboxylase through hormone-induced cAMP production may be restricted to conditions favoring a reduced citrate concentration.

d. Regulation of the Quantity of Enzyme

Activation and inhibition of existing acetyl-CoA carboxylase molecules by the mechanisms outlined above permit cells to respond quickly to sudden changes in various physiological parameters. In addition to this short term response, many tissues can alter the rate of fatty acid synthesis by increasing or decreasing the actual amount of acetyl-CoA carboxylase that they contain. This phenomenon, sometimes called the long-term response, has been studied extensively, particularly in liver.[25,57]

The laboratory of Numa has provided a good insight into the regulatory process. Nishikori et al.[58] compared the role played by increased efficiency of existing acetyl-CoA carboxylase with that of changing enzyme content in rats fasted and then refed a fat-free diet. As illustrated in Figure 10, the rate of acetate incorporation into fatty acids began to increase within 2 hr after feeding. The continuation of this increase throughout the initial 8-hr period can be accounted for by a concurrent drop in the level of inhibitory fatty acyl-CoA derivatives and a sharp rise in the concentration of the activator, citrate. The continued rise in the rate of fatty acid synthesis over the ensuing 40-hr time span, when further activation of preexisting enzyme molecules was unlikely, roughly paralleled the increase in the tissue levels of acetyl-CoA carboxylase measured immunochemically.

Numerous other studies have established firmly this capacity for hepatic acetyl-CoA carboxylase to change in quantity, not only during fasting and refeeding, but also as a result of alloxan diabetes and obesity. The differentiation between altered efficiency and altered amount is best made by treatment of the tissue preparation with antibodies made against purified acetyl-CoA carboxylase.[59,60] This technique is valid for such applications as that shown in Figure 9 because both the active polymeric form of the enzyme and the inactive protomeric form are detected.[60] Carrying the technique a step further, Nakanishi et al.[61] bound ^{125}I-labeled antiacetyl-CoA carboxylase to liver polysomes of rats subjected to various dietary conditions, showing that the relative content of these specific polysomes correlated well with changes in carboxylase activity. More recent studies[62] have confirmed that fluctuating amounts of acetyl-CoA carboxylase in several rat tissues are closely correlated with the abundance of the carboxylase m-RNA.

The high content of acetyl-CoA carboxylase in livers of carbohydrate-fed rats is sharply reduced if the animals are fasted. This results in part from a fall in the quantity of the enzyme. Under these conditions, degradation of acetyl-CoA carboxylase by lysosomal enzymes was

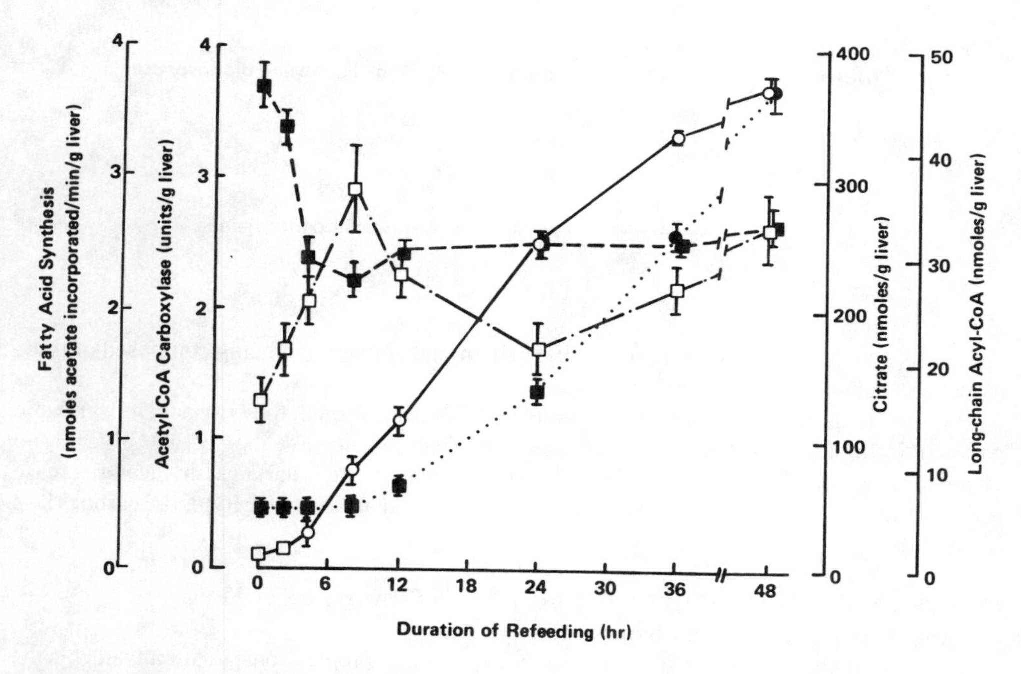

FIGURE 10. Effects of fat-free refeeding on fatty acid synthesis in rat liver slices (○——○) and on the levels of hepatic acetyl-CoA carboxylase (●- - -●), citrate (□- - -□), and long-chain acyl-CoA thioesters (■- - -■). Results are given as means ± SD per gram of wet tissue. For each point, 6 to 8 rats were used. (From Nishikori, K., Iritan, N., and Numa, S., *FEBS Lett.*, 32, 19–21, 1973. With permission.)

shown to be promoted by palmitoyl-CoA, an inhibitor of the enzyme, and diminished by the activator citrate.[63]

It should be emphasized that the fluctuations in acetyl-CoA carboxylase level are by no means unique. The hepatic levels of several other enzymes participating in lipogenesis, such as the malic enzyme, the citrate cleavage enzyme (see Section IV.A), and fatty acid synthetase, are altered in much the same way as lipogenesis waxes and wanes.[64-67] On the other hand, there is less fluctuation in the quantities of such enzymes, including acetyl-CoA carboxylase,[68] in rat lung.

3. Regulation of Acetyl-CoA Carboxylase in Yeast and Higher Plants

The *Saccharomyces cerevisiae* acetyl-CoA carboxylase is similar in some respects to the animal enzyme. It has a comparable specific activity and is sensitive to avidin, yet its subunit structure is apparently different. It exists as a tetramer of mol wt approximately 600,000.[69] The presence of citrate does not induce polymerization, although tricarboxylic acids as well as a number of other common metabolic intermediates seem to activate the enzyme.[70] Citrate-unresponsive forms have also been isolated from yeast.[71]

Long-chain fatty acids reduce the acetyl-CoA carboxylase activity in *S. cerevisiae*, and immunological studies indicated that this is achieved by lowering the absolute amount of the enzyme.[72] Mutants defective in acyl-CoA synthase showed little reduction of acetyl-CoA carboxylase, suggesting that repression of the carboxylase is mediated by long-chain acyl-CoA or a metabolic product thereof.[73]

There has been surprisingly little consensus regarding the molecular organization of acetyl-CoA carboxylase in higher plants. Purification of acetyl-CoA carboxylase from wheat germ or parsley cells in the presence of the proteinase inhibitor phenylmethylsulfonyl fluoride produced polypeptides of 240 and 210 KDa, respectively.[74] It now seems likely that the

$$\text{CoASH} + \text{apo-ACP} \xrightarrow{Mg^{+2}} \text{holo-ACP} + 3',5'\text{-adenosine diphosphate}$$

FIGURE 11.

$$\text{holo-ACP} \xrightarrow{Mn^{+2}} \text{apo-ACP} + 4'\text{-phosphopantetheine}$$

FIGURE 12.

enzyme is a multifunctional protein like that in mammals rather than being composed of three associated subunits as in bacteria.[13]

Post-Beittenmiller et al.[75] have searched for sites of regulation in the fatty acid biosynthetic pathway in higher plants. Based mainly upon the fivefold increase in acetyl-ACP (with no attendant rise in malonyl-ACP) in spinach leaves experiencing a darkness-induced decrease in fatty acid synthesis, it was concluded that the reaction catalyzed by acetyl-CoA carboxylase is rate-limiting.

C. REGULATION OF THE FATTY ACID SYNTHASE SYSTEM

1. Regulation of Fatty Acid Synthase in Bacteria

The enzymatic activity of *E. coli* fatty acid synthase preparations is stimulated by the presence of phosphate compounds, such as glucose-1-phosphate, glucose-6-phosphate, α-glycerophosphate, pyrophosphate, orthophosphate, and most active, fructose-1,6-diphosphate.[76] This effect can help to explain the enhanced rate of lipogenesis observed following carbohydrate uptake. The mechanism of activity stimulation in *E. coli* has not been clarified, but studies with animal systems (see Section 2) suggest that fructose diphosphate competes for a regulatory site or binds at an allosteric site, bringing about a conformational change in the enzyme's structure. There is currently no strong evidence that the allosteric regulation of bacterial fatty acid synthase activity is physiologically important in effecting short-term rate changes in fatty acid synthesis.

Interestingly, there is a very rapid metabolic turnover of the 4′-phosphopantetheine prosthetic groups of *E. coli* fatty acid synthase.[77] An enzyme designated holo-ACP synthase catalyzes the reaction in Figure 11. The functional ACP can in turn be hydrolyzed in a reaction catalyzed by ACP-hydrolase (Figure 12). Once removed, the 4′-phosphopantetheine can be reused for the synthesis of CoA. Whereas the protein moiety of ACP is metabolically stable in growing cells, the 4′-phosphopantetheine prosthetic group is exchanged at approximately four times the rate of net ACP increase. Although this rapid turnover has been implicated in changes in fatty acid synthase activity in animals (Section IV.C.2), its physiological significance in *E. coli* is unclear. The rate of prosthetic group turnover appears not to vary during the life cycle of *E. coli*,[78] but no studies have been reported on turnover in different nutritional states. However, it is known that pantothenate-starved mutant strains of the organism deplete their CoA pool in order to maintain ACP in the holo form.[77]

Certain novel regulatory mechanisms have been identified in the more evolutionarily advanced prokaryote, *Mycobacterium smegmatis* (formerly *M. phlei*).[79] A fatty acid synthase complex found in this organism resembles the multifunctional complex of yeast in having a native mol wt of approximately 2×10^6. However, the *Mycobacterium* complex appears to be composed of six to eight identical subunits of 290,000 mol wt,[80] in contrast to the yeast composition of nonidentical 180,000 and 185,000 mol wt subunits. The enzyme produces a variety of saturated fatty acyl-CoA derivatives, ranging from C_{14} to C_{26} and having a bimodal product pattern of chain lengths peaking quantitatively at C_{16} and C_{24}. An unusual property of the synthase is its pronounced activation by certain polysaccharides found in *M. smegmatis* extracts.[81] The polysaccharides act to lower the K_m for acetyl-CoA by ninefold and for malonyl-CoA by fourfold. The probable *in vivo* regulatory mechanism was set forth by Vance and Bloch.[82] Fatty acid synthesis is promoted by the capacity of the polysaccharides to form a ternary complex with the enzyme-bound C_{24}-CoA product, facilitating release of this long chain compound. In the absence of the polysaccharides, the C_{24}-CoA is retained at the enzyme's termination site, causing a marked reduction in further synthesis. The C_{16}-CoA product, on the other hand, is released easily without aid from the polysaccharides. Formation of the shorter chain (especially C_{16}) fatty acids is favored by a high ratio of acetyl-CoA to malonyl-CoA.[83]

In addition to the above fatty acid synthase, *M. smegmatis* contains a second synthase which is smaller (mol wt ~ 250,000) and is dependent upon ACP. The contrasting properties of the two systems, e.g., the first is sensitive to palmitoyl-CoA inhibition while the second is not, led Odriozola et al.[84] to propose that they complement each other in providing the cellular lipid requirements.

Corynebacterium diphtheriae, an organism about equally advanced as *M. smegmatis*, also has a structurally similar fatty acid synthase.[85,86] It will be interesting to determine at what evolutionary stage the transformation from the simple *E. coli* type of fatty acid synthase to the more complex mammalian type took place.

2. Regulation of Fatty Acid Synthase in Animals

a. Effects of Metabolites

Evidence suggesting allosteric control of fatty acid synthase by fructose-1,6-diphosphate and other phosphorylated sugars has been reported.[76] These sugar derivatives have the effect of overcoming the inhibition that malonyl-CoA exerts on synthases of some tissues. They also decrease the K_m of the enzyme for NADPH. Although the effective levels of the sugar phosphates seem unphysiologically high, they may not be above the concentrations found in certain cellular compartments. The physiological significance of this activation is confused by the failure of other workers[87,88] to confirm the stimulatory effect of hexose diphosphates on purified fatty acid synthases. Further study will be required in order to assess the role of these potential allosteric effectors.

It has been known for some years that fatty acid synthase activity can be inhibited by the product of the enzyme, fatty acyl-CoA. Evidence favoring and opposing a physiological role for product inhibition has been reviewed by Volpe and Vagelos.[47] The general consensus is that while fatty acyl-CoA derivatives do have an inhibitory effect on fatty acid synthase, it is a nonspecific one having little impact in the short-term regulation of fatty acid synthesis. The primary short-term regulatory effect of fatty acyl-CoA is exerted on acetyl-CoA carboxylase (see p. 27).

The turnover of ACP, the prosthetic group of the fatty acid synthase complex, independently of the polypeptide chain as described on page 34 for *E. coli*, also occurs in mammalian tissues. Its rate is accelerated in animals on a fat-free diet and reduced during starvation.[89,90] The factors producing this change in 4′-phosphopantetheine exchange are not known.

The biosynthesis of fatty acid synthase is very markedly affected by an animal's nutritional state. After 1 to 2 days of fasting, the rate of fatty acid synthase formation in rats decreased fourfold, and refeeding fasted rats a high carbohydrate diet stimulated the production of fatty acid synthase by 20- to 25-fold.[91]

The low fatty acid synthase activity in fasted animals could be due to action by an ACP-hydrolase. Roncari[92] demonstrated that a crude extract of fasted rat liver could catalyze the cleavage of ^{14}C-phosphopantetheine from purified fatty acid synthase, whereas extracts of livers from fed rats could not.

TABLE 2
Relative Rates of Synthesis of the Fatty Acid Synthetase in the Livers of Rats in Different Nutritional States[a]

Nutritional state	Mean weight (g)	L-[U-14 C]leucine incorporation: A Fatty acid synthetase (dpm/liver)	B Total soluble protein (dpm/liver)	A/B × 10^5 Relative rates of synthesis
Normal	257	765	3.5×10^5	218
Fasted for 48 hr and subsequently refed a fat-free diet for 72 hr	222	15,900	5.1×10^5	3120
Fasted for 48 hr	204	332	8.3×10^5	40

[a] Rats treated as indicated above were given a single intraperitoneal injection of L-[U-^{14}C]leucine (10 µCi per rat in animals fed a normal diet or refed a fat-free diet and 20 µCi per rat in the fasting group). Four hours later, two rats from each group were killed and the livers pooled. Radioactivity incorporated into the fatty acid synthetase was determined by immunoprecipitation of the DEAE-cellulose-purified fatty acid synthetase.

From Craig, M. C., Nepokroeff, C. M., Lakshmanan, M. R., and Porter, J. W., *Arch. Biochem. Biophys.*, 152, 619–630, 1972. With permission.

Although it is generally agreed that any short-term changes in fatty acid synthase activity are overshadowed by those of acetyl-CoA carboxylase activity, there can be very significant synthase variations over time periods of several hours to days as the result of environmental or genetic factors. The most striking of these is the modification of synthase activity in response to altered nutritional state. Many studies (reviewed by Volpe and Vagelos[47] and by Numa and Yamashita[25]) have confirmed a 10 to 50-fold increase in the specific activity of the hepatic enzyme system following refeeding 48-hr fasted rats with a high carbohydrate, fat-free diet. This rise is apparently due entirely to an increase in the fatty acid synthase content resulting from greatly enhanced formation of the enzyme (Table 2). While the detailed mechanism governing this stimulation is not understood, synthesis appears to be promoted by an intermediate of carbohydrate metabolism at the triose phosphate level or beyond. Induction of synthesis is presumably at the transcriptional level, as judged by the parallel rises in synthase production and cellular content of fatty acid synthase m-RNA.[93]

Under these conditions, the fatty acid synthase activity is clearly rate limiting in lipogenesis, as evidenced by the accumulation of malonyl-CoA.[94] In fasting animals, the synthase content drops due to enhanced degradation[95,96] as well as retarded synthesis. Inhibition of the enzyme by rising levels of polyunsaturated fatty acids spurs degradation of the synthase.[47] Linoleate and possibly other polyunsaturates, but not saturated fatty acids, also sharply decrease fatty acid synthase synthesis in rat liver.

It is probable that these nutritional changes in higher animals are limited to a few specialized tissues. Alterations similar to those described above for liver have been observed in adipose tissue[97] and less noticeably, in intestinal mucosa[98] and lung,[69] but not in brain.[96,99]

b. Effects of Hormones

A variety of evidence suggests that rapid fluctuations in the level of fatty acid synthase are controlled by the action of insulin and catabolic hormones such as glucagon. Paulauskis and Sul[100] showed that insulin administration to diabetic mice increased the fatty acid synthase

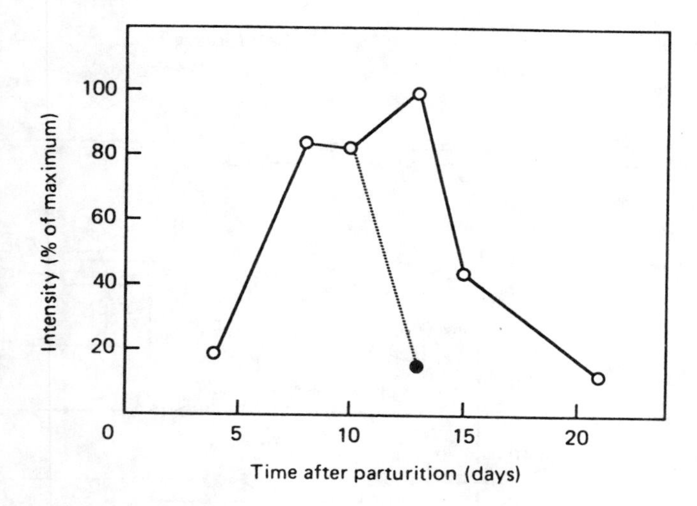

FIGURE 13. Fatty acid synthase mRNA levels during lactation and weaning in rat mammary gland. The continuous line (○) shows the variation in synthase mRNA levels when pups were kept with the lactating females throughout lactation. The dotted line (●) shows the effect of removing the pups at 10 days. (From Braddock, M. and Hardie, D. G., *Biochem. J.*, 249, 603–607, 1988. With permission.)

mRNA level in liver by 4-fold within 1 hour and by a maximum of 19-fold within 6 hours. These workers also demonstrated that dibutyryl-cAMP given to previously fasted normal mice in conjunction with refeeding of a high carbohydrate meal prevented increased transcription of the fatty acid synthase gene that usually follows refeeding.

Similar dramatic changes in fatty acid synthase m-RNA were observed in the mammary gland of lactating rats. The m-RNA levels remained elevated as long as the pups were suckling (Figure 13).[101] The precise mechanism by which gene transcription is regulated under these circumstances remains to be determined.

3. Regulation of Fatty Acid Synthases in Plants

This section might appropriately begin by considering fatty acid synthesis in *Euglena gracilis*, a phytoflagellate which, depending upon its growth conditions, may possess either the properties of a plant or an animal cell. When grown in the dark, *Euglena* resembles an animal cell, both in appearance and in many biochemical characteristics, including the presence of a multifunctional fatty acid synthase system of the animal and yeast type.[102] Exposure of the cells to light results in the development of typical chloroplasts, which contain a second fatty acid synthase system. This latter system appears to be similar to the *Escherichia coli* enzyme in being relatively small and having an absolute requirement for ACP.

Fatty acid synthesis has also been studied in a number of higher plant tissues. A system found in the cytoplasm of developing soybean cotyledons was shown to convert acetyl-ACP and malonyl-ACP to palmitoyl-ACP.[103] Stearoyl-ACP was also produced, but apparently via the action of a separate elongation system. There is as yet little evidence that ACP is covalently bound to a multifunctional fatty acid synthase system of the mammalian or yeast type.[104]

An independent fatty acid synthesizing system is found in the chloroplasts of higher plants. As described in Section IV.A, the substrate for this pathway is drawn directly from the pool of carbon compounds fixed by the photosynthetic apparatus. There is a distinctive acetyl-CoA carboxylase present (Section IV.B.3). The interactions thought to occur between the chloroplast system and the cytosol system are summarized in Figure 14.

A number of studies have shown active regulation of fatty acid synthesis in developing and

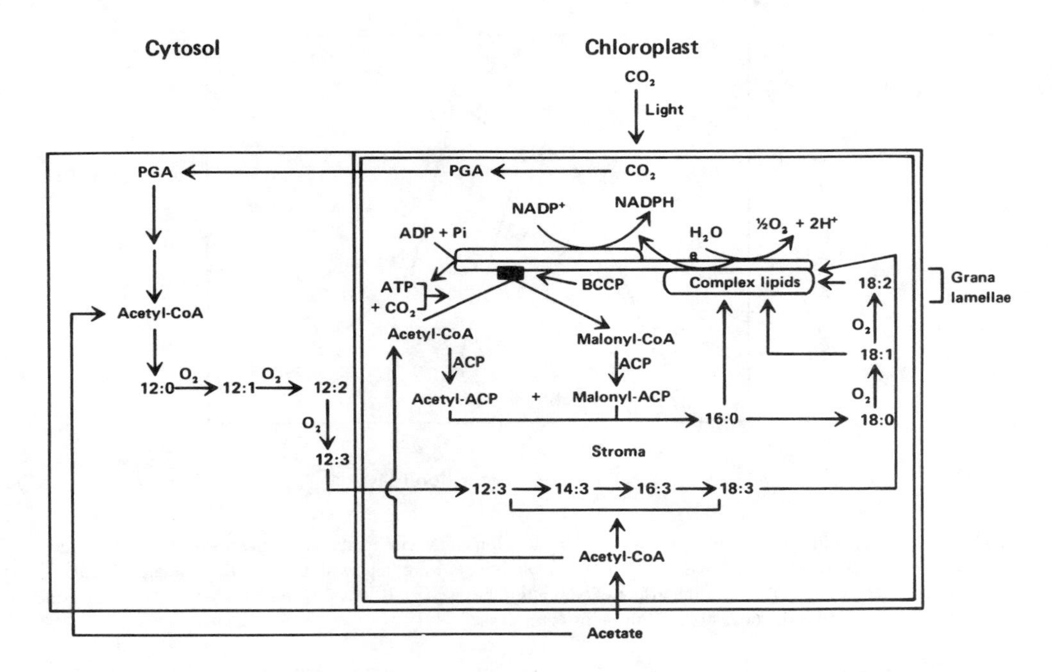

FIGURE 14. Interaction of cytosol and chloroplast in spinach fatty acid biosynthesis. PGA = phosphoglyceric acid. (From Stumpf, P. K., *International Review of Biochemistry*, Vol. 14, Goodwin, T. W., Ed., University Park Press, Baltimore, 1977, 215–237. With permission.)

germinating seeds.[104] During germination, most fatty acids formed *de novo* are utilized for phospholipid synthesis, while developing seeds, on the other hand, produced large amounts of triglycerides. The distinct differences in quantities and types of fatty acids synthesized by seeds of different developmental stages are illustrated in Table 3.

D. REGULATION OF BRANCHED-CHAIN FATTY ACID SYNTHESIS

Some years ago, van Deenen[105] postulated that methyl branched-chain fatty acids, by virtue of their protruding methyl group, would have a fluidizing effect on membranes somewhat analogous to that exerted by the cis-unsaturated fatty acids. Later studies have shown that in some cases branched-chain fatty acids can indeed substitute for the unsaturated fatty acids normally required for growth.[106] One might envision that the concurrent presence in membrane lipids from certain organisms of both unsaturated fatty acids and branched-chain fatty acids permits a sort of fine tuning of fluidity at different levels within the lipid bilayer. *Tetrahymena pyriformis*, strain W, phospholipids normally contain about 5 to 7% branched-chain acids vs. 74% unsaturated acids.[107] Growth of the cells in the presence of isovalerate raised the level of branched-chain derivative to 20 to 25%. Surprisingly, this increase was at the expense of saturated fatty acids rather than unsaturates. Although no direct fluidity measurements were made in the study, it would seem that biosynthesis of branched-chain fatty acids in this instance was regulated by the concentration of isovalerate available to the fatty acid synthesizing system rather than by some physical need for a more fluidizing kind of fatty acid.

Many species of bacteria contain branched-chain fatty acids. In the genus *Bacillus*, there appear to be two fatty acid synthase systems, one producing straight-chain and the other branched-chain acids.[108] The two enzymes differ in specificity only in the first reaction (acyl-CoA to acyl-ACP transacylase), with the synthase leading to branched chain fatty acids

TABLE 3
Comparison of Capacity for Fatty Acid Synthesis in Developing and Germinating Soybean Seeds

Tissue	Total [l-^{14}C]acetate incorporated into fatty acids/hr (mmol/g fresh wt)	Distribution of ^{14}C into fatty acids (%)				
		16:0	18:0	18:1	18:2	18:3
Germinated seed (6 days old)	1.09	60	10	30	0	0
Developing seed (30 days after flowering)	8.4	17	trace	53	30	0

From Stumpf, P. K., *International Review of Biochemistry*, Vol. 14, Goodwin, T. W., Ed., University Park Press, Baltimore, 1977, 215–237. With permission.

preferring a branched-chain-CoA initiator thought to arise through the decarboxylation of branched-chain keto acids derived from valine, leucine, or isoleucine.[109]

Further indications that the extent of fatty acid chain branching is substrate dependent comes from the experiments of Buckner et al.[110] A fatty acid synthase isolated from goose uropygial gland, which normally contains high levels of multiple methyl-branched fatty acids,[111] synthesized palmitic acid when incubated with malonyl- and acetyl-CoA, and made 2,4,6,8-tetramethyldecanoic acid and 2,4,6,8,10-pentamethyldodecanoic acid when incubated with methylmalonyl- and acetyl-CoA. Partially purified fatty acid synthases from rat mammary gland and liver also utilized methylmalonyl-CoA for the synthesis of multiple methyl-branched fatty acids.

In addition to the *de novo* synthesis of branched-chain fatty acids, they are formed in some organisms by the donation of a methyl group to a straight-chain fatty acid, utilizing S-adenosylmethionine. This pathway is described in Chapter 3.

It is highly desirable to conduct experiments in which the physical state of membranes is correlated with their branched-chain fatty acid content. Even in tissues where they are not present at quantitatively significant levels, experiments utilizing fed branched-chain acids might shed light on other aspects of fatty acid metabolism.

REFERENCES

1. **Monod, J., Wyman, J., and Changeux, J.-P.,** On the nature of allosteric transitions: a plausible model, *J. Mol. Biol.*, 12, 88–118, 1965.
2. **Wakil, S. J.,** A malonic acid derivative as an intermediate in fatty acid synthesis, *J. Am. Chem. Soc.*, 80, 6465, 1958.
3. **Lane, M. D., Moss, J., and Polakis, S. E.,** Acetyl coenzyme A carboxylase, *Curr. Top. Cell. Regul.*, 8, 139–195, 1974.
4. **Chang, H.-C., Seidman, I., Teebor, G., and Lane, M. D.,** Liver acetyl CoA carboxylase and fatty acid synthetase: relative activities in the normal state and in hereditary obesity, *Biochem. Biophys. Res. Commun.*, 28, 682–686, 1967.
5. **Prescott, D. J. and Vagelos, P. R.,** Acyl carrier protein, *Adv. Enzymol.*, 36, 269–311, 1972.
6. **Schweizer, E., Kniep, B., Castorph, H., and Holzner, U.,** Pantetheine-free mutants of the yeast fatty-acid-synthetase complex, *Eur. J. Biochem.*, 39, 353–362, 1973.

7. **Schwietz, H., Dietlein, G., Schiltz, E., and Schweizer, E.,** End group analysis of yeast fatty acid synthetase, *Biochim. Biophys. Acta,* 453, 453–458, 1976.
8. **Wakil, S. J.,** Fatty acid synthase, a proficient multifunctional enzyme, *Biochemistry,* 28, 4523–4530, 1989.
9. **Schweizer, E. and Lynen, F.,** Multienzyme complexes. Molecular organization and functional interaction of active sites within the fatty acid synthetase multienzyme complex of yeast, in *Transport by Proteins,* Sund, H. and Blauer, G., Eds., Walter de Gruyter-Verlag, Berlin, 1978, 103–118.
10. **Stoops, J. K., Awad, E. S., Arslanian, M. J., Gunsberg, S., Wakil, S. J., and Oliver, R. M.,** Studies on the yeast fatty acid synthetase. Subunit composition and structural organization of a large multifunctional enzyme complex, *J. Biol. Chem.,* 253, 4464–4475, 1978.
11. **Qureshi, A. A., Lornitzo, F. A., Jenik, R. A., and Porter, J. W.,** Subunits of fatty acid synthetase complexes. Comparative study of enzyme activities and properties of the molecular weight nonidentical subunits of fatty acid synthetase complexes obtained from rat, human, and chicken liver and yeast, *Arch. Biochem. Biophys.,* 177, 364–378, 1976.
12. **Ohlrogge, J. B.,** Biochemistry of plant acyl carrier proteins, in *The Biochemistry of Plants,* Vol. 19, Stumpf, P. K., Ed., Academic Press, Orlando, 1987, 137–157.
13. **Harwood, J. L.,** Fatty acid metabolism, *Annu. Rev. Plant Physiol.,* 39, 101–138, 1988.
14. **Guerra, D. J., Ohlrogge, J. B., and Frentzen, M.,** Activity of acyl carrier protein isoforms in reactions of plant fatty acid metabolism, *Plant Physiol.,* 82, 448– 453, 1986.
15. **Goldberg, I. and Bloch, K.,** Fatty acid synthetases in *Euglena gracilis, J. Biol. Chem.,* 247, 7349–7357, 1972.
16. **Hammes, G. G.,** Fatty acid synthase: elementary steps in catalysis and regulation, *Curr. Top. Cell. Res.,* 26, 311–324, 1985.
17. **Ziegenhorn, J., Niedermeir, R., Nüssler, C., and Lynen, F.,** Charakterisierung der acetyltransferase in der fettsauresynthetase aus hefe, *Eur. J. Biochem.,* 30, 285–300, 1972.
18. **Greenspan, M. D., Birge, C. H., Powell, G., Hancock, W. S., and Vagelos, P. R.,** Enzyme specificity as a factor in regulation of fatty acid chain length in *Escherichia coli, Science,* 170, 1203–1204, 1970.
19. **Aihaud, G. P. and Vagelos, P. R.,** Palmityl-acyl carrier protein as acyl donor for complex lipid biosynthesis in *Escherichia coli, J. Biol. Chem.,* 241, 3866–3869, 1966.
20. **van den Bosch, H. and Vagelos, P. R.,** Fatty acyl-CoA and fatty acyl-acyl carrier protein as acyl donors in the synthesis of lysophosphatidate and phosphatidate in *Escherichia coli, Biochim. Biophys. Acta,* 218, 233–248, 1970.
21. **Aeberhard, E. and Menkes, J. H.,** Biosynthesis of long chain fatty acids by subcellular particles of mature brain, *J. Biol. Chem.,* 243, 3834–3840, 1968.
22. **Khan, A. A. and Kolattukudy, P. E.,** Solubilization of fatty acid synthetase, acyl-CoA reductase, and fatty acyl-CoA alcohol transacylase from the microsomes of *Euglena gracilis, Arch. Biochem. Biophys.,* 170, 400–408, 1975.
23. **Kannangara, C. G. and Stumpf, P. K.,** Fat metabolism in higher plants. LVI. Distribution and nature of biotin in chloroplasts of different plant species, *Arch. Biochem. Biophys.,* 155, 391–399, 1973.
24. **Srere, P. A.,** The enzymology of the formation and breakdown of citrate, *Adv. Enzymol.,* 43, 57–101, 1975.
25. **Numa, S. and Yamashita, S.,** Regulation of lipogenesis in animal tissues, *Curr. Top. Cell. Reg.,* 8, 197–246, 1974.
26. **Caggiano, A. V. and Powell, G. L.,** Regulation of enzymes by fatty acyl coenzyme A. Site specific binding of fatty acyl coenzyme A by citrate synthase — a spin labeling study, *J. Biol. Chem.,* 254, 2800–2806, 1979.
27. **Frenkel, R.,** Regulation and physiological functions of malic enzymes, *Curr. Top. Cell. Reg.,* 9, 157–181, 1975.
28. **Liedvogel, B.,** Lipid precursors in plant cells: the problem of acetyl CoA generation for plastid fatty acid synthesis, in *The Metabolism, Structure and Function of Plant Lipids,* Stumpf, P. K., Mudd, J. B., and Nes, W. D., Eds., Plenum Press, New York, 1987, 509–511.
29. **Brownsey, R. W. and Denton, R. M.,** Acetyl-coenzyme A carboxylase, in *The Enzymes,* 3rd ed., XVIII, Boyer, P. D. and Krebs, E. G., Eds., Academic Press, New York, 1987, 123–146.
30. **Hardie, D. G., Carling, D., and Sim, A. T. R.,** The AMP-activated protein kinase: a multisubstrate regulator of lipid metabolism, *Trends Biochem. Sci.,* 14, 20–23, 1989.
31. **Guchhait, R. B., Moss, J., Sokolski, W., and Lane, M. D.,** The carboxyl transferase component of acetyl CoA carboxylase: structural evidence for intersubunit translocation of the biotin prosthetic group, *Proc. Nat. Acad. Sci. U.S.A.,* 68, 653–657, 1971.
32. **Gregolin, C., Ryder, E., Warner, R. C., Kleinschmidt, A. K., Chang, H. C., and Lane, M. D.,** Liver acetyl coenzyme A carboxylase. II. Further molecular characterization, *J. Biol. Chem.,* 243, 4236–4245, 1968.
33. **Cashel, M.,** Regulation of bacterial ppGpp and pppGpp, *Annu. Rev. Micro.,* 29, 301–318, 1975.
34. **Polakis, S. E., Guchhait, R. B., and Lane, M. D.,** Stringent control of fatty acid synthesis in *Escherichia coli.* Possible regulation of acetyl coenzyme A carboxylase by ppGpp, *J. Biol. Chem.,* 248, 7957–7966, 1973.

35. **Hochstadt-Ozer, J. and Cashel, M.,** The regulation of purine utilization in bacteria. V. Inhibition of purine phosphoribosyltransferase activities and purine uptake in isolated membrane vesicles by guanosine tetraphosphate, *J. Biol. Chem.,* 247, 7067–7072, 1972.
36. **Brady, R. O. and Gurin, S.,** Biosynthesis of fatty acids by cell-free or water-soluble enzyme systems, *J. Biol. Chem.,* 199, 421–431, 1952.
37. **Takai, T., Yokoyama, C., Wada, K., and Tanabe, T.,** Primary structure of chicken liver acetyl-CoA carboxylase deduced from cDNA sequence, *J. Biol. Chem.,* 263, 2651–2657, 1988.
38. **López-Casillas, F., Bai, D. H., Luo, X., Kong, I.-S., Hermodson, M. A., and Kim, K. H.,** Structure of the coding sequence and primary amino acid sequence of acetyl-coenzyme A carboxylase, *Proc. Natl. Acad. Sci. U.S.A.,* 85, 5784–5788, 1988.
39. **Thampy, K. G. and Wakil, S. J.,** Regulation of acetyl-coenzyme A carboxylase, *J. Biol. Chem.,* 263, 6447–6453, 1988.
40. **Gregolin, C., Ryder, E., and Lane, M. D.,** Liver acetyl coenzyme A carboxylase I. Isolation and catalytic properties, *J. Biol. Chem.,* 243, 4227–4235, 1968.
41. **Ryder, E., Gregolin, C., Chang. H.-C., and Lane, M. D.,** Liver acetyl CoA carboxylase: insights into the mechanism of activation by tricarboxylic acids and acetyl CoA, *Proc. Natl. Acad. Sci. U.S.A.,* 57, 1455–1462, 1967.
42. **Lane, M. D. and Moss, J.,** Regulation of fatty acid synthesis in animal tissues, in *Metabolic Regulation,* Vol. 5 of Metabolic Pathways series, 3rd ed., Vogel, H. J., Ed., Academic Press, New York, 1971, 23–54.
43. **Gregolin, C., Ryder, E., Warner, R. C., Kleinschmidt, A. K., and Lane, M. D.,** Liver acetyl-CoA carboxylase: the dissocation-reassociation process and its relation to catalytic activity, *Proc. Natl. Acad. Sci. U.S.A.,* 56, 1751–1761, 1966.
44. **Bortz, W. M. and Lynen, F.,** The inhibition of acetyl-coenzyme A carboxylase by long-chain acyl-coenzyme A derivatives, *Biochem. Zeit.,* 337, 505–509, 1963.
45. **Goodridge, A. G.,** Regulation of the activity of acetyl coenzyme A carboxylase by palmitoyl coenzyme A and citrate, *J. Biol. Chem.,* 247, 6946–6952, 1972.
46. **Ogiwara, H., Tanabe, T., Nikawa, J., and Numa, S.,** Inhibition of rat-liver acetyl-coenzyme-A carboxylase by palmitoyl-coenzyme A., *Eur. J. Biochem.,* 89, 33–41,1978.
47. **Volpe, J. J. and Vagelos, P. R.,** Mechanisms and regulation of biosynthesis of saturated fatty acids, *Physiol. Rev.,* 56, 339–417, 1976.
48. **Berdanier, C. D.,** Role of glucocorticoids in the regulation of lipogenesis, *FASEB J.,* 3, 2179–2183, 1989.
49. **Lee, K. H. and Kim, K. H.,** Regulation of rat liver acetyl coenzyme A carboxylase. Evidence for interconversion between active and inactive forms of enzyme by phosphorylation and dephosphorylation, *J. Biol. Chem.,* 252, 1748–1751, 1977.
50. **Witters, L. A., Kowaloff, E. M., and Avruch, J.,** Glucagon regulation of protein phosphorylation: identification of acetyl coenzyme A carboxylase as a substrate, *J. Biol. Chem.,* 254, 245–248, 1979.
51. **Watkins, P. A., Tarlow, D. M., and Lane, M. D.,** Mechanism for acute control of fatty acid synthesis by glucagon and 3′:5′cyclic AMP in the liver cell, *Proc. Natl. Acad. Sci. U.S.A.,* 74, 1497–1501, 1977.
52. **Davies, S. P., Sim, A. T. R., and Hardie, D. G.,** Location and function of three sites phosphorylated on rat acetyl-CoA carboxylase by the AMP-activated protein kinase, *Eur. J. Biochem.* 187, 183–190, 1990.
53. **Kim, K.-H., López-Casillas, F., Bai, D. H., Luo, X., and Pape, M. E.,** Role of reversible phosphorylation of acetyl-CoA carboxylase in long-chain fatty acid biosynthesis, *FASEB J.,* 3, 2250–2256, 1989.
54. **Chan, C. P., McNall, S. J., Krebs, E. G., and Fischer, E. M.,** Stimulation of protein phosphatase activity by insulin and growth factors in 3T3 cells, *Proc. Natl. Acad. Sci. U.S.A.* 85, 6257–6261, 1988.
55. **Mabrouk, G. M., Helmy, I. M., Thampy, K. G., and Wakil, S. J.,** Acute hormonal control of acetyl-CoA carboxylase. The roles of insulin, glucagon, and epinephrine, *J. Biol. Chem.,* 265, 6330–6338, 1990.
56. **Lent, B. A., Lee, K.-H., and Kim, K. H.,** Regulation of rat liver acetyl-CoA carboxylase, *J. Biol. Chem.,* 253, 8149–8156, 1978.
57. **Romsos, D. R. and Leveille, G. A.,** Dietary regulation of lipid metabolism, in *Modification of Lipid Metabolism,* Perkins, E. G. and Witting, L. A., Eds., Academic Press, New York, 1975, 127–142.
58. **Nishikori, K., Iritani, N., and Numa, S.,** Levels of acetyl coenzyme A carboxylase and its effectors in rat liver after short-term fat-free refeeding, *FEBS Lett.,* 32, 19–21, 1973.
59. **Nakanishi, S. and Numa, S.,** Purification of rat liver acetyl coenzyme A carboxylase and immunochemical studies on its synthesis and degradation, *Eur. J. Biochem.,* 16, 161–173, 1970.
60. **Majerus, P. W. and Kilburn, E.,** Acetyl coenzyme A carboxylase. The roles of synthesis and degradation in regulation of enzyme levels in liver, *J. Biol. Chem.,* 244, 6254–6262, 1969.
61. **Nakanishi, S., Tanabe, T., Horikawa, S., and Numa, S.,** Dietary and hormonal regulation of the content of acetyl coenzyme A carboxylase-synthesizing polysomes in rat liver, *Proc. Natl. Acad. Sci. U.S.A.,* 73, 2304–2307, 1976.

62. **Pape, M. E., Lopez-Casillas, F., and Kim, K. H.,** Physiological regulation of acetyl-CoA carboxylase gene expression: effects of diet, diabetes and lactation on acetyl-CoA carboxylase mRNA, *Arch. Biochem. Biophys.*, 267, 104–109, 1988.
63. **Tanabe, T., Wada, K., Ogiwara, H., and Numa, S.,** Effects of allosteric regulators on proteolysis of rat liver acetyl coenzyme A carboxylase by lysosomal extract, *FEBS Lett.*, 82, 85–88, 1977.
64. **Craig, M. C., Nepokroeff, C. M., Lakshmanan, M. R., and Porter, J. W.,** Effect of dietary change on the rates of synthesis and degradation of rat liver fatty acid synthetase, *Arch. Biochem. Biophys.*, 152, 619–630, 1972.
65. **Volpe, J. J., Lyles, T. O., Roncari, D. A. K., and Vagelos, P. R.,** Fatty acid synthetase of developing brain and liver: content, synthesis, and degradation during development, *J. Biol. Chem.*, 248, 2502–2513, 1973.
66. **Lakshmanan, M. R., Nepokroeff, C. M., and Porter, J. W.,** Control of the synthesis of fatty-acid synthetase in rat liver by insulin, glucagon, and adenosine 3′:5′ cyclic monophosphate, *Proc. Natl. Acad. Sci. U.S.A.*, 69, 3516–3519, 1972.
67. **Gibson, D. M., Lyons, R. T., Scott, D. F., and Muto, Y.,** Synthesis and degradation of the lipogenic enzymes of rat liver, *Adv. Enzyme Regul.*, 10, 187–204, 1972.
68. **Kumar, S.,** Nutritional and hormonal control of lung and liver fatty acid synthesis, *Arch. Biochem. Biophys.*, 183, 625–637, 1977.
69. **Sumper, M. and Riepertinger, C.,** Structural relationship of biotin-containing enzymes. Acetyl-CoA carboxylase and pyruvate carboxylase from yeast, *Eur. J. Biochem.*, 29, 237–248, 1972.
70. **Rasmussen, R. K. and Klein, H. P.,** Regulation of acetyl-CoA carboxylase of *Saccharomyces cerevisiae*, *Biochem. Biophys. Res. Commun.*, 28, 415–419, 1967.
71. **Matsuhashi, M.,** Acetyl-CoA carboxylase from yeast, in *Methods in Enzymology*, Vol. 14, Lowenstein, J. M., Ed., Academic Press, New York, 1969, 3–8.
72. **Kamiryo, T. and Numa, S.,** Reduction of the acetyl coenzyme A carboxylase content of *Saccharomyces cerevisiae* by exogenous fatty acids, *FEBS Lett.*, 38, 29–32, 1973.
73. **Kamiryo, T., Parthasarathz, S., and Numa, S.,** Evidence that acetyl coenzyme A synthetase activity is required for repression of yeast acetyl coenzyme A carboxylase by exogenous fatty acids, *Proc. Natl. Acad. Sci. U.S.A.*, 73, 386–390, 1976.
74. **Egin-Bühler, B., Loyal, R., and Ebel, J.,** Comparison of acetyl CoA carboxylases from parsley cell cultures and wheat germ, *Arch. Biochem. Biophys.*, 203, 90–100, 1980.
75. **Post-Beittenmiller, D., Jaworski, J. G., and Ohlrogge, J. B.,** *In vivo* pools of free and acylated acyl carrier proteins in spinach, *J. Biol. Chem.*, 266, 1858–1865, 1991.
76. **Wakil, S. J., Goldman, J. K., Williamson, I. P., and Toomey, R. E.,** Stimulation of fatty acid biosynthesis by phosphorylated sugars, *Proc. Natl. Acad. Sci. U.S.A.*, 55, 880–887, 1966.
77. **Powell, G. L., Elovson, J., and Vagelos, P. R.,** Acyl carrier protein XII. Synthesis and turnover of the prosthetic group of acyl carrier protein *in vivo*, *J. Biol. Chem.*, 244, 5616–5624, 1969.
78. **Bauza, M. T., deLoach, J. R., Aguanno, J. J., and Larrabee, A. R.,** Acyl carrier protein prosthetic group exchange and phospholipid synthesis in synchronized cultures of a pantothenate auxotroph of *Escherichia coli*, *Arch. Biochem. Biophys.*, 174, 344–349, 1976.
79. **Bloch, K.,** Control mechanisms for fatty acid synthesis in *Mycobacterium smegmatis*, *Adv. Enzymol.*, 45, 1–84, 1977.
80. **Wood, W. I., Peterson, D. O., and Bloch, K.,** Subunit structure of *Mycobacterium smegmatis* fatty acid synthetase, *J. Biol. Chem.*, 253, 2650–2656, 1978.
81. **Ilton, M., Jevans, A. W., McCarthy, E. D., Vance, D., White, H. B., II, and Bloch, K.,** Fatty acid synthetase activity in *Mycobacterium phlei:* regulation by polysaccharides, *Proc. Natl. Acad. Sci. U.S.A.*, 68, 87–91, 1971.
82. **Vance, D. and Bloch, K.,** Control mechanisms in the synthesis of saturated fatty acids, *Annu. Rev. Biochem.*, 46, 263–298, 1977.
83. **Flick, P. K. and Bloch, K.,** *In vitro* alteration of the product distribution of the fatty acid synthetase from *Mycobacterium phlei*, *J. Biol. Chem.*, 249, 1031–1036, 1974.
84. **Odriozola, J. M., Ramos, J. A., and Bloch, K.,** Fatty acid synthetase activity in *Mycobacterium smegmatis*. Characterization of the acyl carrier protein-dependent elongating system, *Biochim. Biophys. Acta*, 488, 207–217, 1977.
85. **Knoche, H., Esders, T. W., Koths, K., and Bloch, K.,** Palmityl coenzyme A inhibition of fatty acid synthesis. Relief by bovine serum albumin and mycobacterial polysaccharides, *J. Biol. Chem.*, 248, 2317–2322, 1973.
86. **Knoche, H. W. and Koths, K. E.,** Characterization of a fatty acid synthetase from *Corynebacterium diphtheriae*, *J. Biol. Chem.*, 248, 3517–3519, 1973.

87. **Porter, J. W., Kumar, S., and Dugan, R. E.,** Synthesis of fatty acids by enzymes of avian and mammalian species, *Prog. Biochem. Pharmacol.,* 6, 1–101, 1971.
88. **Smith, S. and Abraham, S.,** Fatty acid synthetase from lactating rat mammary gland. I. Isolation and properties, *J. Biol. Chem.,* 245, 3209–3217, 1970.
89. **Tweto, J. and Larrabee, A. R.,** The effect of fasting on synthesis and 4′–phosphopantetheine exchange in rat liver fatty acid synthetase, *J. Biol. Chem.,* 247, 4900–4904, 1972.
90. **Volpe, J. J. and Vagelos, P. R.,** Fatty acid synthetase of mammalian brain, liver, and adipose tissue. Regulation by prosthetic group turnover, *Biochim. Biophys. Acta,* 326, 293–304, 1973.
91. **Lakshmanan, M. R., Nepokroeff, C. M., and Porter, J. W.,** Control of the synthesis of fatty-acid synthetase in rat liver by insulin, glucagon, and adenosine 3′:5′-cyclic monophosphate, *Proc. Natl. Acad. Sci. U.S.A.,* 69, 3516–3519, 1972.
92. **Roncari, D. A.,** Mammalian fatty acid synthetase. II. Modification of purified human liver complex activity, *Can. J. Biochem.,* 53, 135–142, 1975.
93. **Craig, M. C., Nepokroeff, C. M., Lakshmanan, M. R., and Porter, J. W.,** Effect of dietary change on the rate of synthesis and degradation of rat liver fatty acid synthetase, *Arch. Biochem. Biophys.,* 152, 619–630, 1972.
94. **Nepokroeff, C. M. and Porter, J. W.,** Translation and characterization of the fatty acid synthetase messenger RNA, *J. Biol. Chem.,* 253, 2279–2283, 1978.
95. **Guynn, R. W., Veloso, D., and Veech, R. I.,** The concentration of malonyl-coenzyme A and the control of fatty acid synthesis *in vivo, J. Biol. Chem.,* 247, 7325–7331, 1972.
96. **Volpe, J. J., Lyles, T. O., Roncari, D. A. K., and Vagelos, P. R.,** Fatty acid synthetase of developing brain and liver. Content, synthesis, and degradation during development, *J. Biol. Chem.,* 248, 2502–2513, 1973.
97. **Saggerson, E. D. and Greenbaum, A. L.,** The regulation of triglyceride synthesis and fatty acid synthesis in rat epididymal adipose tissue. Effects of altered dietary and hormonal conditions, *Biochem. J.,* 119, 221–242, 1970.
98. **Zakim, D. and Ho, W.,** The acetyl-CoA carboxylase and fatty acid synthetase activities of rat intestinal mucosa, *Biochim. Biophys. Acta,* 222, 558–559, 1970.
99. **Volpe, J. J. and Kishimoto, Y.,** Fatty acid synthetase of brain: development, influence of nutritional and hormonal factors and comparison with liver enzyme, *J. Neurochem.,* 19, 737–753, 1972.
100. **Paulauskis, J. D. and Sul, H. S.,** Hormonal regulation of mouse fatty acid synthase gene transcription in liver, *J. Biol. Chem.,* 264, 574–577, 1989.
101. **Braddock, M. and Hardie, D. G.,** Cloning of cDNA to rat mammary-gland fatty acid synthase mRNA, *Biochem. J.,* 249, 603–607, 1988.
102. **Delo, J., Ernst-Fonberg, M. L., and Bloch, K.,** Fatty acid synthetases from *Euglena gracilis, Arch. Biochem. Biophys.,* 143, 384–391, 1971.
103. **Porra, R. J. and Stumpf, P. K.,** Lipid biosynthesis in developing and germinating soybean cotyledons. The formation of palmitate and stearate by chopped tissue and supernatant preparations, *Arch. Biochem. Biophys.,* 176, 53–62, 1977.
104. **Stumpf, P. K.,** Lipid biosynthesis in higher plants, *International Review of Biochemistry,* Vol. 14, Goodwin, T. W., Ed., University Park Press, Baltimore, 1977, 215–237.
105. **van Deenen, L. L. M.,** Phospholipids and biomembranes, *Prog. Chem. Fats Other Lipids,* 8, 1–127, 1965.
106. **Silbert, D. F., Ladenson, R. C., and Honegger, J. L.,** The unsaturated fatty acid requirement in *Escherichia coli.* Temperature dependence and total replacement by branched-chain fatty acids, *Biochim. Biophys. Acta,* 311, 349–361, 1973.
107. **Conner, R. L. and Reilly, A. E.,** The effect of isovalerate supplementation on growth and fatty acid composition of *Tetrahymena pyriformis* W, *Biochim. Biophys. Acta,* 398, 209–216, 1975.
108. **Kaneda, T.,** Fatty acids of the genus *Bacillus:* an example of branched-chain preference, *Bacteriol. Rev.,* 41, 391–418, 1977.
109. **Oku, H. and Kaneda, T.,** Biosynthesis of branched-chain fatty acids in *Bacillis subtilis, J. Biol. Chem.,* 263, 18386–18396, 1988.
110. **Buckner, J. S., Kolattukudy, P. E., and Rogers, L.,** Synthesis of multimethyl-branched fatty acids by avian and mammalian fatty acid synthetase and its regulation by malonyl-CoA decarboxylase in the uropygial gland, *Arch. Biochem. Biophys.,* 186, 152–163, 1978.
111. **Buckner, J. S. and Kolattukudy, P. E.,** Lipid biosynthesis in the sebaceous glands: synthesis of multi-branched fatty acids from methylmalonyl-coenzyme A in cell-free preparations from the uropygial gland of goose, *Biochemistry,* 14, 1774–1782, 1975.

Chapter 3

FATTY ACID MODIFICATION

I. INTRODUCTION

We have seen in Chapter 2 that the principal product of *de novo* fatty acid synthesis in most organisms is palmitic acid. Yet palmitic acid comprises only 10 to 20% of the total fatty acids found in a typical membrane. Obviously, extensive alteration of the palmitate is necessary before it can satisfy the structural needs of the cell.

Phospholipids and glycolipids make up the bulk of membrane structural lipids in animals and plants. Although several dozen species of fatty acids have been identified as components of these lipids, the major constituents are nearly always those listed in Table 1. The table clearly shows that the principal modifications required for newly synthesized palmitate are elongation to a C_{18} or C_{20} chain length and insertion of one or more double bonds.

The very fact that the distribution of fatty acid species in a particular kind of cell is characteristic and unchanging under any fixed set of conditions is proof enough that this alteration of fatty acids is strictly regulated. Remembering the all important influence of chain length and unsaturation on membrane fluidity (Chapter 1), it is easy to understand why so much attention is currently being devoted to the enzymatic regulation of these properties.

Also considered in the chapter are selected modifications giving rise to fatty acids that are important structural elements of only certain classes of organisms. For example, cyclopropane fatty acids can account for over 50% of the total fatty acids of some bacterial species,[1] and very long chain (C_{24} to C_{36}) fatty acids are conspicuous in some lipids of vertebrate retina.[2]

II. FATTY ACID ACTIVATION

Most enzymatic reactions of fatty acids require them to first be converted to acyl CoA thioesters. This is achieved by the action of fatty acyl CoA synthetase (fatty acid:CoA ligase, EC 6.2.1.3), an enzyme widely distributed in the endoplasmic reticulum, the outer mitochondrial membrane, and the peroxisomal membrane. Despite early evidence to the contrary, work in the early 1980s indicated that the acyl-CoA synthetase enzymes from these three fractions of rat liver are indistinguishable.[3] However, differences in substrate specificity and other enzymatic properties have more recently suggested that fatty acyl CoA synthetases of the three fractions are indeed distinct, with stearoyl CoA synthetase being present in mitochondria, microsomes, and peroxisomes, while lignoceroyl CoA ($C_{24:0}$ CoA) synthetase is present only in microsomes and peroxisomes.[4] Fatty acyl CoA synthetases having broad and narrow (favoring arachidonic acid) specificities have also been reported from murine T lymphocytes,[5] supporting the concept that these enzymes may exert a strong influence on the pattern of intracellular fatty acid utilization by providing a distinctive mixture of acyl CoAs for each organelle. The rate of fatty acid activation *in vivo* is also likely to be governed by the availability of free fatty acids and CoASH as well as by acyl-CoA feedback inhibition.

III. FATTY ACID ELONGATION

Cells typically contain large amounts of 18-carbon fatty acids, and animal cells usually have significant concentrations of 20-carbon fatty acids as well. Even longer chain acids are major components of certain specialized tissues, e.g., myelin of mammalian brain contains high levels of 22- and 24-carbon components. It is generally considered that most of these

TABLE 1
The Most Common Naturally Occurring Fatty Acids

Systemic name	Shorthand	Common name	Structural formula
n-Hexadecanoic	16:0	Palmitic	$CH_3(CH_2)_{14}COOH$
n-Octadecanoic	18:0	Stearic	$CH_3(CH_2)_{16}COOH$
cis-9-Hexadecenoic	Δ9-16:1	Palmitoleic	$CH_3(CH_2)_5CH{=}CH(CH_2)_7COOH$
cis-9-Octadecenoic	Δ9-18:1	Oleic	$CH_3(CH_2)_7CH{=}CH(CH_2)_7COOH$
cis-11-Octadecenoic	Δ11-18:1	Vaccenic	$CH_3(CH_2)_5CH{=}CH(CH_2)_9COOH$
cis,cis-9,12-Octadecadienoic	Δ9,12-18:2	Linoleic	$CH_3(CH_2)_4CH{=}CHCH_2CH{=}CH(CH_2)_7\ COOH$
All *cis*-9,12,15-octadecatrienoic	Δ9,12,15-18:3	α-Linolenic	$CH_3CH_2\ CH{=}CHCH_2CH{=}CHCH_2CH{=}CH(CH_2)_7COOH$
All *cis*-6,9,12-octadecatrienoic	Δ6,9,12-18:3	γ-Linolenic	$CH_3(CH_2)_4CH{=}CHCH_2CH{=}CHCH_2CH{=}CH(CH_2)_4COOH$
All *cis*-5,8,11,14-eicosatetraenoic	Δ5,8,11,14-20:4	Arachidonic	$CH_3(CH_2)_4CH{=}CHCH_2CH{=}CHCH_2CH{=}CHCH_2CH{=}CH(CH_3)_3COOH$

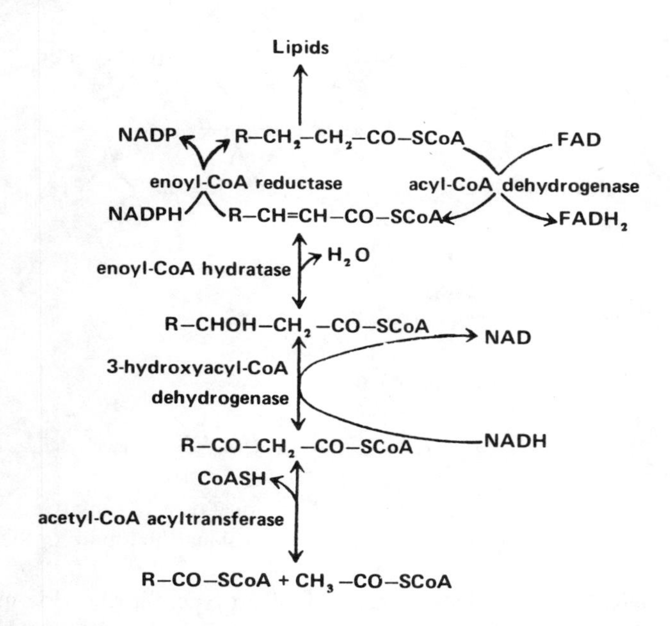

FIGURE 1. Fatty acid elongation (from bottom to top) and β-oxidation (from top to bottom).

longer chain fatty acids are derived either from dietary sources or from elongation of palmitoyl-CoA (and perhaps some stearoyl-CoA) formed by the cytoplasmic fatty acid synthetase system. Elongation can also extend the length of dietary fatty acids, as exemplified by the ready conversion of ingested linoleic acid (18:2) to arachidonic acid (20:4).

Two enzymatic systems are known to carry out elongation reactions in eukaryotic cells. The first is located in mitochondria and the second is associated with microsomal membranes. They will each be considered in turn and will be compared to the equivalent process in prokaryotes.

A. MITOCHONDRIAL ELONGATION

1. The Pathway

Elongation of fatty acids in mitochondria involves reversing some of the reactions of the well-known ß-oxidation pathway for fatty acid oxidation. Thus, ß-oxidation (Figure 1, from top to bottom) and elongation (Figure 1, from bottom to top) both make use of acetyl-CoA acyltransferase (thiolase), 3-hydroxyacyl-CoA dehydrogenase, and enoyl-CoA hydratase. The final step of fatty acid elongation utilizes an enzyme, enoyl-CoA reductase, which is not a part of the ß-oxidation system.

The elongation process takes place in the mitochondrial matrix. In liver and kidney, it operates efficiently only in the presence of both NADH and NADPH, whereas mitochondria of heart muscle, skeletal muscle, and aortic intima require only NADH. Optimal activity of the pathway is found with octanoyl-CoA in the case of kidney cortex and decanoyl-CoA in liver.[6] Elongation of palmitoyl-CoA and stearoyl-CoA, once thought to be the principal substrates, is minimal because they are inhibitory (at least *in vitro*) to the component enzymes 3-hydroxyacyl-CoA dehydrogenase and enoyl-CoA reductase.[7]

2. Regulation of Mitochondrial Elongation

The discovery of low enzymatic elongation activity toward C_{16} and C_{18} derivatives has pointed up the need to reevaluate the physiological role of mitochondrial fatty acid elongation.

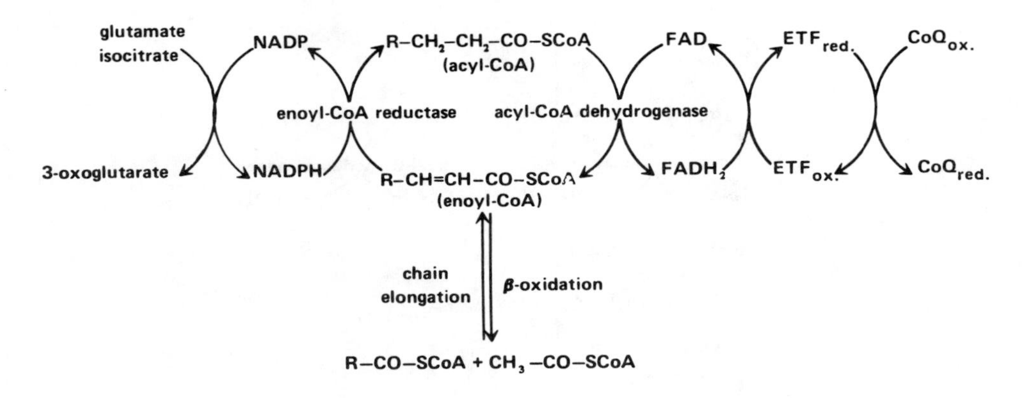

FIGURE 2. Mechanism of fatty acid elongation stimulation by glutamate and isocitrate.

The postulated function of elongating palmitoyl-CoA must now be tentatively assigned to the microsomal elongation system (see below) and a new explanation found for the mitochondrial enzymes.

Hinsch et al.[7] have proposed that elongation in heart-type mitochondria may function to store reducing equivalents (NADH) and acetate units in situations such as anoxia, which favor high ratios of reduced-to-oxidized pyridine nucleotides. It is indeed known that high NADH to NAD ratios stimulate fatty acid synthesis in heart mitochondria.[8] On the other hand, the liver type of elongation, with its requirement for NADPH, could lead to a transfer of hydrogen from NADPH-generating compounds, such as glutamate and citrate, to the mitochondrial respiratory chain via a cycling of substrate between enoyl-CoA reductase and acyl CoA dehydrogenase (Figure 2), thus regenerating NADP and bypassing the early part of the respiratory chain. This type of coupling is supported by the observed stimulatory effect of glutamate and isocitrate on chain elongation.[9] However, the primary physiological function(s) of mitochondrial fatty acid chain elongation must be clarified by further experimentation.

B. MICROSOMAL ELONGATION

1. The Pathway

Rat hepatic microsomal fatty acid elongation utilizes malonyl CoA to lengthen acyl CoA chains in a process involving sequential action by four enzymes: (1) condensing enzyme, (2) ß-ketoacyl-CoA reductase, (3) ß-hydroxyacyl-CoA dehydrase, and (4) *trans*-2-enoyl-CoA reductase. Through competition studies with different fatty acyl CoAs, Prasad et al.[10] concluded that at least three different condensing enzymes are present, one utilizing saturated acyl CoAs, a second for monounsaturated acyl-CoAs, and a third for polyunsaturated acyl-CoAs. The remaining three enzymes of the pathway are probably active towards all elongation intermediates.[11]

Fatty acid elongation is active in plant cells, particularly in those species that accumulate extracellular wax esters on epidermal surfaces.[12] The process involves microsomal enzymes generally using stearoyl-CoA as substrate but capable of extending the acyl chain length to C_{26} in the well studied leek system. The elongation products are also acyl-CoA derivatives.

2. Regulation of Microsomal Elongation

Elongation activity and specificity vary markedly from one cell type to another. Because of the high levels of C_{22} and C_{24} fatty acids in the sphingolipids of nervous system tissues, brain microsomes have been examined in some detail. An effort has been made to determine why the production of C_{22} and C_{24} fatty acids increases sharply at the onset of myelination.[13]

Whereas microsomal acyl-CoA hydrolase activity declined during myelination in such a fashion as to create high levels of palmitoyl- and stearoyl-CoA for elongation, elevated levels of these substrates achieved by other means prior to the onset of myelination did not stimulate elongation.

Considerable insight into the sequence of events regulating fatty acid elongation in the central nervous system emerged from studies with quaking mice, which carry an inherited deficiency in myelination. Elongation of C_{16} acyl-CoA to C_{18} acyl-CoA and of C_{18} acyl-CoA to C_{20} acyl-CoA was much less active than in control mice.[14] This explained nicely the buildup of tissue C_{20} fatty acids instead of the normal C_{24} species. Further evidence is needed in order to decide whether a specific enzyme or enzyme system is required for the addition of each C_2 unit. It has been established that the changing enzyme activity in microsomes of the mutant mice is not matched by changes in the comparable mitochondrial activites.[15]

C. ELONGATION IN PROKARYOTES

Relatively little is known regarding fatty acid elongation in prokaryotes. An analysis of *in vivo* chain-length specificity has been conducted using *Acholeplasma laidlawii* B.[16] Optimal elongation of straight-chain saturated fatty acids was found with C_9 to C_{13} substrates, while slightly longer chains were preferred among branched-chain and unsaturated fatty acids. One or two C_2 units were added to the acids.

If a relatively low-melting, exogenous fatty acid, which was not itself a substrate for elongation, was incorporated into lipids of *Acholeplasma*, elongation of the other fatty acid species was stimulated. The incorporation of high-melting acids had the opposite effect. The authors concluded that the modulation of chain elongation was caused not by membrane fluidity alterations, but rather by competition of the added fatty acids with elongation substrates and products for stereospecific acylation of the 1- and 2-positions of *sn*-glycerol-3-phosphate. Thus a long-chain, high-melting fatty acid would preferentially occupy the 1-position of the glycerol moeity, creating a demand for shorter-chain acids for the 2-position.

IV. FATTY ACID DESATURATION

The introduction of a double bond into the hydrocarbon chain of a fatty acid seems such a straightforward, uncomplicated reaction. Considering the variety of electron acceptors available in nature, this dehydrogenation reaction might be expected to proceed by an easily characterized, universal mechanism subject to clearly defined controls. Nothing could be further from the truth. There are at least three distinct pathways of major importance for fatty acid desaturation. Details of the enzymology and energy transfer are still poorly understood in all three pathways. And extensive work on control mechanisms only began a few years ago.

Our meager state of understanding can be attributed in part to the fact that fatty acid desaturation, like so many other reactions of lipid metabolism, involves a two-phase system — substrates are provided from or through an aqueous phase to an enzyme situated in a nonaqueous membrane environment. An entire new arsenal of techniques is still under development for tracing substrates and reactants through such systems as this and a variety of other metabolic schemes involving membrane synthesis. The especially crucial role of unsaturated fatty acids in regulating membrane fluidity requires that more sophisticated efforts be made to probe the hydrophobic interior of those membranes that apparently house the black box controlling desaturation.

Considering our grossly incomplete understanding of the molecular mechanisms regulating fatty acid desaturation, it is surprising what a wealth of detailed information we have acquired on the outcome of the desaturation process. The more important details from early research are presented in a comprehensive review by Gurr.[17] The salient features of each pathway are

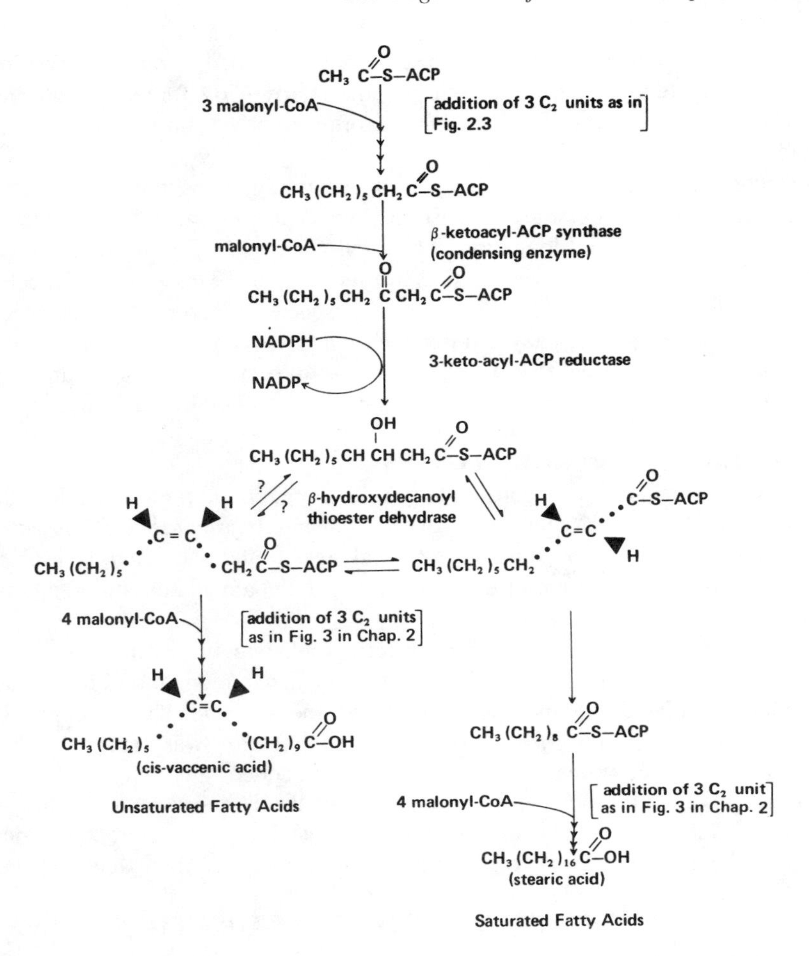

FIGURE 3. The biosynthesis of saturated and unsaturated fatty acids in *Escherichia coli.*

presented below in four separate installments for the convenience of readers primarily interested in a particular organism. Apart from desaturation in anaerobic bacteria, which employs a totally unique mechanism, the differences in the pathways in aerobic bacteria, animals, and plants are matters of detail — the basic reaction is the same in all three cases. However, there appear to be enough significant differences in the regulation of desaturation by the three classes of organisms to warrant treating them separately.

A. DESATURATION IN ANAEROBIC BACTERIA

1. The Pathway

Much of the pathway for synthesis of bacterial monounsaturated fatty acids is similar to that described in Chapter 2, Section 11, for the formation of saturated fatty acids. The major difference lies in the existence of a branch point at which a commitment is made as to whether the fatty acid will be completed with or without the presence of a double bond. Figure 3 illustrates the pathway as it operates in *Escherichia coli*. The branch point in this species comes during dehydration of the C_{10} derivative by an enzyme termed ß-hydroxydecanoyl-ACP dehydrase.[18] The enzyme dehydrates its substrate to produce *trans*-2-decenoyl-ACP by a mechanism analogous to that involved in earlier dehydration steps. But unlike the other

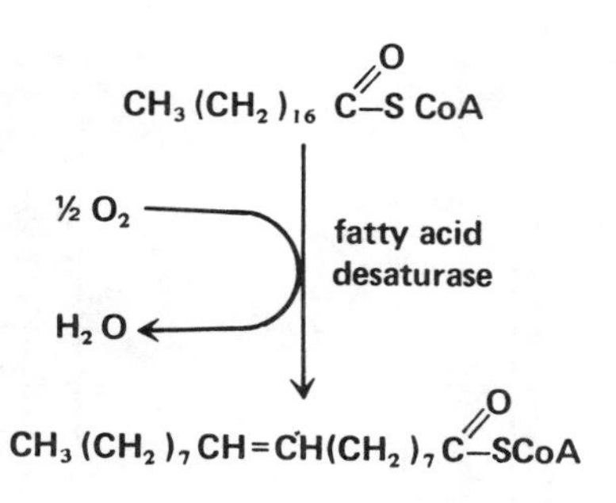

FIGURE 4. Fatty acid desaturation in aerobic bacteria.

dehydrases, it also has the capability of isomerizing the *trans*-2 unsaturated product to *cis*-3-decenoyl-ACP.

Further utilization of these unsaturated intermediates is governed by two distinct 3-ketoacyl-ACP synthases.[19] Synthase I but not synthase II can catalyze a critical step in the pathway to unsaturated fatty acids. Although the identity of this step has not been confirmed, it may be the elongation of *cis*-3-decenoyl-ACP, thereby producing the unsaturated intermediate which upon further elongation yields palmitoleoyl-ACP. On the other hand, only synthase II can elongate palmitoleoyl-ACP to *cis*-vaccenoyl-ACP, the final unsaturated fatty acid (Figure 3).

2. Regulation of Anaerobic Desaturation

The increased level of unsaturated fatty acids present in lipids of *E. coli* grown at low temperatures, e.g., 27°C, appears to stem mainly from an increase in the activity of 3-ketoacyl-ACP synthase II relative to that of 3-ketoacyl-ACP synthase I. Because synthase II is less susceptible than synthase I to cold inactivation, elongation of its preferred substrate, palmitoleoyl-ACP, to *cis*-vaccenyl-ACP continues, leading to an increase in average fatty acid chain length as well as unsaturation.[20] It is this change in the activity of existing enzymes rather than *de novo* enzyme synthesis that permits adaptation to low temperature.

B. DESATURATION IN AEROBIC BACTERIA

1. The Pathway

Aerobic bacteria introduce unsaturated bonds into preformed saturated long chain fatty acyl-CoA derivatives by an oxygen-dependent, iron-requiring process[21] (Figure 4) apparently similar to that found in higher plants and animals. The major differences are that with rare exceptions,[22,23] only a single double bond is inserted, and the positional specificity is quite varied from species to species. For example, *Mycobacterium phlei* desaturates palmitate mainly at the 10 position, while *Bacillus megaterium* converts both palmitate and stearate to the *cis*-5 unsaturated derivatives. Probably the most common products are palmitoleic and oleic acids (both *cis*-9-isomers).

2. Regulation of Aerobic Desaturation

Fulco and co-workers carried out extensive studies on the temperature-induced changes in fatty acid composition in *Bacillus megaterium.* The findings, as summarized by Fulco,[24] provide strong evidence for a direct effect of temperature on the Δ^5-desaturase enzyme and its rate of synthesis. The essentials of the system are outlined in Figure 5. Desaturation seems to be affected by three factors:

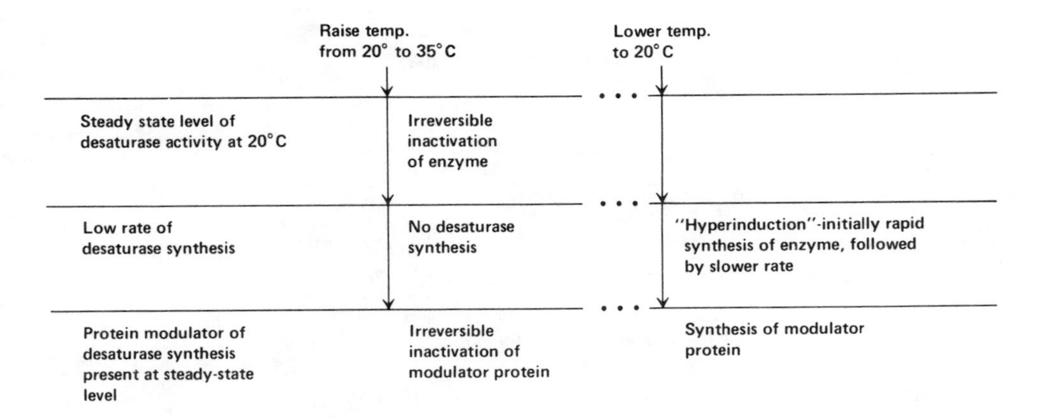

FIGURE 5. Outline of fatty acid desaturation control in *Bacillus megaterium*. (Modified from Fujii, D. K. and Fulco, A. J., *J. Biol. Chem.*, 252, 3660–3670, 1977. With permission.)

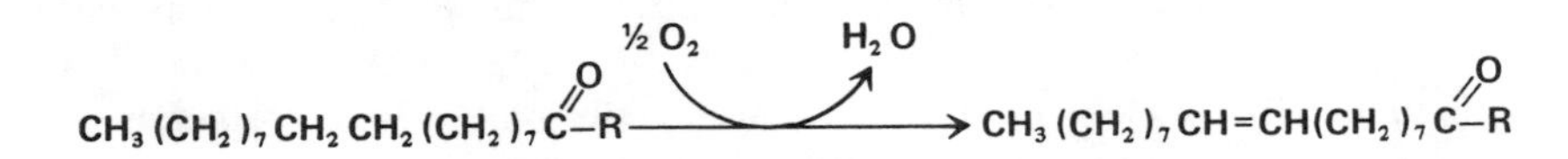

FIGURE 6. Generalized fatty acid desaturation in eukaryotic cells.

1. There is an irreversible loss of desaturase activity at high temperature.
2. Desaturase synthesis stops at high temperature, but then resumes at an unexpectedly rapid rate when cells are shifted from a high to a low temperature.
3. A variety of kinetic studies and inhibition experiments suggests the presence of a protein modulator capable of regulating the transcription of the desaturase gene by governing the synthesis of desaturase mRNA.

The striking hyperinduction phenomenon, which permits cells to form unsaturated fatty acids rapidly in response to a fall in temperature, is thought to result from a lag in the development of functional modulator complexes. These active complexes are envisioned as oligomers, whose formation depends upon first achieving a high cellular concentration of the inactive monomers.

C. DESATURATION IN AEROBIC EUKARYOTES

1. The Pathway of Double Bond Insertion

Fatty acid desaturation in eukaryotic cells is basically similar to that carried out in aerobic bacteria. However, the much broader specificity for double bond number and sequence of addition present complications that have not yet been clarified experimentally. Information concerning desaturation of fatty acids in a wide variety of animals and plants has been thoroughly analyzed by Gurr.[17] Although Gurr rightly emphasizes the frequently overlooked similarities among the various types of organisms, there appear to be significant differences between the pathways of desaturation in animals and plants, particularly as regards the regulation of desaturation. This section outlines those aspects of the pathway that all aerobic animals and plants have in common. The more specialized properties of the pathway will be described in later sections as they are applicable to the process of regulation.

The desaturation reaction can be formulated in its most general way, as follows (Figure 6). The form of the substrate is not specified because the enzymes appear capable of desaturating acyl-CoA derivatives, acyl-ACP derivatives, and even phospholipid-bound acyl groups, de-

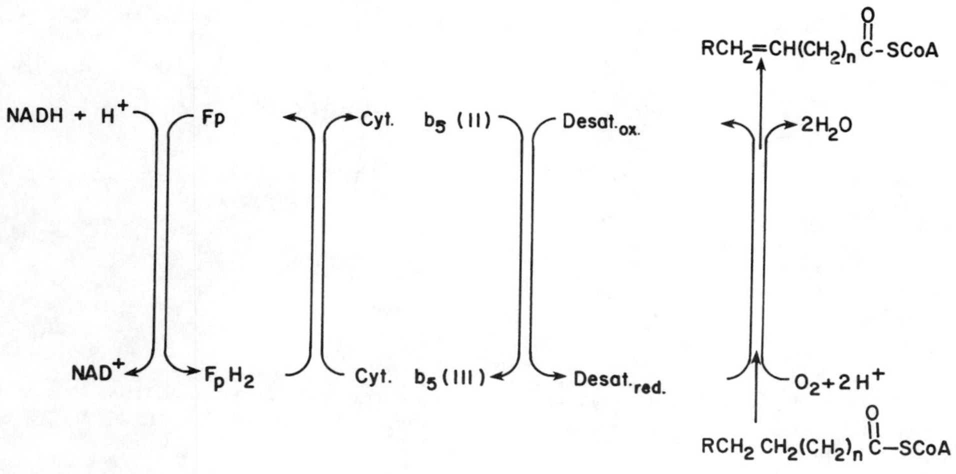

FIGURE 7. The electron transport system providing electrons for the insertion of the Δ^9 double bond into stearoyl-CoA.

pending upon the particular system under consideration. All systems do have an absolute requirement for O_2, and no effective artificial electron acceptors have been discovered. The natural electron donors are reduced pyridine nucleotides. In many systems NADH is most effective, but in some cases there is a strong preference for NADPH.[17]

Electrons supplied by the reduced pyridine nucleotides are transferred to the desaturase active site via an electron transfer chain. In rat liver microsomes, the electron flow appears to pass through a flavoprotein and cytochrome b_5 before becoming available for use by the desaturase (Figure 7). Cytochrome P_{450}, although present in the preparations, is not involved in this chain. Purified cytochrome b_5 is an amphipathic protein of mol wt 16,700.[25] The active site is on the hydrophilic head oriented towards the cytoplasmic side of the endoplasmic reticulum, while the protein is affixed to the membrane by a hydrophobic tail some 44 amino acid residues in length. It appears that both the microsomal NADH-cytochrome b_5 reductase and the NADPH-cytochrome b_5 reductase are also bound to the endoplasmic reticulum in much the same fashion as cytochrome b_5.[26] Stearoyl CoA desaturase has been purified from rat liver microsomes by Strittmatter et al.,[27] and its gene has been isolated and sequenced.[28] The desaturase is a 53,000 dalton single polypeptide containing one atom of nonheme iron. The molecule is very hydrophobic, containing 62% nonpolar amino acid residues, and requires phospholipid as well as O_2, stearoyl-CoA, NADH, and the two electron transport proteins for activity. Cytochrome b_5 has also been implicated in the Δ^6-desaturation of linoleic acid to γ-linolenic acid.[29]

Spectrophotometric measurements showed that the reduction level of cytochrome b_5 in microsomes could be decreased by the addition of stearoyl-CoA. However the decrease due to stearoyl-CoA desaturation was much less rapid than the rate of cytochrome b_5 reduction in the presence of NADH. It thus appears that the transport chain has the capacity to furnish electrons for other microsomal reactions besides fatty acid desaturation per se.

The mechanism whereby hydrogen is extracted from the substrate has not been unequivocally determined. However, kinetic isotope effects observed with deuterium labeled fatty acids and tritium labeled fatty acids in preparations from *Chlorella*, liver, and bacteria[30-32] suggest that desaturation involves the concerted extraction of a pair of hydrogen atoms rather than the participation of some form of oxygen-containing intermediate.

Fatty acid desaturation in plants requires plant ferredoxin, ferredoxin:NADP$^+$ reductase,

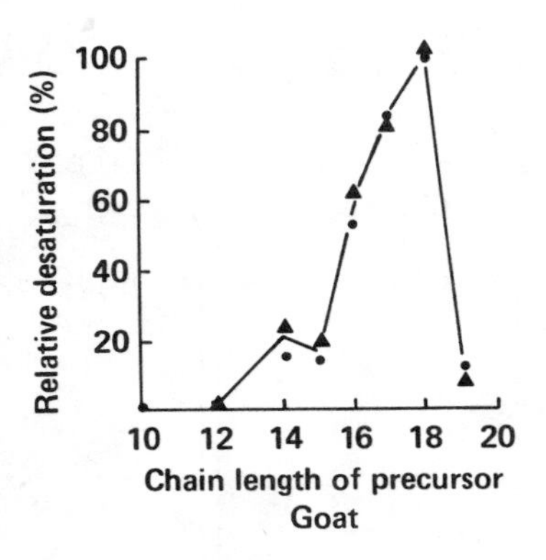

FIGURE 8. Effect of chain length on fatty acid desaturation in microsomes from goat liver. (From Gurr, M. I., Robinson, M. P., James, A. T., Morris, L. J., and Howling, D., *Biochim. Biophys. Acta,* 280, 415–421, 1972. With permission.)

NADPH, and oxygen.[33,34] The stearoyl-ACP desaturase systems of both *Euglena*[33] and developing safflower seeds[34] are inhibited by KCN, but not by CO. Because neither ferredoxin nor NADPH oxidase are affected by KCN, it was concluded[34] that the cyanide-sensitive factor lies between those two components and could be the fatty acid desaturase itself.

2. Positional Specificity of Fatty Acid Desaturases

It is generally considered that animal and higher plant cells contain several distinct desaturases, each with its own characteristic positional specificity. Recent studies with desaturase mutants of the plant *Arabidopsis thaliana* [35,36] have clearly shown this to be true. The most extensively studied desaturase is the enzyme responsible for the formation of nonoenoic fatty acids. The most widespread of these desaturases acting on saturated precursors inserts a *cis*-double bond at the Δ^9 position (between carbons 9 and 10, counting from the carboxyl end). This specificity is maintained although the chain length of the substrate may vary from C_{10} to C_{22}, suggesting that the fatty acid is positioned at the active site of the enzyme through binding of its carboxyl group. Fatty acids of certain chain lengths are preferred substrates, with the highest reactivity belonging to the C_{18} precursor in most animals and plants (Figures 8 and 9).

In some species, e.g., *Chlorella*, a separate desaturase capable of inserting double bonds at the Δ^7 position has been described.[37] The recent use of more sensitive analytical procedures has revealed that monoenes having double bonds in positions other than Δ^9 are more widespread than previously thought.[38]

The specificity of the desaturase capable of inserting a second double bond into a monoene chain has been tested in several species, but perhaps most thoroughly in *Chlorella*.[39] A wide variety of monoenes differing in chain length and double bond position were administered to the cells. In those substrates that were desaturated further, the second double bond was, with only one exception, placed on the side of the original double bond opposite the carboxyl group so as to produce a methylene-interrupted diene. Formation of this new bond generally required that the initial double bond have a *cis* configuration and be either at the Δ^9 position or at a position 9 carbon atoms from the methyl end of the chain (Table 2). The double bond of the most prevalent natural substrate, oleic acid, possesses both of the above-positional characteristics, and not surprisingly was the most reactive compound tested. The exception to the rule was *cis*-12-octadecenoic acid, which was efficiently reduced to Δ9, 12-octadecadienoate.

The ability to form a diene by insertion of a second double bond on the methyl terminal

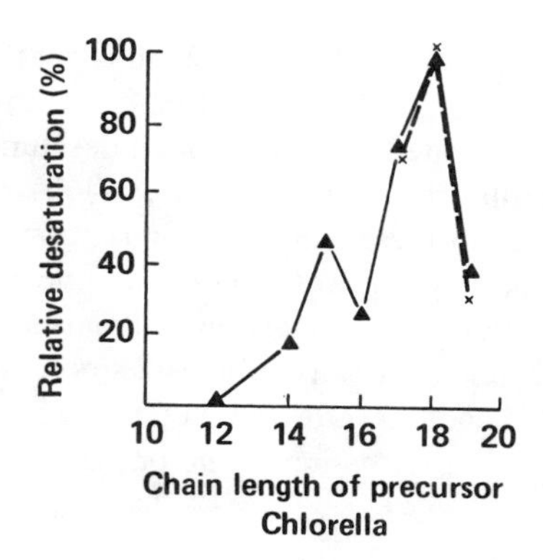

FIGURE 9. Effect of chain length on fatty acid desaturation in homogenates from *Chorella pyrenoidosa*. The symbols ▲ and × represent two sets of experiments. (From Gurr, M. I., Robinson, M. P., James. A. T., Morris, L. J., and Howling, D., *Biochim. Biophys. Acta,* 280, 415–421, 1972. With permission.)

TABLE 2
The Conversion of Monoenoic to Dienoic Fatty Acids by *Chlorella vulgaris*

Precursor	% Direct conversion[a]	Double bond position
Group I		
cis-9-Hexadecenoic (*n*-7)[b]	21.0	9-10, 12-13
cis-9-Hepadecenoic (*n*-8)	37.4	9-10, 12-13
cis-9-Octadecenoic (*n*-9)	79.0	9-10, 12-13
cis-9-Nonadecenoic (*n*-10)	25.0	9-10, 12-13
Group II		
cis-7-Hexadecenoic (*n*-9)	36.0	7-8, 10-11 (*n*-6, *n*-9)
cis-8-Heptadecenoic (*n*-9)	26.0	8-9, 11-12 (*n*-6, *n*-9)
cis-9-Octadecenoic (*n*-9)	79.0	9-10, 12-13 (*n*-6, *n*-9)
cis-10-Nonadecenoic (*n*-9)	24.1	10-11, 13-14 (*n*-6, *n*-9)
Group III		
cis-7-Octadecenoic (*n*-11)	< 1	~~
cis-8-Octadecenoic (*n*-10)	< 1	~~
cis-10-Octadecenoic (*n*-8)	< 1	~~
cis-11-Octadecenoic (*n*-7)	< 1	~~
cis-12-Octadecenoic (*n*-6)	48.0	9-10, 12-13
Group IV		
trans-9-Octadecenoic (*n*-9)	< 1	~~
9-Octadecynoic	< 1	~~
cis-9,18-Nonadecadienoic (*n*-9)[c]		

[a] In some cases breakdown and resynthesis *de novo* occurred.
[b] Presumed precursor from *cis*-l8 nonadecenoic acid. The figure of 72.7% represents radioactivity in the monoene and diene fraction. Individual figures are not available.
[c] Presumed precursor from palmitic acid.

(From Howling, D., Morris, L. J., Gurr, M. I., and James, A. T., *Biochim. Biophys. Acta,* 260, 10–19, 1972. With permission.)

side of the initial unsaturated bond is typical of plants, but is not found in higher animals. Most dienoic acids (principally linoleic acid, Δ9,12-18:2) found in animal tissues are of plant origin. These dienes can be further desaturated by some animal desaturases to form γ-linolenic acid (Δ6,9,12-18:3) or, after elongation, arachidonic acid (Δ5,8,11,14-20:4).[40] While the desaturases yielding polyunsaturates are almost certainly different from the enzyme responsible for monoenoic fatty acid formation, they appear to share the same microsomal electron transport system.[41] Note that in these acids, the additional double bonds are added towards the carboxyl end. Plants, as stated above, characteristically synthesize more highly unsaturated fatty acids by continuing to insert double bonds towards the methyl end, forming such structures as α-linolenic acid (Δ9,12,15-18:3). The formation of the principal polyunsaturated fatty acids is outlined in Figure 10.

3. Regulation of Fatty Acid Desaturation in Animal Cells

a. Nature of the Substrate

The ability of microsomal systems prepared from animal tissues to catalyze the desaturation of CoA derivatives of saturated and unsaturated fatty acids has been well established for some years.[42] Free fatty acids are not utilized as substrates by the desaturases. However, despite early indications to the contrary,[43] there is widespread evidence that phospholipid-bound fatty acyl chains can also be desaturated.[44] These phospholipid substrates for desaturation have been demonstrated in protozoa, yeast, higher plants, and rat liver as well as other tissues. One of the most common reactions is the desaturation of phosphatidylcholine- or phosphatidylethanolamine-bound oleate to linoleate. This Δ^{12} desaturation appears to utilize the same cytochrome b_5-dependent electron transport system of the endoplasmic reticulum that supplies electrons for the Δ^9 desaturation of stearoyl-CoA (Figure 7). In rat liver, desaturation can also occur at the Δ^5 and Δ^6 positions of phospholipid-bound oleate and eicosatrienoate, respectively. This reaction proceeds more effectively than on CoA esters of the same fatty acids.

b. Fate of the Reaction Products

Whatever the preferred *in vivo* substrate may be, the desaturation products are found almost entirely in phospholipid form in most tissues. In fact, it is this rapid appearance in phospholipids which makes it difficult to establish the true substrate. Even *in vitro* systems frequently transfer the majority of their desaturated fatty acyl chains to phospholipids through action of the acyl transferases present.[17]

One advantage gained by the rapid disposal of any desaturase-generated unsaturated fatty acyl-CoA is that pronounced inhibition of the enzymes by an accumulation of CoA-bound products[17] can be thus avoided. On the other hand, dissemination of fatty acyl-CoA derivatives into phospholipids may be so fast in some cases as to deplete not only the pool of products, but also the supply of substrate, thus reducing the rate of desaturation. Available in the cytosol of many animal cells as a potential buffer for maintaining an optimal concentration of free fatty acids and their CoA esters is a family of abundant 14- to 15-kDa binding proteins.[45] These proteins generally have a stronger affinity for free fatty acids than fatty acyl CoAs, but a recently discovered binding protein of rat and bovine liver prefers the fatty acyl CoA derivatives.[46] While these proteins have been implicated in relieving the inhibition of acetyl-CoA carboxylase activity,[47] they have not yet been shown to influence desaturation.

c. Regulation via Control of Cofactor Levels

One must consider the possibility that fatty acid desaturation is controlled by limiting the concentration of essential cofactors, or in the case of oxygen, cosubstrate. The degree of microsomal fatty acid unsaturation in the yeast *Candida lipolytica* is increased under conditions featuring higher aeration.[48] Oxygen has also been proposed to be the rate-limiting factor

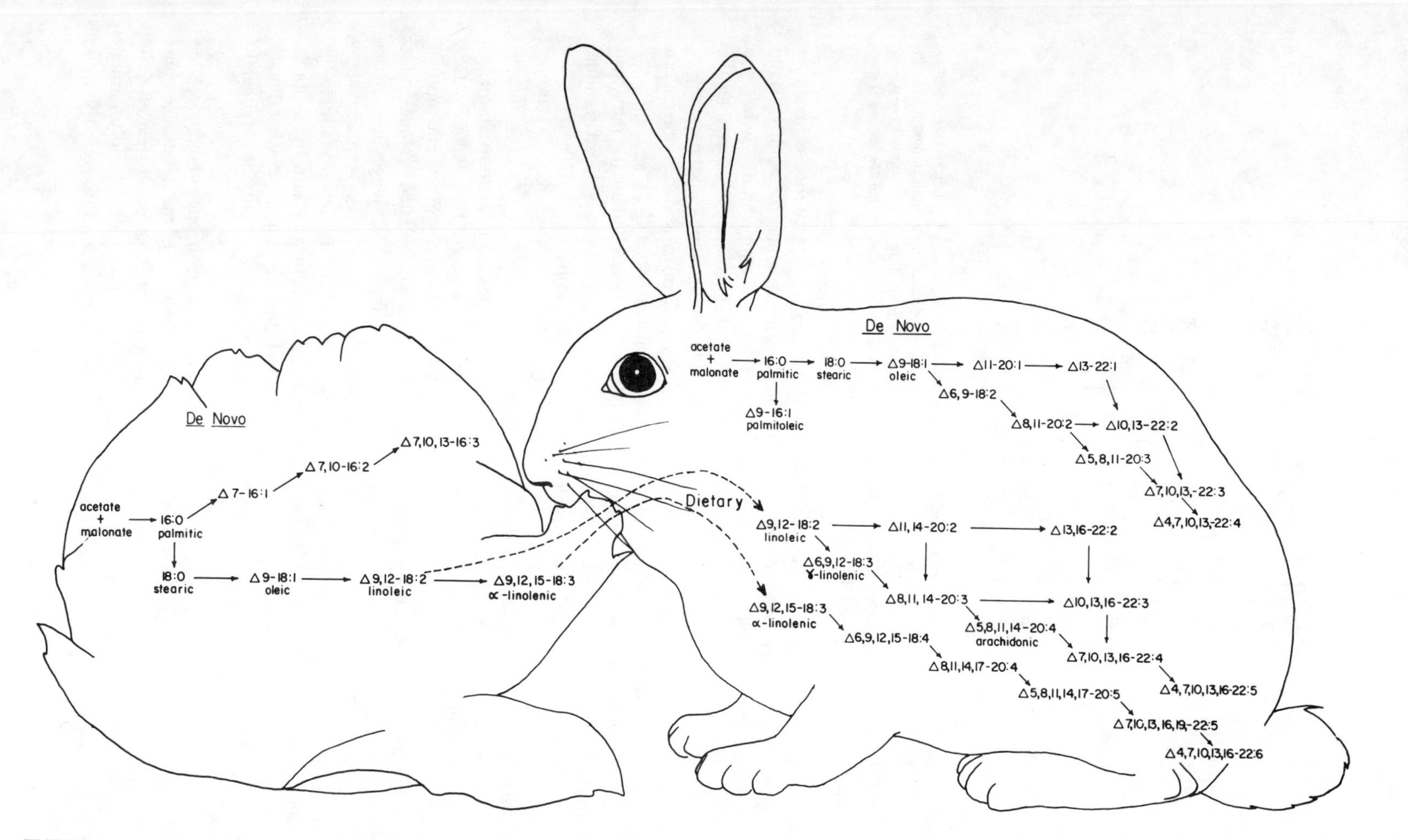

FIGURE 10. Important pathways of unsaturated fatty acid biosynthesis in plants and animals. Note the conversion of dietary 18:2 from plant sources to γ-18:3 and 20:4 in animals, and the further desaturation of plant-derived α-18:3 by animals.

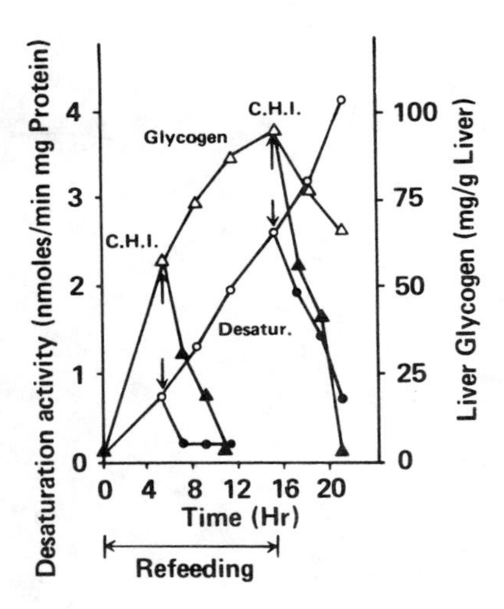

FIGURE 11. Effect of cycloheximide on the induction of desaturase activity in refed rats. Cycloheximide (0.2 mg/100 g body weight) was injected intraperitoneally at 5 or 16 hr after initiation of refeeding as indicated by arrows in the figure. Glycogen content in control liver (—△—), liver glycogen content in the drug-treated animals (—▲—), desaturation activity in control animals (—○—), desaturation activity in the drug-treated animals (—●—). (From Oshino, N. and Sato, R., *Arch. Biochem. Biophys.*, 149, 369–377, 1972, With permission.)

in desaturation by certain higher plant cells,[49] but there is not direct evidence for this in animal tissues. While it is true that poikilothermic animals, such as the frog and the goldfish, have a higher degree of fatty acid unsaturation when grown at low temperatures, where the solubility of oxygen in water is greater, it is not excluded that parameters other than the concentration of dissolved oxygen are primarily responsible for the altered lipid pattern (see Section 3.d). The unicellular animal cell, *Tetrahymena pyriformis*, possessed the same pattern of fatty acids when grown over a wide range of oxygen concentrations.[50]

Most animal systems reveal a preference of NADH as an electron donor for fatty acid desaturation. Limitation of the rate of desaturation due to insufficient reduced pyridine nucleotides does not appear to be a physiologically significant problem.

d. Regulation via Changes in Desaturase Activity

There is a striking decrease in the activity of stearoyl-CoA desaturase activity in microsomes isolated from the liver of fasted rats as compared with microsomes from fed animals.[51] These findings complemented the earlier report of low desaturase activity in diabetic rats and its restoration to normal by insulin treatment.[52] Oshino and Sato[51] showed that cycloheximide blocked the increase of desaturase activity in refed rats (Figure 11), suggesting to the investigators that this was a true enzyme induction rather than merely being an activation stemming from the increased synthesis of fatty acids. The authors did not investigate the interesting possibility that parts of the fatty acid desaturase system may turn over rapidly enough to decrease significantly in amount when synthesis is blocked, no matter what dietary regime is invoked. Rat liver stearoyl-CoA desaturase has a half life of only approximately 4 hours.[53]

There was evidence that the protein component most sensitive to changes in dietary status was the cyanide sensitive factor, which participates as a component of the desaturase complex.[54] Prasad and Joshi[55] were subsequently able to show that the depressed stearoyl-CoA desaturation in microsomes from diabetic rats could be restored to normal simply through addition of the purified terminal desaturase. These latter workers also discovered that the feeding of fructose, which can be utilized by diabetic rats, restored full desaturase activity. This points to some glycolytic intermediate as the inducer of desaturase synthesis.

Jeffcoat and James[56,57] observed that while a high carbohydrate diet stimulated fatty acid desaturase activity (as well as fatty acid synthetase activity) in isolated rat liver microsomes, a fat supplemented diet, particularly if it contained unsaturated fatty acids, reduced desaturase activity. The control of Δ^6 and Δ^5 desaturases was less sensitive to dietary changes than was that of Δ^9 desaturase. It was suggested that unsaturated fatty acids might repress induction of the desaturase enzyme.

An alternative explanation is that the sizable amount of the fed unsaturated fatty acids that was incorporated into hepatic membranes (e.g., see Hopkins and West[58]) changed the physical properties of the membrane lipid bilayer in such a way as to partially and reversibly inactivate the desaturase. This latter mechanism was not supported by experiments in which microsomes isolated from high-desaturase-activity rats (fed a low-fat diet) and low-desaturase-activity rats (fed a safflower oil diet) were each delipidated and reconstituted with a common lipid.[59] In this case, the desaturase activity was restored to nearly the same very different levels measured in the respective starting preparations. It is now generally agreed that the fluctuation of rat liver stearoyl-CoA desaturase activity in response to changes in dietary status or hormone levels comes largely through alterations in the amount of desaturase present. As mentioned above, sharp declines in stearoyl-CoA desaturase levels could in theory be generated through the rapid metabolic turnover of the enzyme coupled with reduced synthesis. Conversely, dietary induction by feeding carbohydrates to fasted animals has been shown to effect a 30 to 60 fold increase in stearoyl-CoA desaturase m-RNA levels over a period of 20 hr.[60]

A large decrease in Δ^9-desaturase activity was noted in yeast cells fed unsaturated fatty acids.[61] The decrease was traced to reduced levels of desaturase m-RNA, but the mechanism underlying the reduced transcription was not determined.

Extensive studies of fatty acid desaturase regulation have been conducted with the protozoan *Tetrahymena pyriformis*. Under most experimental circumstances, *Tetrahymena* synthesizes fatty acids solely for use in membrane phospholipids. Therefore, there is no need to differentiate between events regulating that pathway and the pathway leading to the production of storage fats. As described on page 206, the fatty acids of phospholipids from *Tetrahymena* growing at low temperatures are much more highly unsaturated than those of cells grown at a higher temperature.[62-64] A detailed analysis of the control of fatty acid desaturation in this organism has been undertaken by the laboratories of Nozawa and Thompson,[65] using a thermotolerant strain that grows well over a broad temperature range. Through the use of electron spin resonance[64] and freeze fracture electron microscopy[66] for estimating the fluidity of the various intracellular membranes, it was determined that the increased desaturase activity induced by shifting cells to lower temperatures was correlated with a more viscous lipid milieu in the lipid bilayers. Related experiments established that several factors that can alter membrane fluidity result in modification of desaturase activity.[67,68] Thus fluidizing additives, such as exogenous polyunsaturated fatty acids or general anesthetic, inhibited one or more desaturases while rigidifying factors activated the desaturases.

The changes in *Tetrahymena* desaturase activity were found to be quite independent of the level of dissolved O_2.[49] But evidence exists for at least two mechanisms of desaturase regulation, namely, through modulating the activity of existing enzyme molecules and through changing the absolute number of enzyme molecules. These two modes of control appear to operate in tandem to help readjust *Tetrahymena* membrane fluidity perturbed by environmental stress.[69]

On the basis of experiments with *Tetrahymena*, it was postulated[70] that low temperature-induced rigidification of membrane lipid molecules might cause conformational changes in the membrane embedded desaturase molecules, leading them to retain a higher degree of enzymatic activity upon chilling than do other lipid biosynthetic enzymes. Under such conditions the level of fatty acid unsaturation would gradually rise. As the increasing unsaturation "refluidized" the endoplasmic reticulum, the physical constraints favoring high desaturase activity would be relaxed until a new, lower level of activity was reached. Membrane fluidity would thereby be self-regulating. Data best interpreted in this way have been obtained using several other organisms as well.[71]

Membrane fluidity can also be regulated in the same *Tetrahymena* cells through the induced synthesis of fatty acid desaturase. In low temperature-stressed *Tetrahymena* cycloheximide prevented the three- to fourfold increase in palmitoyl-CoA and stearoyl-CoA desaturase and blocked smaller increases in components of the electron transport chain coupled to desaturation.[72] These changes in fatty acyl-CoA desaturase levels may be determined in part by a membrane fluidity-induced reduction in the normally rapid metabolic turnover of these proteins.[69] Increased amounts of stearoyl-CoA desaturase have also been noted in avian liver following a reduction of membrane fluidity by estradiol injection.[73] On the other hand, little evidence is available for the induced synthesis of monoenoic or polyenoic fatty acid desaturases.

4. Regulation of Fatty Acid Desaturation in Plant Cells

a. Nature of the Substrate

The production of unsaturated fatty acids is a more complex process in higher plants than in animals. Desaturation occurs both in the cytosol and in the chloroplasts, and many properties of the two systems are different. A combination of direct and indirect evidence indicates that the chloroplast pathway for desaturating stearate requires the stearoyl-ACP derivative as substrate. This was shown directly in spinach chloroplasts[33] and by inference in *Euglena*.[74,75] In the latter case, *Euglena* grown with organic nutrients in the dark so that they lacked functional chloroplasts could desaturate only stearoyl-CoA, whereas light-grown cells required stearoyl-ACP.

Cotyledons of developing soybean seeds contained, in their cytosol, a desaturase that acted on stearoyl-ACP, but not stearoyl-CoA or free stearic acid.[76] The activity was no longer present in the seeds following germination.

Differences also appear in the synthesis of polyunsaturated fatty acids. Again, many of the first indications came from study of *Euglena*, with its mixture of plant and animal characteristics. Dark-grown cells contained fatty acids typical of animals, e.g., arachidonic acid, while light-grown, chloroplast-containing cells were more typical of plants in having linoleic and α-linolenic acids.[77,78] The plant-like pattern of fatty acids in the photoautotrophic *Euglena* was consistent with the finding of large amounts of galactosyl diacylglycerols, whereas the dark-grown cells contained primarily phospholipids.

By feeding *Euglena* a synthetic radioactive fatty acid postulated to be an intermediate only in the animal-like pathway of the organism, Gurr and colleagues[17] demonstrated that components of this biosynthetic sequence do not enter the other pathway in chloroplasts. The radioactive substrate and its metabolic products were incorporated into phospholipids, but not glycolipids.

Research over the past few years[79] has confirmed that unsaturated fatty acids of both higher plants and algae are synthesized by two separate pathways, one in the chloroplast and the other in the endoplasmic reticulum. The biosynthetic scheme is outlined in Figure 12. The initial product of *de novo* fatty acid synthesis, palmitoyl-ACP, arises via a soluble fatty acid synthase system in the chloroplast stroma (see Chapter 2). Much of the palmitoyl-ACP is further elongated to stearoyl-ACP by a stromal elongase, and this latter compound is largely desaturated by another soluble stromal enzyme, stearoyl-ACP desaturase.[80]

The fate of the two principal fatty acid products, palmitoyl-ACP and oleoyl-ACP, very much depends upon the species of plant in question. Some plants, such as spinach, *Arabidopsis thaliana*, and the green alga *Chlamydomonas reinhardtii*, synthesize a substantial proportion of their abundant chloroplast lipids *in situ* through pathways involving the stepwise desaturation of oleic acid to linoleic and linolenic acids and palmitic acid to hexadecatrienoic and sometimes hexadecatetraenoic acids. The double bonds are introduced into fatty acids esterified to phosphatidylglycerol, galactolipids, and sulfolipids synthesized within the chloroplast by the "prokaryotic pathway", so named because properties of the biosynthetic enzymes resemble those of bacterial lipid-forming enzymes.[79]

Other portions of the newly-synthesized palmitoyl-ACP and oleoyl-ACP are hydrolyzed to free acids[81] and then converted to CoA derivatives in the chloroplast envelope for export to the endoplasmic reticulum. Further desaturation of the oleate chain takes place during phospholipid synthesis via the so-called "eukaryotic pathway".[79] In *Chlorella vulgaris* [82] and pumpkin leaves[83] the favored substrate for desaturation in the E.R. is oleate esterified to phosphatidylcholine. But the principal site of linoleate to linolenate desaturation is less certain. Evidence (reviewed by Jaworski[80]) has been presented for linoleate desaturation both on endoplasmic reticulum-associated phosphatidylcholine and on chloroplast-associated galactolipids. The relative contribution of each site is not easy to establish because most plants transport large quantities of linoleate and linolenate from microsomal membranes back into the chloroplast for use in glycolipid synthesis (Figure 12).

Due to the extreme lability of the membrane-bound oleate and linoleate desaturases, it has not been possible to purify the enzymes from E.R. and chloroplasts and compare their properties. Somerville and associates have approached the question from a genetic point of view by analyzing fatty acid desaturase mutants of *Arabidopsis*. Two mutants deficient in the desaturation of linoleate were studied in detail. The fadD mutant contained a defective ω-3 desaturase resulting in lowered linolenate concentrations in membranes of the chloroplast as well as other organelles.[84] This desaturase was thought to be localized in the chloroplast, since levels of the characteristic chloroplast fatty acid hexadecatrienoate were also sharply reduced. Decreased levels of linolenate outside the chloroplast were attributed to a diminution of the usual flux of chloroplast-derived linolenate into other cellular compartments to supplement the linolenate formed by a microsomal ω-3 desaturase.

A second *Arabidopsis* mutant, fadC, was deficient in the desaturation of oleate within the chloroplast.[36] Oleate accumulated in all the chloroplast lipid classes and, to a lesser extent, in the other cellular phosphoglycerides, again suggesting an enhanced export of chloroplast-produced oleate in the mutant cells.

b. Regulatory Mechanisms in Plants

Just as is the case with animals, plants clearly possess the ability to regulate the pattern of unsaturated fatty acids that they synthesize. This ability is most dramatically revealed in studies of plant acclimation to changing environmental factors such as temperature[85] or salinity[86] (see pp. 210 and 213).

The mechanisms for regulating fatty acid desaturase activity under these various conditions of stress are not yet well understood. Harris and James[87] proposed a very simple system for controlling desaturation in nonphotosynthetic tissues such as seeds and bulbs. Here the supply of dissolved oxygen can be low enough to limit desaturation. At high temperatures, the decreased solubility of oxygen in water reduced desaturation significantly, providing the more highly saturated mixture of fatty acids needed for synthesizing membranes of optimal fluidity

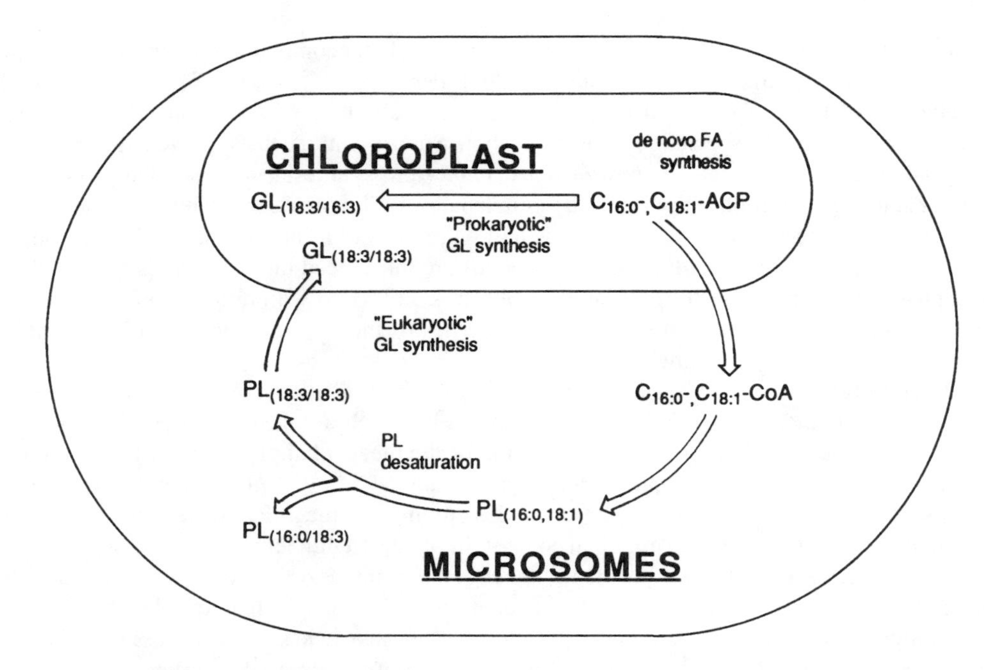

FIGURE 12. Formation of unsaturated fatty acids via the "prokaryotic" and the "eukaryotic" pathways in plants.

at the elevated temperature. The degree of fatty acid unsaturation in nonphotosynthetic, cultured sycamore cells was also affected more by the concentration of dissolved O_2 than by temperature changes per se.[88] This effect of temperature is less noticeable in green tissues, which presumably maintain a higher concentration of dissolved oxygen by virtue of their capacity to generate it photosynthetically. It is conceivable that a gradient in oxygen tension might exist even at the subcellular level. Wilson and Crawford[85] observed that temperature change produced a marked alteration in the degree of saturation in leaf phospholipid fatty acids while the glycolipids, which constitute the bulk of the chloroplast lipids, did not change their fatty acid pattern.

Other arguments have been advanced to explain the generally observed decrease in lipid unsaturation found in plants grown at higher temperatures. Thus Browse and Slack[89] proposed that the rate of oleic acid biosynthesis in safflower cotyledons developing at high temperature exceeded the capacity of further desaturation reactions to keep pace.

Lower plants may have their own separate regulatory schemes for fatty acid desaturation. The yeast, *Torulopsis utilis*, has been clearly shown to possess greater stearoyl-CoA desaturase activity when grown at low temperatures.[90,91] It was not possible to confirm whether the observed differences were caused mainly by variations in desaturase quantity, specific activity, or in the availability of the O_2 cosubstrate.

V. METABOLISM OF CYCLOPROPANE FATTY ACIDS

Cyclopropane fatty acids are common in both gram-positive and gram-negative bacteria. Their biochemistry was recently reviewed by Cronan and Rock.[18] The *cis*-cyclopropane ring is formed by action of the enzyme cyclopropane synthetase on phospholipid-bound unsaturated fatty acids (Figure 13). All of the more common bacterial phospholipids can serve as acceptors of the methyl group from S-adenosylmethionine, and the cyclopropane ring can be inserted into fatty acids bound at either the 1 or the 2 position of the glycerol moiety.[92]

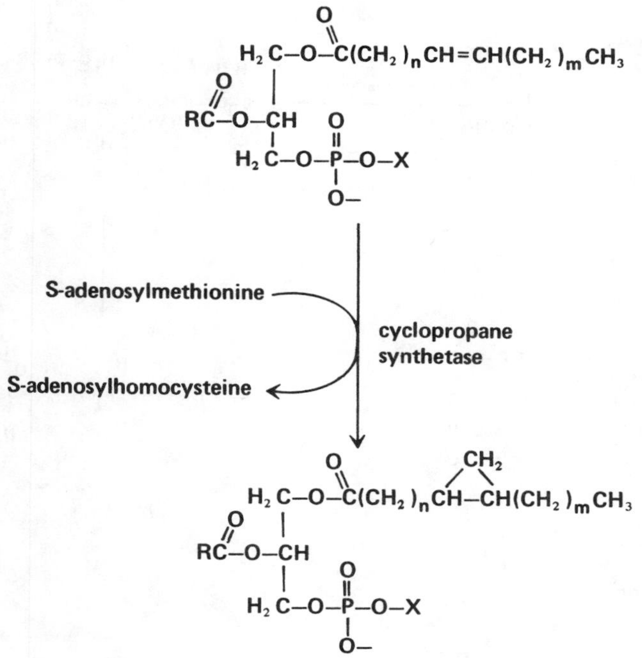

$$
\begin{array}{l}
H_2C{-}O{-}\overset{O}{\overset{\|}{C}}(CH_2)_nCH{=}CH(CH_2)_mCH_3 \\
RC(=O){-}O{-}CH \\
H_2C{-}O{-}P(=O)(O^-){-}O{-}X
\end{array}
$$

S-adenosylmethionine → S-adenosylhomocysteine

cyclopropane synthetase

$$
\begin{array}{l}
H_2C{-}O{-}\overset{O}{\overset{\|}{C}}(CH_2)_nCH{-}CH(CH_2)_mCH_3 \quad (CH_2 \text{ bridging the two CH}) \\
RC(=O){-}O{-}CH \\
H_2C{-}O{-}P(=O)(O^-){-}O{-}X
\end{array}
$$

FIGURE 13.

Logarithmic phase *Escherichia coli* cells contain only traces of cyclopropane fatty acids, but as the culture enters the stationary phase, phospholipid-bound monoenes are transformed to the cyclopropane derivative in large numbers. Using mutants of *E. coli*, Cronan et al.[93] demonstrated that the final concentration of cyclopropane fatty acids was determined by the level of monoenoic fatty acids attained during logarithmic growth. The cellular concentration of S-adenosylmethionine was not a controlling factor. *E. coli* strains lacking the ability to form cyclopropane fatty acids and those modified to contain additional cyclopropane fatty acid synthase genes (thereby producing the cyclic acids throughout logarithmic growth) all grow normally under various environmental conditions. There is as yet no unique function known for the cyclopropane fatty acids, nor has the mechanism triggering their abrupt synthesis from the unsaturated precursors been identified.

VI. CONVERSION OF STRAIGHT-CHAIN TO BRANCHED-CHAIN FATTY ACIDS

Branched-chain fatty acids arise through two quite different biosynthetic pathways. The iso- and anteiso-fatty acids, bearing a methyl group on the methylene carbon one or two positions removed, respectively, from the terminal methyl group, are formed when fatty acid synthesis is initiated with a branched-chain substrate rather than acetyl-CoA (see Chapter 2). In the mycobacteria, the methyl branch appears in a different position, e.g., in tuberculostearic acid (10-methylstearic acid). The methyl group in this case is added to a phospholipid-bound oleic acid chain.

Many details of this reaction were elucidated through the efforts of Law and associates.[94] The basic reaction was formulated to proceed as follows:

A cytosolic enzyme from *Mycobacterium phlei* catalyzes the alkylenation of a C_{16} or C_{18}

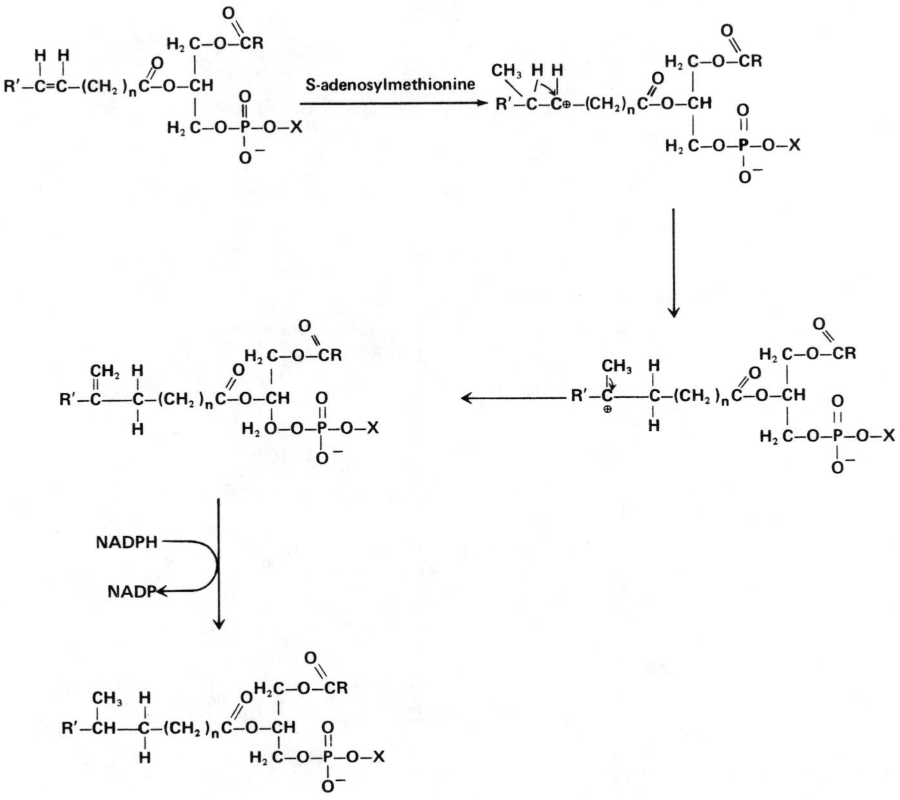

S-adenosylmethionine

NADPH

NADP

FIGURE 14.

Δ^9 monoolefinic fatty acid bound to either the *sn*-1 or *sn*-2 position of phosphatidylglycerol, phosphatidylinositol, or phosphatidylethanolamine (Figure 14). The methylene group is subsequently reduced to a methyl group with the aid of NADPH. The branched-chain compound tuberculostearic acid increases in cells entering the stationary phase of growth. The regulatory basis for the increase is not known.

VII. BIOHYDROGENATION OF UNSATURATED FATTY ACIDS

Polyunsaturated fatty acids of plant lipids are known to be hydrogenated enzymatically by microorganisms found in the digestive system of ruminant animals. Details of biohydrogenation have been discussed by Moore[95] and by Garton.[96] The impact of this process can be appreciated by considering that as much as 90% of dietary unsaturated fatty acids can be hydrogenated in the goat's rumen.[97]

It appears that dietary lipids are hydrolyzed to free the esterified fatty acids before enzymatic hydrogenation commences. Whereas reduction of the unsaturated bonds takes place within the anaerobic confines of the rumen, which has been shown to contain gaseous hydrogen, the reductant is suspected to be a more biologically familiar agent, possibly a flavin system.[98]

The system for fatty acid biohydrogenation has been studied extensively using both mixed rumen microorganisms and the isolated rumen bacterium *Butyrivibrio fibrisolvens*. The prob-

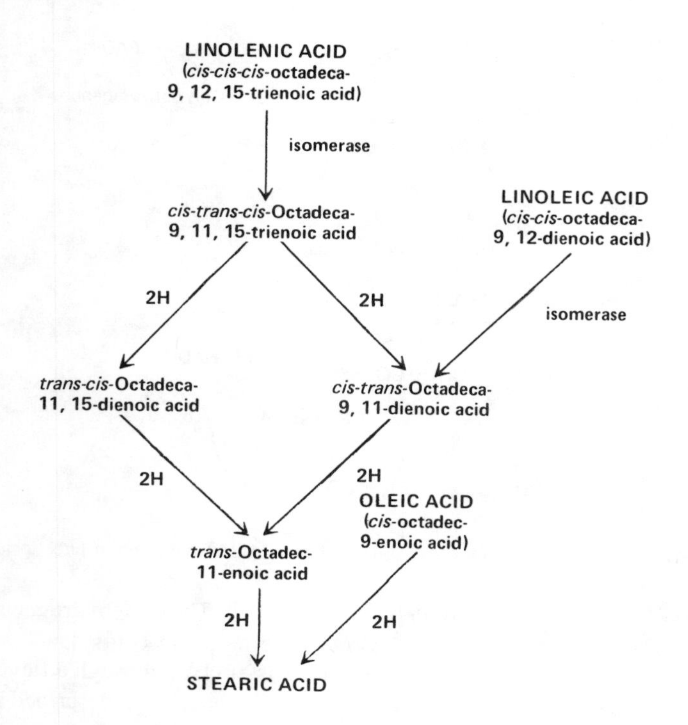

FIGURE 15. Probable principal pathways involved in the ruminal biohydrogenation of C_{18} unsaturated fatty acids. (From Garton, G. A., *Biochemistry of Lipids II*, Vol. 14, Goodwin, T. W., Ed., University Park Press, Baltimore, 1977, 337–370. With permission.)

able pathway is illustrated in Figure 15. The *B. fibrosolvens* reductase was purified and shown to require α-tocopherolquinol as a reductant.[99] The enzymes are membrane-bound and, enigmatically, appear to serve no useful purpose for the microorganisms, at least in the case of *B. fibrisolvens*, since the hydrogenated fatty acids produced are not incorporated into cellular lipids.

Biohydrogenation is generally thought of as an obligately anaerobic process, but Kepler and Tove[100] reported aerobic reduction of oleate to stearate by a strain of *Bacillus cereus* obtained from the rumen of a sheep. Interestingly, the hydrogenation system was not present if the organisms were grown at 20°C but was fully active in 37°C grown cells. Inhibitors of protein synthesis blocked induction of the activity following a temperature increase. The regulatory significance, if any, of this finding is obscure.

VIII. FATTY ACID OXIDATION

A. β-OXIDATION OF SATURATED FATTY ACIDS

The degradation of long-chain fatty acids to acetyl-CoA by the so-called β-oxidation process occurs in both mitochondria and peroxisomes of animal cells, while in plants most of the activity is found either in peroxisomes or the functionally similar glyoxysomes. The contribution of the peroxisomes to total cellular β-oxidation is variable, since peroxisomal enzymes are inducible by certain substrates and hypolipidemic drugs, leading to increases of 5 to 10 fold in flux through the pathway. Glyoxysomes are also much more prevalent at certain stages of plant development, for example, in germinating seeds.

The mitochondrial β-oxidation scheme (Figure 16) is essentially the reverse of the reactions utilized for fatty acid elongation (Figure 1). In mitochondria the initial step is catalyzed

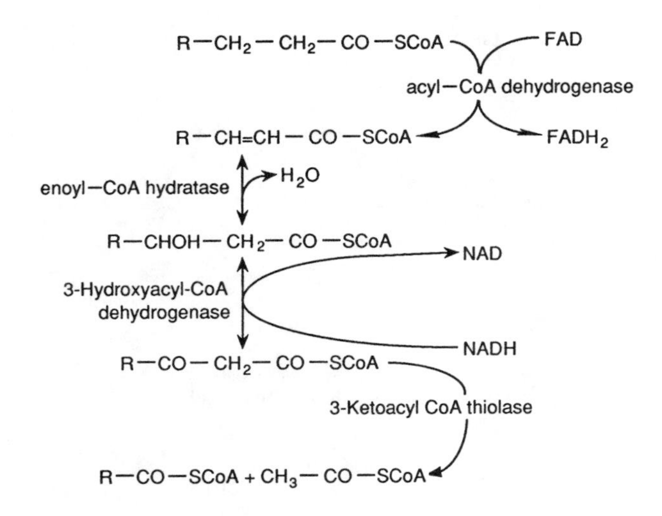

FIGURE 16. The enzymes involved in mitochondrial β-oxidation of fatty acids.

by one of three different acyl-CoA dehydrogenases.[101] These dehydrogenate long-chain, medium-chain, and short-chain fatty acyl CoAs. The reducing equivalents removed during dehydrogenation are transferred into the electron transfer chain through a flavoprotein. Properties of the other enzymes of mitochondrial β-oxidation have been described by Bremer and Osmundsen.[102]

The reactions of β-oxidation in peroxisomes resemble those in mitochondria; however, a different group of enzymes is operational. Acyl-CoA dehydrogenases are replaced by an acyl-CoA oxidase coupled to the conversion of O_2 to H_2O_2 (Figure 17). The oxidase reaction appears to be the rate-limiting step in the pathway.[103] The peroxisomal β-oxidation system is only active with octanoyl CoA and longer acyl-CoAs. An observed inability of mitochondria to degrade very long-chain fatty acids (longer than C_{22}) is probably due to the absence in mitochondria (but not peroxisomes) of an acyl-CoA synthetase capable of activating the longer chain length acids.[104] Peroxisomes do actively oxidize the very long chain fatty acids, but often through only a few round of β-oxidation, producing shortened fatty acids better suited for the energy conserving mitochondrial β-oxidation process.

The two pathways of β-oxidation in animal cells are affected quite differently by the presence of carnitine. Whereas peroxisomes can take up free fatty acids and convert them directly to acyl-CoA derivatives, the mitochondrial inner membrane is poorly permeable to free fatty acids. Passage through the inner membrane requires that the acyl moiety be transferred from CoA to carnitine. Two types of oligomeric carnitine palmitoyltransferase participate in this inward movement.[105] One, carnitine palmitoyltransferase (CPT_o), whose exact localization is still being questioned, is accessible to fatty acyl-CoA entering the mitochondrion. A second and nearly identical form of the carnitine palmitoyltransferase (CPT_i) is oriented inward toward the mitochondrial matrix. As indicated in Figure 18, acyl-CoA and carnitine are converted to acylcarnitine by CPT_o in a reversible reaction. The acylcarnitine product is then transferred across the inner mitochondrial membrane by CPT_i and hydrolyzed to release acyl-CoA into the matrix, where it is subject to β-oxidation.

This carnitine-mediated transport process offers opportunities for metabolic control of fatty acid oxidation. CPT_o is subject to inhibition by malonyl-CoA at several binding sites, and acyl-CoA is a competitive inhibitor of malonyl-CoA binding at some of these sites. Thus the

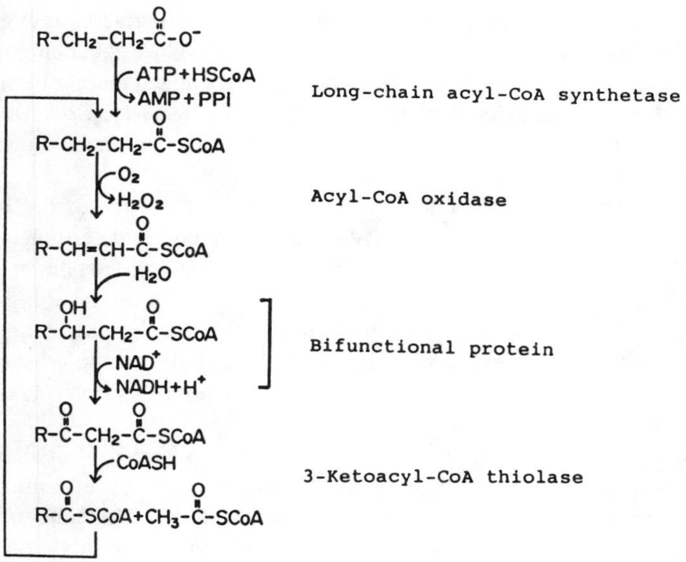

$$R\text{-}CH_2\text{-}CH_2\text{-}\overset{O}{\overset{\|}{C}}\text{-}O^-$$

$$\xrightarrow[AMP + PPi]{ATP + HSCoA}$$ Long-chain acyl-CoA synthetase

$$R\text{-}CH_2\text{-}CH_2\text{-}\overset{O}{\overset{\|}{C}}\text{-}SCoA$$

$$\xrightarrow[H_2O_2]{O_2}$$ Acyl-CoA oxidase

$$R\text{-}CH{=}CH\text{-}\overset{O}{\overset{\|}{C}}\text{-}SCoA$$

$$\xrightarrow{H_2O}$$

$$R\text{-}\overset{OH}{\overset{|}{C}H}\text{-}CH_2\text{-}\overset{O}{\overset{\|}{C}}\text{-}SCoA$$ Bifunctional protein

$$\xrightarrow[NADH + H^+]{NAD^+}$$

$$R\text{-}\overset{O}{\overset{\|}{C}}\text{-}CH_2\text{-}\overset{O}{\overset{\|}{C}}\text{-}SCoA$$

$$\xrightarrow{CoASH}$$ 3-Ketoacyl-CoA thiolase

$$R\text{-}\overset{O}{\overset{\|}{C}}\text{-}SCoA + CH_3\text{-}\overset{O}{\overset{\|}{C}}\text{-}SCoA$$

FIGURE 17. The β-oxidation sequence in peroxisomes. (From Hashimoto, T., *Peroxisomes in Biology and Medicine,* Fahimi, H. D. and Sies, H., Eds., Springer-Verlag, Heidelburg, 1987, 97–104. With permission.)

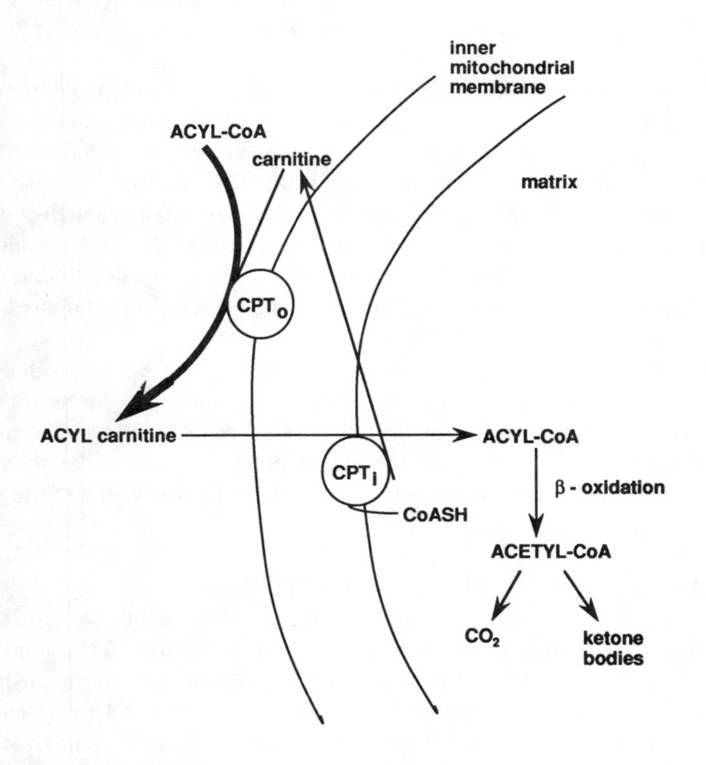

FIGURE 18. The carnitine-mediated system for transporting fatty acids across the mitochondrial inner membrane.

ratio of free palmitoyl-CoA to malonyl-CoA can determine the rate of acyl group transfer to the site of β-oxidation, with high proportions of palmitoyl-CoA enhancing oxidation.

Most cells also contain a carnitine acetyltransferase, localized mainly in mitochondria and peroxisomes.[105] In heart mitochondria this enzyme can shuttle acetyl-CoA from the mitochondrial matrix to the cytosol. It also appears to aid in the efflux of chain-shortened fatty acids from peroxisomes.

Mitochondrial β-oxidation of fatty acids is capable of supplying the body with large amounts of catabolic energy. The activity of the pathway is determined largely by the concentration of free fatty acids supplied to the cell from storage depots or dietary sources. This supply is controlled hormonally, mainly through shifts in the insulin/glucagon ratio or through changing levels of vasopressin, thyroid hormones, etc.[102] It is seemingly incidental to this often massive energy-related flux of fatty acids through β-oxidation that chain shortening and selective fatty acid degradation helps to tailor the pool of fatty acids utilized for the metabolism of membrane lipids.

In gram negative bacteria such as *Escherichia coli*, β-oxidation of fatty acids proceeds via the same chemical steps as have been described for the pathway in animal mitochondria. However, the enzymatic activities are organized differently.[106] The conversion of free fatty acid to acyl-CoA is catalyzed by a cytosolic acyl-CoA synthetase having a broad substrate specificity for medium and long chain fatty acids. The initial β-oxidation enzyme, acyl-CoA dehydrogenase, is cytosolic, as is the two protein multienzyme complex that contains the remaining five enzymatic activities. The synthesis of the β-oxidation enzymes is controlled by the *fad* regulon. Long chain fatty acids but not medium chain fatty acids can induce the expression of these genes,[107] apparently by acting through a repressor coded for by the *fad* R^+ gene.

β-Oxidation in plants has been studied most extensively in seeds, which in many species contain large reserves of fat that is rapidly mobilized during germination. The β-oxidation process occurs in peroxisome-like glyoxysomes rather than mitochondria and utilizes enzymes that resemble those found in animal peroxisomes (see Figure 17). Also present within the glyoxysomes are the enzymes of the glyoxylate cycle. β-oxidation and the glyoxylate cycle function well together because the latter pathway utilizes the acetyl-CoA produced through β-oxidation and resupplies the former pathway with the CoASH needed to continue fatty acid activation.[108] The net products of this cooperative arrangement, succinate and NADH, leave the glyoxysome for use elsewhere in the cell.

Control of β-oxidation in plant cells is exerted at the gene expression level. Enzymes of the β-oxidation pathway are induced during late seed ripening and again during seed germination. Glyoxysomes are highly enriched in the oil-rich cotyledons of seedlings. The molecular basis for the preferential synthesis of glyoxysomal components is not understood. As seedling growth proceeds and coyledons green, glyoxysomes are gradually transformed into typical peroxisomes.[109]

B. β-OXIDATION OF UNSATURATED FATTY ACIDS

β-oxidation operates on unsaturated fatty acids as described above for saturated acids until the initial double bond is approached. Depending upon its position, the double bond is either isomerized or reduced to yield an intermediate having unsaturation in the position necessary for β-oxidation to proceed normally.[110] Both types of double bond treatment are required in the oxidation of linoleoyl-CoA (Figure 19). In rat liver peroxisomes the double bond isomerization is catalyzed by a multifunctional protein having Δ^3, Δ^2-enoyl CoA isomerase activity.[111]

Recent studies of oleic acid catabolism in rat liver[112] have suggested that the expected cycle

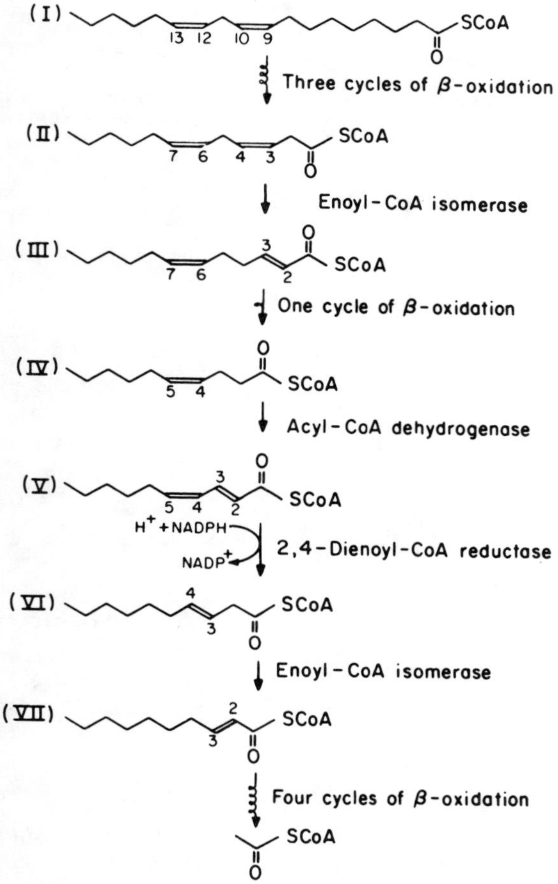

FIGURE 19. β-oxidation of unsaturated fatty acids. (From Schulz, H., *Biochemistry of Lipids and Membranes,* Vance, E. E. and Vance, J. E., Eds., Benjamin/Cummings, Menlo Park, 1985, 116–142. With permission.)

of β-oxidation converting *cis*-5-tetradecenoyl-CoA to *cis*-3-dodecenoyl-CoA does not in fact occur, probably because of an unfavorable equilibrium. With this avenue for oleate β-oxidation nonfunctional, *cis*-5-tetradecenoyl-CoA is instead reduced to tetradecanoyl-CoA by a NADPH dependent pathway, yielding a saturated product that then proceeds through the conventional steps of β-oxidation (Figure 20).

C. α-OXIDATION

An alternative mechanism for fatty acid catabolism, found primarily in plants, is α-oxidation. The discovery of this pathway in 1952 and much of its subsequent characterization is attributable to Stumpf and co-workers.[113] Our present understanding of the reaction sequence is summarized in Figure 21. A free fatty acid is first converted to a free radical derivative in an O_2-triggered reaction that simultaneously produces a hydroperoxy compound. This compound then reacts with the free radical fatty acid derivative to yield a very unstable D-α-hydroperoxyl fatty acid, which can either be decarboxylated and oxidized to form a fatty acid one carbon atom shorter, or be reduced to form a D-α-hydroxy fatty acid.

The enzymes catalyzing α-oxidation are found loosely associated with intracellular membranes of plant tissues. The capacity for removing one carbon atom from a fatty acid rather

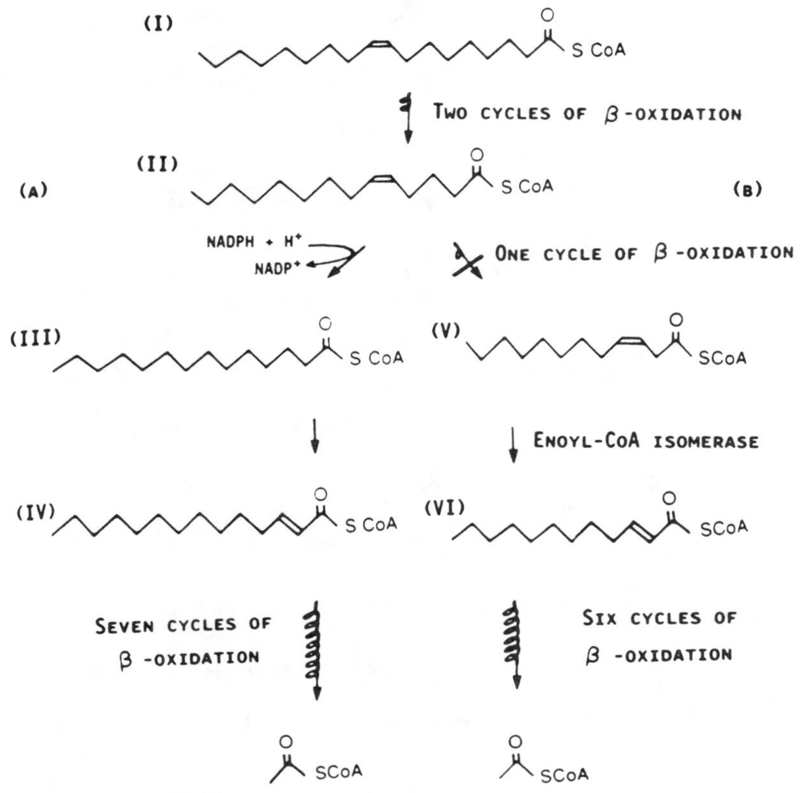

(I) S CoA

TWO CYCLES OF β-OXIDATION

(II) S CoA

(A) (B)

NADPH + H+ NADP+

ONE CYCLE OF β-OXIDATION

(III) S CoA (V) SCoA

ENOYL-CoA ISOMERASE

(IV) S CoA (VI) SCoA

SEVEN CYCLES OF β-OXIDATION

SIX CYCLES OF β-OXIDATION

SCoA SCoA

FIGURE 20. A direct, NADPH-mediated reduction (route A) allows an unsaturated intermediate to bypass an unfavorable cycle of β-oxidation (route B). (From Tserng, H.-Y. and Jin, S.-J., *J. Biol. Chem.*, 266, 11614–11620, 1991. With permission.)

than the usual two permits the double bond or other substituents of certain fatty acids to be repositioned in such a way as to be suitable substrates for the β-oxidation sequence.

Relatively large amounts of α-hydroxy fatty acids, e.g., cerebronic acid (1-hydroxy C_{24}) are found in mammalian brain as components of cerebrosides and sulfatides. These derivatives are formed from straight chain fatty acids by the action of a mixed function oxygenase requiring O_2 and NADPH. The enzyme prefers C_{20} to C_{26} fatty acids.[114] Details of the reaction mechanism are unknown. Ciliary membranes of the protozoan *Tetrahymena pyriformis* are also enriched in α-hydroxy fatty acid-containing sphingolipids.[115] In this case the substrate, palmitate, is apparently hydroxylated only after it is incorporated into an intact sphingolipid.

A different α-oxidation system is found in mammalian kidney and liver. Its principal function appears to be the oxidation of phytanic acid (3,7,11,15-tetramethyl palmitic acid), the isoprenoid product formed from the phytol tail of dietary chlorophyll. This and other such fatty acids branched in the 3 position are not susceptible to β-oxidation per se. However, following α-hydroxylation and decarboxylation, the resulting 2 methyl product pristanic acid can be degraded by β-oxidation, presumably yielding propionyl-CoA as a reaction product (Figure 22).[116,117] Singh et al.[118] have reported that the isoprenoid-derived fatty acids pristanic and heptadecanoic acids are preferentially oxidized in the peroxisomes of skin fibroblasts.

D. LIPID PEROXIDATION

A methylene carbon atom of an unsaturated lipid is susceptible to hydrogen atom abstrac-

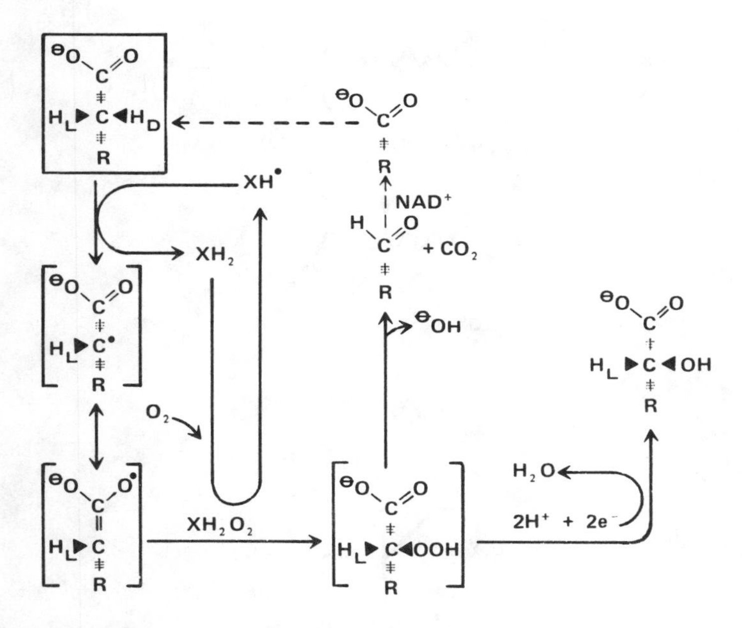

FIGURE 21. A proposed mechanism for α-oxidation of fatty acids A cofactor such as flavin semiquinone (XH°) extracts a hydrogen atom from the substrate fatty acid (upper left), forming XH_2 and a free radical fatty acid derivative. The free radical then reacts with the subsequently oxygenated XH_2 to form the D-2-hydroperoxy-fatty acid, thereby regenerating XH°. The D-2-hydroperoxy-fatty acid can either decarboxylate to form CO_2 and a fatty aldehyde or be reduced to form the D-2-hydroxy-fatty acid. (From Shine, W. E. and Stumpf, P. K., *Arch. Biochem. Biophys.*, 162, 147–157, 1974. With permission.)

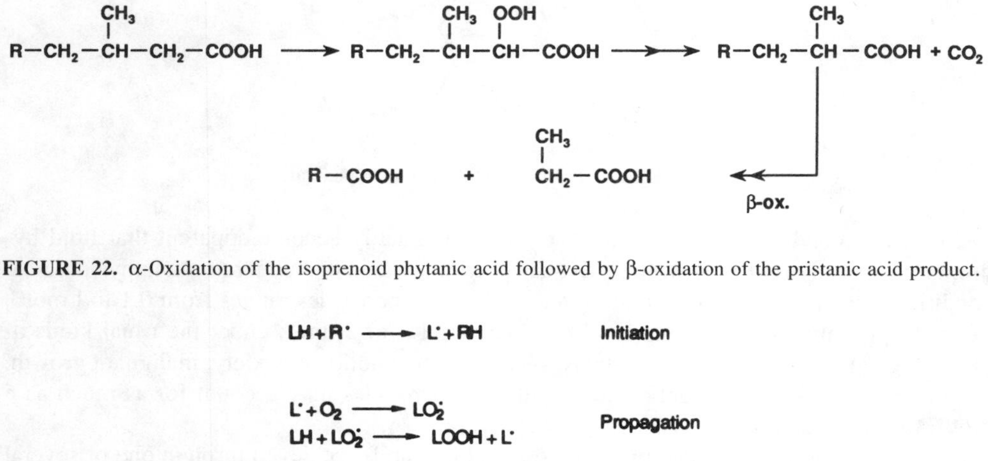

$$R{-}CH_2{-}CH(CH_3){-}CH_2{-}COOH \longrightarrow R{-}CH_2{-}CH(CH_3){-}CH(OOH){-}COOH \longrightarrow R{-}CH_2{-}CH(CH_3){-}COOH + CO_2$$

$$\xrightarrow{\beta\text{-ox.}} R{-}COOH \; + \; CH_3{-}CH_2{-}COOH$$

FIGURE 22. α-Oxidation of the isoprenoid phytanic acid followed by β-oxidation of the pristanic acid product.

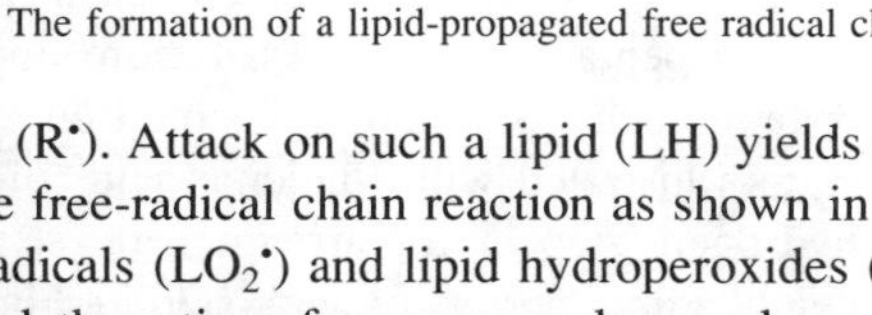

$$LH + R^\cdot \longrightarrow L^\cdot + RH \qquad \text{Initiation}$$

$$L^\cdot + O_2 \longrightarrow LO_2^\cdot$$
$$LH + LO_2^\cdot \longrightarrow LOOH + L^\cdot \qquad \text{Propagation}$$

FIGURE 23. The formation of a lipid-propagated free radical chain reaction.

tion by many free radicals ($R^\cdot$). Attack on such a lipid (LH) yields a carbon-centered radical ($L^\cdot$) that can propagate the free-radical chain reaction as shown in Figure 23.

Formation of peroxy radicals ($LO_2^\cdot$) and lipid hydroperoxides (LOOH) can occur either nonenzymatically or through the action of enzymes such as cyclooxygenase and lipoxygenase. Lipid hydroperoxides are generally, although not always[119] formed from free fatty acids and are only indirectly a part of membrane lipid metabolism. However, their further reactions will be briefly outlined here. After a number of years during which most lipid peroxidation

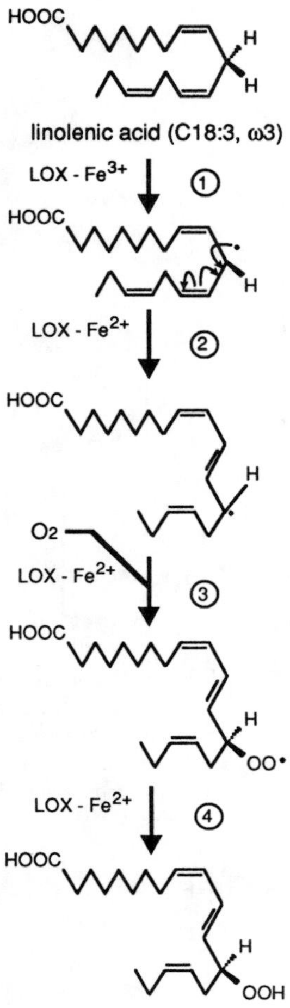

FIGURE 24. The formation of lipid hydroperoxides.

research was conducted by food chemists, it has gradually become apparent that lipid hydroperoxides can occur *in vivo*. In fact, it has been calculated that under normal physiological conditions the steady state concentration of lipid hydroperoxides ranges from 0.1 to 1 mol% of total lipid fatty acids.[120] This level can increase 1.5 to 2 times under the usual kinds of physiological stress. Under severe stress, such as acute emotional anxiety, malignant growth, X-irradiation, or vitamin E deficiency, lipid hydroperoxides may account for as much as 5 mol% of tissue fatty acids.

The enzymatic formation of lipid hydroperoxides can be achieved through one of several closely related lipoxygenases (E.C. 1.13.11.12). These enzymes catalyze the peroxidation of molecules containing *cis, cis*-1,4-pentadiene moieties. Before lipoxygenase can function it must be activated, i. e., the one molecule of bound nonheme iron must be oxidized to Fe^{3+} by a free radical. The reaction, as illustrated with linolenic acid (Figure 24), proceeds by the stereospecific removal of hydrogen (step 1), a rearrangement (step 2), and a reaction of the resulting radical with O_2 to yield a lipid peroxy radical (step 3). Finally this radical is reduced to a hydroperoxide (step 4) with concurrent oxidation of the lipoxygenase back to the active Fe^{3+} form.

In plant tissues lipoxygenase acts primarily on either the 9- or 13-position of linoleate and linolenate.[121] In contrast, lipid peroxidation in animal cells more commonly involves arachidonate and is catalyzed by one of four lipoxygenases, yielding various hydroperoxy fatty acid products. Some of the better known pathways leading to important eicosanoids[122] are illustrated in Figure 25.

The eicosanoids include various forms of prostaglandins (PG), thromboxanes (TX), and leukotrienes (LT), some of which are listed in Figure 26. The physiological actions of these individual compounds are very much a function of their structures, which are predetermined to a large extent by the nature of their fatty acid precursor. Eicosanoids derived from arachidonic acid are in many cases physiologically inactive or antagonistic to those of the equivalent class arising from eicosapentaenoic acid. Thus TXA_2 is a potent aggregator of platelets while TXA_3 is much less active in this respect. There is keen interest in studying the effects of diet, especially foods rich in eicosapentaenoic and other n-3 acids, on the balance of eicosanoid classes in humans.[123]

Relatively little is known regarding the physiological control of lipoxygenase action. Lipoxygenases are generally considered to be cytosolic enzymes. However, Rouzer and Kargman[124] observed that elevated Ca^{2+} levels in A23187-treated leukocytes led to a translocation of 35% of the cells' 5-lipoxygenase protein to membranes. The Ca^{2+}-induced activation was also accompanied by a decrease in overall 5-lipoxygenase activity, possibly due to suicide inactivation. The suggestion that enzyme translocation is important in stimulating the 5-lipoxygenase-catalyzed early reaction in leukotriene synthesis is supported by the finding that drugs specific for the inhibition of leukotriene synthesis prevent translocation of 5-lipoxygenase to membranes (although they do not inhibit 5-lipoxygenase activity per se[125]).

Leukocyte 5-lipoxygenase activity can apparently be inhibited by 15-lipoxygenase and its arachidonic acid product 15-hydroperoxyeicosatetraenoic acid.[126] The physiological significance of this in terms of regulating the balance of tissue eicosanoids is unknown.

IX. REDUCTION OF FATTY ACIDS TO FATTY ALCOHOLS

The membrane phospholipids containing ether-linked hydrocarbon side chains derive them from fatty alcohols (Chapter 4). The fatty alcohols may come from the catabolism of glycerol ethers, sphingosine bases (Chapter 4), or from direct reduction of long-chain fatty acids.

Fatty acids, as the CoA derivatives, are reduced to alcohols in a two-step process. The initial step converts the fatty acyl-CoA to free aldehyde. This is followed by a reduced pyridine nucleotide-requiring reduction of the fatty aldehyde to a fatty alcohol by a soluble enzyme.[127,128] A lack of chain-length specificity by the NADPH-dependent fatty aldehyde reductase of bovine cardiac muscle[127] led the authors to propose that the rather characteristic pattern of ether chain lengths formed is determined at the initial acyl-CoA reductase step.

The reduction of fatty acids to fatty alcohols seems particularly prominent in nervous tissue, heart, and neoplasms, all of which are rich in ether-containing lipids. It has been suggested that a high level of reduced pyridine nucleotides in a tissue may be a controlling factor in alcohol, and consequently, ether-lipid biosynthesis.[129]

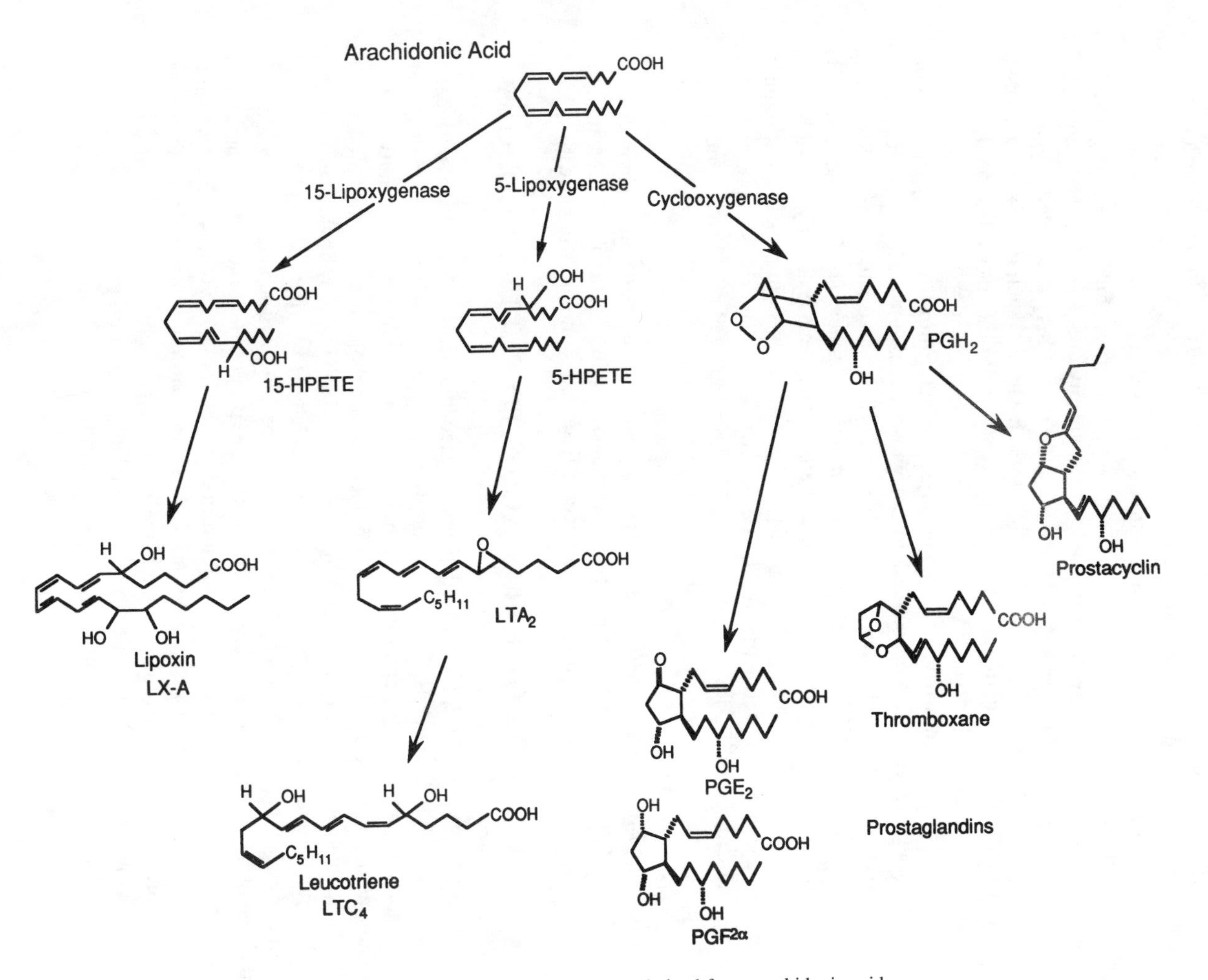

FIGURE 25. Major eicosanoid classes derived from arachidonic acid.

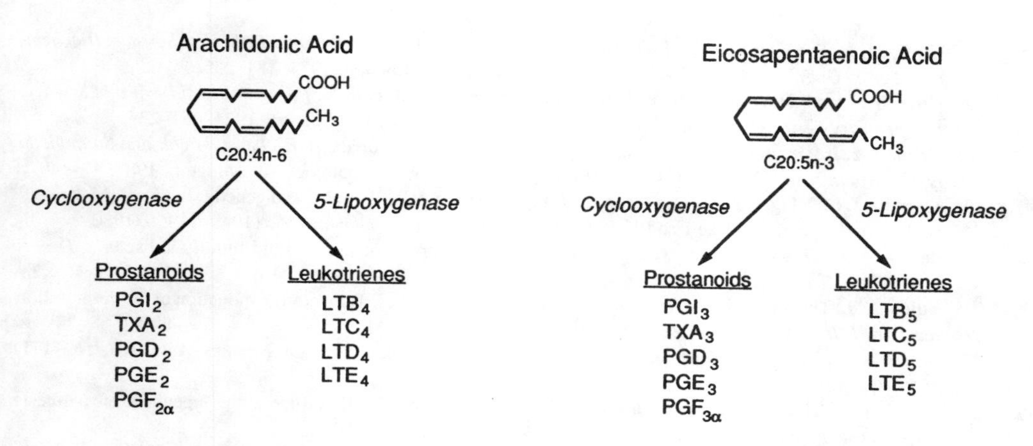

FIGURE 26. Comparison of eicosanoids formed from arachidonic acid and eicosapentaenoic acid.

REFERENCES

1. **Finnerty, W. R. and Makula, R. A.,** Microbial lipid metabolism, *CRC Crit. Rev. Microbiol.*, 4, 1–40, 1975.
2. **Aveldano, M. I.,** A novel group of very long chain polyenoic fatty acids in dipolyunsaturated phosphatidylcholines from vertebrate retina, *J. Biol. Chem.* , 267, 1172–1179, 1987.
3. **Miyazawa, S., Hashimoto, T., and Yokota, S.,** Identity of long-chain acyl-coenzyme A synthetase of microsomes, mitochondria, and peroxisomes in rat liver, *J. Biochem. (Tokyo)* 98, 723–733, 1985.
4. **Singh, H. and Poulos, A.,** Distinct long chain and very long chain fatty acyl CoA synthetases in rat liver peroxisomes and microsomes, *Arch. Biochem. Biophys.*, 266, 486–495, 1988.
5. **Taylor, A. A., Sprecher, H., and Russell, J. H.,** Characterization of an arachidonic acid-selective acyl-CoA synthetase from murine T lymphocytes, *Biochim. Biophys. Acta* 833, 229–238, 1985.
6. **Hinsch, W. and Seubert, W.,** On the mechanism of malonyl-CoA-independent fatty-acid synthesis. Characterization of the mitochondrial chain-elongating system of rat liver and pig-kidney cortex, *Eur. J. Biochem.*, 53, 437–447, 1975.
7. **Hinsch, W., Klages, C., and Seubert, W.,** On the mechanism of malonyl-CoA independent fatty-acid synthesis. Different properties of the mitochondrial chain elongation and enoyl-CoA reductase in various tissues, *Eur. J. Biochem.*, 64, 45–55, 1976.
8. **Whereat, A. F., Orishimo, M. W., Nelson, J., and Phillips, S. J.,** The location of different synthetic systems for fatty acids in inner and outer mitochondrial membranes from rabbit heart. *J. Biol. Chem.*, 244, 6498–6506, 1969.
9. **Podak, E. R.,** Ph.D. thesis, Medical Faculty, Gottingen, 1971.
10. **Prasad, M. R., Nagi, M. N., Ghesquier, D., Cook, L., and Cinti, D. L.,** Evidence for multiple condensing enzymes in rat hepatic microsomes catalyzing the condensation of saturated, monounsaturated, and polyunsaturated acyl coenzyme A, *J. Biol. Chem.*, 261, 8213–8217, 1986.
11. **Nagi, M. N., Cook, L., Prasad, M. R., and Cinti, D. L.,** Do rat hepatic microsomes contain multiple NADPH-supported fatty acid chain elongation pathways or a single pathway?, *Biochem. Biophys. Res. Commun.*, 140, 74–80, 1986.
12. **Cassagne, C., Lessire, R., Bessoule, J. J., and Moreau, P.,** Plant elongases, in *The Metabolism, Structure and Function of Plant Lipids,* Stumpf, P. K., Mudd, J. B., and Nes, W. D., Eds., Plenum Press, New York, 1987, 481–489,
13. **Brophy, P. J. and Vance, D. E.,** The synthesis and hydrolysis of long-chain fatty acyl-coenzyme A thioesters by soluble and microsomal fractions from the brain of the developing rat, *Biochem. J.*, 160, 247–251, 1976.
14. **Goldberg, I., Shechter, I., and Bloch, K.,** Fatty acyl-coenzyme A elongation in brain of normal and quaking mice, *Science,* 182, 497–499, 1973.
15. **Murad, S. and Kishimoto, Y.,** Chain elongation of fatty acid in brain: a comparison of mitochondrial and microsomal enzyme activities, *Arch. Biochem. Biophys.*, 185, 300–306, 1978.
16. **Saito, Y., Silvius, J. F., and McElhaney, R. N.,** Membrane lipid biosynthesis in *Acholeplasma laidlawii* B: elongation of medium- and long-chain exogenous fatty acids in growing cells, *J. Bacteriol.*, 133, 66–74, 1978.

17. **Gurr, M. I.,** The biosynthesis of unsaturated fatty acids, in *MTP International Review of Science, Biochemistry Series One,* Vol. 4, Goodwin, T. W., Ed., Butterworths, London, 1974, 181–235.
17a. **Gurr, M. I. and Bloch, K.,** unpublished data. Quoted in Gurr, M. I., in *MTP International Review of Science, Biochemistry Series One,* Vol. 4, Goodwin, T. W., Ed., Butterworths, London, 1974, 181–235.
18. **Cronan, J. E., Jr. and Rock, C. O.,** Biosynthesis of membrane lipids, in *Escherichia coli* and *Salmonella typhimurium,* Vol. 1, Neidhardt, F., Ed., American Society for Microbiology, Washington, 1987, 474–497.
19. **Garwin, J. L., Klages, A. L., and Cronan, J. E., Jr.,** Structural enzymatic, and genetic studies of β-ketoacyl-acyl carrier protein synthase I and II of *Escherichia coli, J. Biol. Chem.,* 255, 11949–11956, 1980.
20. **de Mendoza, D. and Cronan, J. E., Jr.,** Thermal regulation of membrane lipid fluidity in bacteria, *Trends in Biochem. Sci.,* 8, 49–52, 1983.
21. **Fulco, A. J.,** The biosynthesis of unsaturated fatty acids by bacilli. I. Temperature induction of the desaturation reaction, *J. Biol. Chem.,* 244, 889–895, 1969.
22. **Fulco, A. J.,** Bacterial biosynthesis of polyunsaturated fatty acids, *Biochim. Biophys. Acta,* 187, 169–171, 1969.
23. **Fulco, A. J.,** The biosynthesis of unsaturated fatty acids by bacilli. II. Temperature-dependent biosynthesis of polyunsaturated fatty acids, *J. Biol. Chem.,* 245, 2885–2990, 1970.
24. **Fulco, A. J.,** Regulation and pathways of membrane lipid biosynthesis by bacilli, *Membrane Fluidity,* Kates, M. and Manson, L. A., Eds. Plenum Press, New York, 1984, 303–327.
25. **Spatz, L. and Strittmatter, P.,** A form of cytochrome b_5 that contains an additional hydrophobic sequence of 40 amino acid residues, *Proc. Natl. Acad. Sci. U.S.A.,* 68, 1042–1046, 1971.
26. **dePierre, J. W. and Ernster, L.,** Enzyme topology of intracellular membranes, *Annu. Rev. Biochem.,* 46, 201–262, 1977.
27. **Strittmatter, P., Spatz, L., Corcoran, D., Rogers, M. J., Setlow, B., and Redline, R.,** Purification and properties of rat liver microsomal stearyl coenzyme A desaturase, *Proc. Natl. Acad. Sci. U.S.A.,* 71, 4565–4569, 1974.
28. **Thiede, M. A., Ozols, J., and Strittmatter, P.,** Construction and sequence of cDNA for rat liver stearyl coenzyme A desaturase, *J. Biol. Chem.,* 261, 13230–13235, 1986.
29. **Okayasu, T., Ono, T., Shinojima, K., and Imai, Y.,** Involvement of cytochrome b_5 in the oxidative desaturation of linoleic acid to γ-linoleic acid in rat liver microsomes, *Lipids,* 12, 267–271, 1977.
30. **Morris, L. J., Harris, R. V., Kelly, W., and James, A. T.,** The stereochemistry of desaturations of long-chain fatty acids in *Chlorella vulgaris, Biochem. J.,* 109, 673–678, 1968.
31. **Morris. L. J.,** Mechanisms and stereochemistry in fatty acid metabolism, *Biochem. J.,* 118, 681–693, 1970.
32. **Johnson, A. R. and Gurr, M. I.,** Isotope effects in the desaturation of stearic to oleic acid, *Lipids,* 6, 78–84, 1971.
33. **Nagai, J. and Bloch, H.,** Enzymatic desaturation of stearyl acyl carrier protein, *J. Biol. Chem.,* 243, 4626–4633, 1968.
34. **Jaworksi, J. G. and Stumpf, P. K.** Properties of a soluble stearyl-acyl carrier protein dèsaturase from maturing *Carthamus tinctorius, Arch. Biochem. Biophys.,* 162,158–165, 1974.
35. **Grantz, D. A. and Zeiger, E.,** Stomatal responses to light and leaf-air water vapor pressure difference show similar kenetics in sugarcane and soybean, *Plant Physiol.,* 81, 865–868, 1986.
36. **Browse, J., Kunst, L., Anderson, S., Hugly, S., and Somerville, C.,** A mutant of *Arabidopsis* deficient in the chloroplast 16:1/18:1 desaturase, *Plant Physiol.,* 90, 522–529, 1989.
37. **Howling, D., Morris, L. J., and James, A. T.,** The influence of chain length on the dehydrogenation of saturated fatty acids, *Biochem. Biophys. Acta,* 152, 224–226, 1968.
38. **Wood, R. and Wiegand, R. D.,** Hepatoma, host liver, and normal rat liver lipids: distribution of isomeric monoene fatty acids in individual lipid classes, *Lipids,* 10, 746–749, 1975.
39. **Howling, D., Morris, L. J., Gurr, M. I., and James, A. T.,** The specificity of fatty acid desaturases and hydroxylases. The dehydrogenation and hydroxylation of monoenoic acids, *Biochem. Biophys. Acta,* 260, 10–19, 1972.
40. **Gurr, M. I. and James, A. T.,** *Lipid Biochemistry,* 2nd ed., Chapman and Hall, London, 1975.
41. **Brenner, R. R.,** The desaturation step in the animal biosynthesis of polyunsaturated fatty acids, *Lipids,* 6, 567–575, 1971.
42. **Mead, J. F.,** Synthesis and metabolism of polyunsaturated acids, *Fed. Proc.,* 20, 952–955, 1961.
43. **Stoffel, W.,** Uber biosynthese und biologischen abbau hochungesattigter fettsauren, *Naturwissenschaften,* 53, 621–630, 1966.
44. **Kates, M., Pugh, E. L., and Ferrante, G.,** Regulation of membrane fluidity by lipid desaturases, in *Membrane Fluidity,* Kates, M. and Manson, L. A., Eds. Plenum Press, New York, 1984, 379–395.
45. **Clarke, S. D. and Armstrong, M. K.,** Cellular lipid binding proteins: expression, function, and nutritional regulation, *FASEB J.,* 3, 2480–2487, 1989.

46. **Rasmussen, J. T., Borchers, T., and Knudsen, J.,** Comparison of the binding affinities of acyl-CoA-binding protein and fatty-acid binding protein for long-chain acyl-CoA esters, *Biochem J.,* 265, 849–855, 1990.
47. **Lunzer, M. A., Manning, J. A., and Ockner, R. K.,** Inhibition of rat liver acetyl coenzyme A carboxylase by long chain acyl coenzyme A and fatty acid, *J. Biol. Chem.,* 252, 5483–5487, 1977.
48. **Ferrante, G., Ohno, Y., and Kates, M.,** Influence of temperature and growth phase on desaturase activity of the mesophilic yeast *Candida lipolytica, Can. J. Biochem.,* 61, 171–177, 1983.
49. **Rebeille, F., Bligny, R., and Douce, R.,** Oxygen and temperature effects on the fatty acid composition of sycamore cells (*Acer pseudoplatanus*), in *Biogenesis and Function of Plant Lipids,* Mazliak, P., Benveniste, P., Costes, C., and Douce, R., Eds., Elsevier, Amsterdam, 1980, 203–206.
50. **Skriver, L. and Thompson, G. A., Jr.,** Environmental effects on *Tetrahymena* membranes. Temperature-induced changes in membrane fatty acid unsaturation are independent of the molecular oxygen concentration, *Biochim. Biophys. Acta,* 431, 180–188, 1976.
51. **Oshino, N. and Sato, R.,** The dietary control of the microsomal stearyl CoA desaturation enzyme system in rat liver, *Arch. Biochem. Biophys.,* 149, 369–377, 1972.
52. **Gelhorn, A. and Benjamin, W.,** The effect of insulin on monosaturated fatty acid synthesis in diabetic rats. The stability of the informational RNA and of the enzyme system concerned with fatty acid desaturation, *Biochim. Biophys. Acta,* 116, 460–466, 1966.
53. **Gelhorn, A. and Benjamin, W.,** The intracellular localization of an enzymatic defect of lipid metabolism in diabetic rats, *Biochim. Biophys. Acta,* 84, 167–175, 1964.
54. **Oshino, N.,** The dynamic behavior during dietary induction of the terminal enzyme (cyanide-sensitive factor) of the stearyl CoA desaturation system of rat liver microsomes, *Arch. Biochem. Biophys.,* 149, 378–387, 1972.
55. **Prasad, M. R. and Joshi, V. C.,** Regulation of rat hepatic stearoyl coenzyme A desaturase. The roles of insulin and carbohydrate, *J. Biol. Chem.,* 254, 997–999, 1979.
56. **Jeffcoat, R. and James, A. T.,** Interrelation between the dietary regulation of fatty acid synthesis and the fatty acyl-CoA desaturases, *Lipids,* 12, 469–474, 1977.
57. **Jeffcoat, R. and James, A. T.,** The control of stearoyl-CoA desaturase by dietary linoleic acid, *FEBS Lett.,* 85, 114–118, 1978.
58. **Hopkins, G. J. and West, C. E.,** Diet-induced changes in the fatty acid composition of mouse hepatocyte plasma membranes, *Lipids,* 12, 327–334, 1977.
59. **Holloway, C. T. and Holloway, P. W.,** Stearyl coenzyme A desaturase activity in mouse liver microsomes of varying lipid composition, *Arch. Biochem. Biophys.,* 167, 496–504, 1975.
60. **Thiede, M. A. and Strittmatter, P.,** The induction and characterization of rat liver stearyl-CoA desaturase mRNA, *J. Biol. Chem.,* 260, 14459–14463, 1985.
61. **Bossie, M. A. and Martin, C. E.,** Nutritional regulation of yeast Δ9 fatty acid desaturase activity, *J. Bact.,* 171, 6409–6413, 1989.
62. **Erwin, J. and Bloch, K.,** Lipid metabolism of ciliated protozoa, *J. Biol. Chem.,* 238, 1618–1624, 1963.
63. **Wunderlich, F., Speth, V., Batz, W., and Kleinig, H.,** Membranes of *Tetrahymena,* III. The effect of temperature on membrane core structures and fatty acid composition of *Tetrahymena* cells, *Biochim. Biophys. Acta,* 298, 39–49, 1973.
64. **Nozawa, Y., Iida, H., Fukushima, H., Ohki, K., and Ohnishi, S.,** Studies on *Tetrahymena* membranes. Temperature-induced alteration in fatty acid composition of various membrane fractions in *Tetrahymena pyriformis* and its effect on membrane fluidity as inferred by spin-label study, *Biochim. Biophys. Acta,* 367, 134–147, 1974.
65. **Thompson, G. A., Jr. and Nozawa, Y.,** The regulation of membrane fluidity in *Tetrahymena,* in *Membrane Fluidity,* Kates, M. and Manson, L. A., Eds., Plenum Press, New York, 1984, 397–432.
66. **Martin, C. E., Hiramitsu, K., Kitajima, Y., Nozawa, Y., Skriver, L., and Thompson, G. A., Jr.,** Molecular control of membrane properties during temperature acclimation. Fatty acid desaturase regulation of membrane fluidity in acclimating *Tetrahymena* cells, *Biochemistry,* 15, 5218–5227, 1976.
67. **Thompson, G. A., Jr. and Nozawa, Y.,** *Tetrahymena:* a system for studying dynamic membrane alterations within the eukaryotic cell, *Biochim. Biophys. Acta,* 472, 55–92, 1977.
68. **Nandini-Kishore, S. G., Kitajima, Y., and Thompson, G. A., Jr.,** Membrane fluidizing effects of the general anesthetic methoxyflurane elicit an acclimation response in *Tetrahymena, Biochim. Biophys. Acta,* 471, 157–161, 1977.
69. **Kasai, R., Yamada, T., Hasegawa, I., Muto, Y., Yoshioka, S., Nakamaru, T., and Nozawa, Y.,** Regulatory mechanism of desaturation activity in cold acclimation of *Tetrahymena pyriformis,* with special reference to quick cryoadaptation, *Biochim. Biophys. Acta,* 836, 397–401, 1985.
70. **Kasai, R., Kitajima, Y., Martin, C. E., Nozawa, Y., Skriver, L., and Thompson, G. A., Jr.,** Molecular control of membrane properties during temperature acclimation. Membrane fluidity regulation of fatty acid desaturase action?, *Biochemistry,* 15, 5228–5233, 1976.

71. **Cossins, A. R. and Sinensky, M.,** Adaptation of membranes to temperature, pressure, and exogenous lipids, in *Physiology of Membrane Fluidity*, Shinitzky, M., Ed., CRC Press Inc., Boca Raton, 1984, 1–20.
72. **Umeki, S., Fukushima, H., Watanabe T., and Nozawa, Y.,** Temperature acclimation mechanisms in *Tetrahymena pyriformis*: effects of decreased temperature on microsomal electron transport., *Biochem. Internat.*, 4, 101–107, 1982.
73. **Lippiello, P. M., Holloway, C. T., Garfield, S. A., and Holloway, P. W.,** The effects of estradiol on stearyl-CoA desaturase activity and microsomal membrane properties in rooster liver, *J. Biol. Chem.*, 254, 2004–2009, 1979.
74. **Nagai, J. and Bloch, K.,** Synthesis of oleic acid by *Euglena gracilis*, *J. Biol. Chem.*, 240, PC3702–3703, 1965.
75. **Nagai, J. and Bloch, K.,** Enzymatic desaturation of stearyl acyl carrier protein, *J. Biol. Chem.*, 241, 1925–1927, 1966.
76. **Stumpf, P. K. and Porra, R. J.,** Lipid biosynthesis in developing and germinating soybean cotyledons. The formation of oleate by a soluble stearyl acyl carrier protein desaturase, *Arch. Biochem. Biophys.*, 176, 63–70, 1976.
77. **Erwin, J. and Bloch, K.,** The α-linolenic acid content of some photosynthetic microorganisms, *Biochem. Biophys. Res. Commun.*, 9, 103–108, 1962.
78. **Rosenberg, A., Pecker, M., and Moschides, E.,** Fatty acids of the pellicles and plastids of light-grown and dark-grown cells of *Euglena gracilis*, *Biochemistry*, 4, 680–685, 1965.
79. **Roughan, P. G. and Slack, C. R.,** Cellular organization of glycerolipid metabolism, *Ann. Rev. Plant Phys.*, 33, 97–132, 1982.
80. **Jaworski, J. G.,** Biosynthesis of monoenoic and polyenoic fatty acids, in *The Biochemistry of Plants*, Vol. 9, Stumpf, P. K. and Conn, E. E., Eds. Academic Press, Orlando, 1987, 159–174.
81. **Ohlrogge, J. B., Shine, W. E., and Stumpf, P. K.,** Fat metabolism in higher plants. Characterization of plant acyl-ACP and acyl-CoA hydrolases, *Arch. Biochem. Biophys.*, 189, 382–391, 1978.
82. **Gurr, M. I., Robinson, M. P., and James, A. T.,** The mechanism of formation of polyunsaturated fatty acids by photosynthetic tissue, *Eur. J. Biochem.*, 9, 70–78, 1969.
83. **Roughan, P. G.,** Turnover of the glycerolipids of pumpkin leaves, *Biochem. J.*, 117, 1–8, 1970.
84. **Grantz, D. A. and Zeiger, E.,** Stomatal responses to light and leaf-air water vapor pressure difference show similar kinetics in sugarcane and soybean, *Plant Physiol.*, 81, 865, 1986.
85. **Wilson, J. M. and Crawford, R. M. M.,** The acclimation of plants to chilling temperatures in relation to the fatty acid composition of leaf polar lipids, *New Phytol.*, 73, 805–820, 1974.
86. **Kuiper, P. J. C.,** Role of lipids in water and ion transport, in *Recent Advances in the Chemistry and Biochemistry of Plant Lipids*, Galliard, T. and Mercer, E. I., Eds. Academic Press, London, 1975, 359–386.
87. **Harris, P. and James, A. T.,** The effect of low temperatures on fatty acid biosynthesis in plants, *Biochem. J.*, 112, 325–330, 1969.
88. **Rebeille, F., Bligny, R., and Douce, P.,** Oxygen and temperature effects on the fatty acid composition in sycamore cells (*Acer pseudoplatanus* L.), *Biochim. Biophys. Acta*, 620, 1–9, 1980.
89. **Browse, J. and Slack, R.,** The effects of temperature and oxygen on the rates of fatty acid synthesis and oleate desaturation in safflower *(Carthamus tinctorius)* seed, *Biochim. Biophys. Acta*, 753, 145–152, 1983.
90. **Meyer, F. and Bloch, K.,** Effects of temperature on the enzymatic synthesis of unsaturated fatty acids in *Torulopsis utilis*, *Biochim. Biophys. Acta*, 77, 671–673, 1963.
91. **Brown, C. M. and Rose, A. H.,** Fatty acid composition of *Candida utilis* as affected by growth temperature and dissolved oxygen tension, *J. Bacteriol.*, 99, 371–378, 1969.
92. **Law, J. H.,** Biosynthesis of cyclopropane rings, *Acc. Chem. Res.*, 4, 199–203, 1971.
93. **Cronan, J. E., Jr., Nunn, W. D., and Batchelor, J. G.,** Studies on the biosynthesis of cyclopropane fatty acids in *Escherichia coli*, *Biochim. Biophys. Acta*, 348, 63–75, 1974.
94. **Akamatsu, Y. and Law, J. H.,** Enzymatic alkylenation of phospholipid fatty acid chains by extracts of *Mycobacterium phlei*, *J. Biol. Chem.*, 245, 701–708, 1970.
95. **Moore, J. H.,** Lipid biochemistry, from forage to milk, in *Industrial Aspects of Biochemistry*, Spencer, B., Ed., North Holland, Amsterdam, 1974, 835–863.
96. **Garton, G. A.,** Fatty acid metabolism in ruminants, in *Biochemistry of Lipids II*, Vol. 14, Goodwin, T. W., Ed., University Park Press, Baltimore, 1977, 337–370.
97. **Bickerstaffe, R. and Johnson, A. R.,** The effect of intravenous infusions of sterculic acid on milk fat synthesis, *Br. J. Nutr.*, 27, 561–570, 1972.
98. **Hunter, W. J., Baker, F. C., Rosenfeld, I. S., Keyser, J. B., and Tove, S. B.,** Biohydrogenation of unsaturated fatty acids, *J. Biol. Chem.*, 251, 2241–2247, 1976.
99. **Hughes, P. E., Hunter, W. J., and Tove, S. B.,** Biohydrogenation of unsaturated fatty acids. Purification and properties of *cis*-9, *trans*-11-octadecadienoate reductase, *J. Biol. Chem.*, 257, 3643–3649, 1982.
100. **Kepler, C. R. and Tove, S. B.,** Induction of biohydrogenation of oleic acid in *Bacillus cereus* by increase in temperature, *Biochem. Biophys. Res. Commun.*, 52, 1434–1439, 1973.

101. **Hashimoto, T.,** Comparison of enzymes of lipid β-oxidation in peroxisomes and mitochondria, in *Peroxisomes in Biology and Medicine,* Fahimi, H. D. and Sies, H., Ed., Springer-Verlag, Heidelburg, 1987, 97–104.
102. **Bremer, J. and Osmundsen, H.,** Fatty acid oxidation and its regulation, in *Fatty Acid Metabolism and Its Regulation,* Numa, S., Ed., Elsevier, New York, 1984, 113–154.
103. **Reubsaet, F. A. G., Veerkamp, J. M., Bukkens, S. G. F., Trijbels, J. M. F., and Monnens, L. A. H.,** Acyl-CoA oxidase activity and peroxisomal fatty acid oxidation in rat tissues, *Biochim. Biophys. Acta,* 958, 434–442, 1988.
104. **Singh, H., Derwas, N., and Poulos, A.,** β-oxidation of very-long-chain fatty acids and their coenzyme A derivatives by human skin fibroblasts, *Arch. Biochem. Biophys.,* 254, 526–533, 1987.
105. **Bieber, L. L.,** Carnitine, *Annu. Rev. Biochem.,* 57, 261–283, 1988.
106. **Nunn, W. D.,** A molecular view of fatty acid catabolism in *Escherichia coli, Microbiol. Rev.,* 50, 179–192, 1986.
107. **Weeks, G., Shapiro, M., Burns, R. O., and Wakil, S. J.,** Control of fatty acid metabolism. I. Induction of the enzymes of fatty acid oxidation in *Escherichia coli, J. Bact.,* 97, 827–836, 1969.
108. **Kindl, H.,** β-oxidation of fatty acids by specific organelles, in *The Biochemistry of Plants,* Vol. 9, Stumpf, P. K. and Conn, E. E., Eds., Academic Press, Orlando, 1987, 31–52.
109. **Eising, R. and Gerhardt, B.,** Catalase degradation in sunflower coyledons during peroxisome transition from glyoxysomal to leaf peroxisomal function, *Plant Physiol.,* 84, 225–232, 1987.
110. **Schulz, H.,** Oxidation of fatty acids, in *Biochemistry of Lipids and Membranes,* Vance, D. E. and Vance, J. E., Eds., Benjamin/Cummings, Menlo Park, 1985, 116–142.
111. **Palosaari, P. M. and Hiltunen, J. K.,** Peroxisomal bifunctional protein from rat liver is a trifunctional enzyme possessing 2-enoyl-CoA hydratase, 3-hydroxyacyl-CoA dehydrogenase, and Δ^3, Δ^2-enoyl-CoA isomerase activities, *J. Biol. Chem.,* 265, 2446–2449, 1990.
112. **Tserng, K.-Y. and Jin, S.-J.,** NADPH-dependent reductive metabolism of *cis*-5 unsaturated fatty acids. A revised pathway for β-oxidation of oleic acid, *J. Biol. Chem.,* 266, 11614–11620, 1991.
113. **Stumpf, P. K.,** Membrane-bound enzymes in plant lipid metabolism, in *The Enzymes of Biological Membranes,* Vol. 2, Martonosi, A., Ed., Plenum Press, New York, 1976, 145–159.
114. **Tatsumi, K., Murad, S., and Kishimoto, Y.,** Mechanism and stereospecificity of α-hydroxylation of lignoceric acid in rat brain, *Arch. Biochem. Biophys.,* 171, 87–92, 1975.
115. **Kaya, K., Ramesha, C. S., and Thompson, G. A., Jr.,** On the formation of α-hydroxy fatty acids, *J. Biol. Chem.,* 259, 3548–3553, 1984.
116. **Stokke, O.,** The degradation of a branched chain fatty acid by alterations between α- and β-oxidations, *Biochim. Biophys. Acta,* 176, 54–59, 1969.
117. **Tsai, S. C., Avigan, J., and Steinberg, D.,** Studies on the α-oxidation of a phytanic acid by rat liver mitochondria, *J. Biol. Chem.,* 244, 2682–2692, 1969.
118. **Singh, H., Usher, S., Johnson, D., and Poulos, A.,** A comparative study of straight chain and branched chain fatty acid oxidation in skin fibroblasts from patients with peroxisomal disorders, *J. Lipid Res.,* 31, 217–225, 1990.
119. **Schewe, T., Rapoport, S. M., and Kuhn, H.,** Enzymology and physiology of reticulocyte lipoxygenases: comparison with other lipoxygenases, *Adv. Enzymol.,* 58, 191–272, 1986.
120. **Kagan, V. E.,** *Lipid Peroxidation in Biomembranes,* CRC Press, Boca Raton, 1988, pp. 181.
121. **Hildebrand, D. F.,** Lipoxygenases, *Physiol. Plant.,* 76, 249–253, 1989.
122. **Needleman, P., Turk, J., Jakschik, B. A., Morrison, A. R., and Lefkowith, J. B.,** Arachidonic acid metabolism, *Annu. Rev. Biochem.,* 55, 69–102, 1986.
123. **Weaver, B. J. and Holub, B. J.,** Health effects and metabolism of dietary eicosapentaenoic acid, *Prog. Food and Nutr. Sci.,* 12, 111–150, 1988.
124. **Rouzer, C. A. and Kargman, S.,** Translocation of 5-lipoxygenase to the membrane in human leukocytes challenged with ionophore A23187, *J. Biol. Chem.,* 263, 10980–10988, 1988.
125. **Rouzer, C. A., Ford-Hutchinson, A. W., Morton, H. E., and Gillard, J. W.,** MK886, a potent and specific leukotriene biosynthesis inhibitor blocks and reverses the membrane association of 5-lipoxygenase in ionophore-challenged leukocytes, *J. Biol. Chem.,* 265, 1436–1442, 1990.
126. **Cashman, J. R., Lambert, C., and Sigal, E.,** Inhibition of human leukocyte 5-lipoxygenase by 15-HPETE and related eicosanoids, *Biochem. Biophys. Res. Commun.,* 155, 38–44, 1988.
127. **Kawalek, J. C. and Gilbertson, J. R.,** Enzymic reduction of free fatty aldehydes in bovine cardiac muscle, *Biochem. Biophys. Res. Commun.,* 51, 1027–1033, 1973.
128. **Lee, T.-C. and Snyder, F.,** Cancer cells, in *Lipid Metabolism in Mammals,* Vol. 2, Snyder, F., Ed. Plenum Press, New York, 1977, 293–310.
129. **Snyder, F.,** The enzymatic pathways of ether-linked lipids and their precursors, in *Ether Lipids: Chemistry and Biology,* Academic Press, New York, 1972, 121–156.

Chapter 4

PHOSPHOLIPIDS

I. INTRODUCTION

Phospholipids are by far the major lipid components of most membranes. The principal exceptions are nervous tissue, with high levels of cerebrosides and gangliosides, and plant chloroplast membranes, which are rich in diacylglycolipids (see Chapter 7). Different phospholipid classes are always found in fixed proportions in a given membrane type while another functionally distinct membrane in the same cell might be quite different in its phospholipid distribution.[1] Whereas the lipid composition of a particular membrane is invariant while the cell's environment remains constant, polar moieties as well as fatty acids may change during adaptation to environmental stress (see Chapter 10). I shall not review in detail each step of the individual biosynthetic and catabolic pathways; these have been elaborated elsewhere.[2] Instead, I shall present them in outline form and discuss only those references that relate to control aspects.

One must constantly remind oneself that the biosynthetic pathways of the various phospholipids are so closely interrelated that a perturbation in one pathway nearly always has repercussions on others. Most of the structural phospholipids are derived from the same precursor pools, and interconversions of one functional phospholipid into another are commonplace. A summary of anastomosing biogenetic interrelationships is presented in Figure 1.[3]

Many membranes have at least trace amounts of all the phospholipids listed in Figure 1. Not only is each phospholipid always present as a specific percentage of the total, but each maintains its own characteristic distribution of fatty acyl and alkyl side chains. We shall therefore consider regulatory factors controlling both relative rates of the synthetic enzymes and their selectivity for certain molecular species occurring in the available precursor pools. Furthermore, it is essential to include in the overall formulation those degradative enzymes that deplete precursor pools in a manner capable of altering the precise composition of the pools.

With the exceptions noted, I shall generally consider that the metabolic pools of prime biochemical interest in eukaryotic cells are those available to the microsomal membranes (endoplasmic reticulum), for most of the cellular phospholipids are synthesized here. However, the rapid exchange of phospholipids between the microsomes and other functionally different membranes renders it impossible to view these pools as being discrete. While the constant intermixing of molecules wreaks havoc upon the observer's efforts to monitor regulatory activities, it suits beautifully the needs of the cell to redistribute lipids made or retailored in this microsomal clearing house. The important regulatory factors surrounding intracellular movement of phospholipids are discussed in Chapter 8.

In eukaryotic cells some phospholipid synthesis definitely occurs inside the mitochondrion and the chloroplast. The latter organelle is the major site of phosphatidylglycerol biosynthesis in green plants. There have been several reports that phosphatidic acid, phosphatidylglycerol, and diphosphatidylglycerol can be synthesized in mammalian mitochondria.[5] Yeast mitochondria appear to synthesize these lipids and others, including phosphatidylethanolamine, phosphatidylinositol, and phosphatidylserine.[6] Despite stringent precautions to prevent cross contamination of mitochondria with microsomes and vice versa, it is extremely difficult to be certain of a membrane-bound enzyme's intracellular location. The rate of enzymatic activity measured *in vitro* is affected by such a wide variety of factors, e.g., substrate levels and

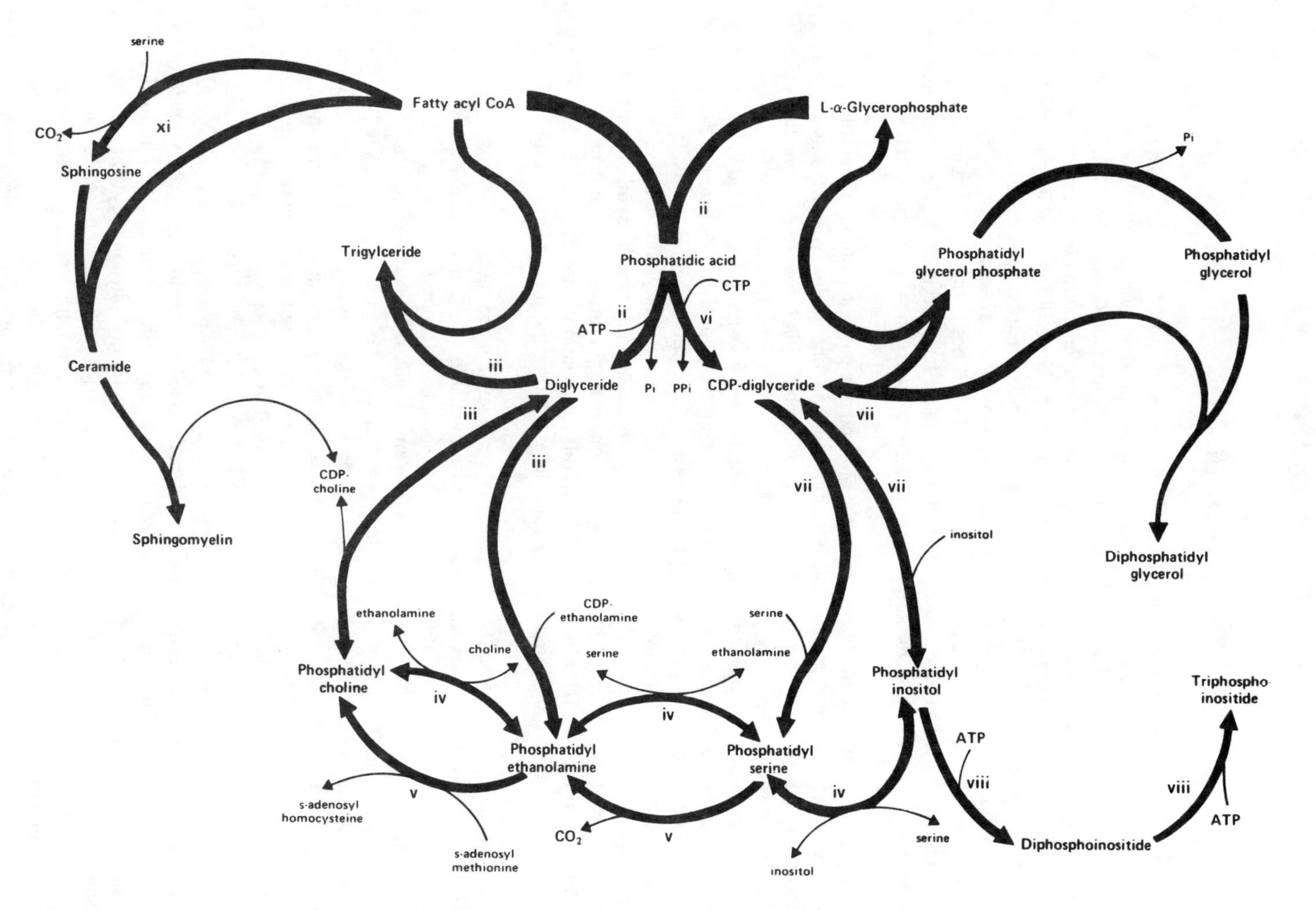

FIGURE 1. A summary chart showing interrelated pathways of phospholipid biosynthesis. (Modified from Thompson, G. A., Jr., *Form and Function of Phospholipids*, Ansell, G. B., Dawson, R. M. C., and Hawthorne, J. N., Eds., Elsevier, Amsterdam, 1973, 67. With permission.)

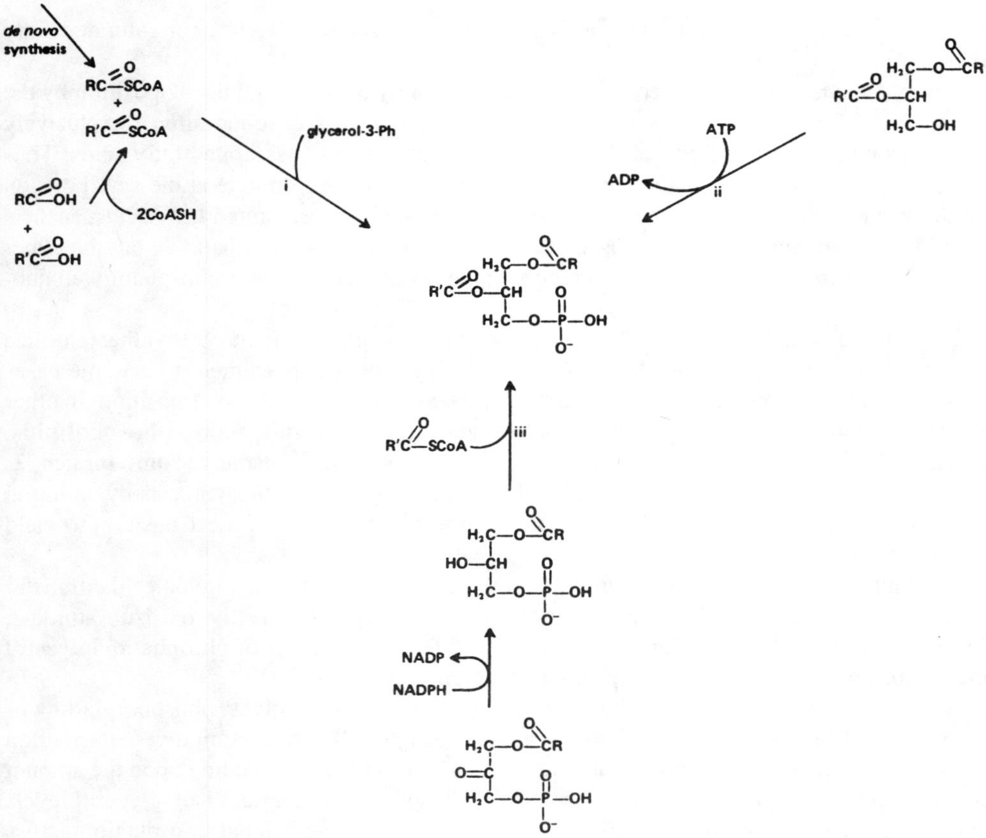

FIGURE 2.

accessibility, cofactor concentrations, and product acceptor sites, that a final resolution of the question will be years in coming.

The purification and characterization of phospholipid biosynthetic enzymes are severely hampered by the fact that techniques for working with membrane-bound proteins are still poorly developed. The recent development of new detergent solubilization and reconstitution procedures has allowed considerable progress in this area.[6] The first steps towards establishing the molecular genetics of these enzymes are now being taken, mainly using yeast and *E. coli*.

II. REGULATION OF THE PHOSPHATIDIC ACID POOL

Phosphatidic acid can arise within the cell via several pathways (Figure 2). Its major source in most cells is through the acylation of *sn*-glycerol-3-phosphate in microsomes. Acyl-CoA derivatives of mixed chain length and degree of unsaturation are provided through *de novo* synthesis (Chapters 2 and 3) or through activation of free fatty acids by microsomal (and perhaps mitochondrial) acyl-CoA synthetases. In animal cells the initial step of phosphatidic acid synthesis involves the formation of l-acyl-*sn*-glycerol-3-phosphate (lysophosphatidic acid) through action by acyl-CoA: glycerol-3-phosphate acyltransferase reactions (reaction i, Figure 2). This enzyme is located primarily in the microsomal membranes although some mammalian tissues, especially liver, show considerable activity by a different acyltransferase in the mitochondrial fraction.[7] The acylation shows a strong preference for saturated fatty acids.

Once formed, l-acyl-*sn*-glycerol-3-phosphate is rapidly acylated at the *sn*-2 position by the enzyme acyl-CoA:l-acyl-*sn*-glycerol-3-phosphate acyltransferase, found almost exclusively in microsomes. The specificity of this reaction strongly favors unsaturated fatty acids. Thus phosphatidic acid formed *de novo* has a striking enrichment of palmitate at the *sn*-1 position and oleic and linoleic acids at the *sn*-2 position. More highly unsaturated fatty acids, such as arachidonic acid, appear not to be used in the direct synthesis of phosphatidate but may later be exchanged into some of its products, e.g., phosphatidylcholine and phosphatidylethanolamine.

In bacterial cells, the acyltransferases responsible for phosphatidate biosynthesis utilize acyl-ACPs as substrates rather than acyl-CoAs.[8] The fatty acyl placement is in some cases similar to that in animal cells, with a saturated acid being attached at the *sn*-1 position. In other bacteria, such as *Brevibacterium ammoniagenes*, one of the major phospholipids, phosphatidylglycerol, exhibits a reverse positioning of fatty acids, namely, 1-unsaturated, 2-saturated.[9] This pattern is thought to be established during *de novo* biosynthesis by an initial formation of 2-acyl-*sn*-glycerol-3-phosphate followed by acylation of the 1 position to yield phosphatidate.

The pathway of phosphatidate synthesis in the chloroplasts of plants involves the transfer of acyl chains from acyl-ACP as in bacteria and also leads predominantly to a 1-unsaturated, 2-saturated product.[10] This is subsequently utilized for the synthesis of phosphatidylglycerol and, in the so-called 16:3 plants (see p. 161), glycolipids.

An alternative pathway for phosphatidic acid synthesis involves phosphorylation of diacylglycerol by diacylglycerol kinase (reaction ii, Figure 2). The quantitative contribution of this reaction to the overall phosphatidate pool can vary greatly, depending upon the amount of diacylglycerol produced by interconnecting pathways. For example, diacylglycerol levels can fluctuate sharply during hormone- or growth factor-mediated signal transduction across the plasma membrane[11](see also p. 92), and sphingomyelin biogenesis yields diacylglycerol as a byproduct (see p. 101).

The substrate specificity of diacylglycerol kinase appears to depend upon its location in the cell. The membrane-bound enzyme of Swiss 3T3 cells shows a marked preference for 1-stearoyl-2-arachidonoyl-diacylglycerol, leading investigators to conclude that it is involved in the synthesis of physiologically important arachidonoyl-enriched molecular species of phosphatidylinositol.[12] The enzyme diacylglycerol kinase can be translocated to membranes from the cytosol in fibroblasts, possibly as a result of increased substrate levels in the membranes.[13] Based on *in vitro* studies with protein kinase C, it has been proposed that the translocation of soluble diacylglycerol kinase to membranes is enhanced following its phosphorylation.[14] In other examples, e.g., Swiss 3T3 cells,[15] no translocation of diacylglycerol kinase was observed, but platelet-derived growth factor enhanced phosphatidate synthesis by the membrane-bound kinase, apparently by elevating the level of diacylglycerol substrate.

A relatively minor alternative route for phosphatidic acid formation involves the acylation and subsequent reduction of dihydroxyacetone phosphate (reaction iii, Figure 2). The acylation is catalyzed by the same microsomal acyltransferase that acts on *sn*-glycerol-3-phosphate.[16] Acyldihydroxyacetone phosphate is a major metabolic product in peroxisomes, where it is utilized largely for the formation of ether lipids (see Section XII).

Utilization of phosphatidic acid involves one of two principal reactions: its dephosphorylation to diglyceride by the enzyme phosphatidate phosphohydrolase (reaction ii, Figure 1) or its activation to CDP diacylglycerol by the enzyme cytidine 5′-triphosphate:phosphatidate cytidylyltransferase (reaction vi, Figure 1). The routing of intermediates through this major

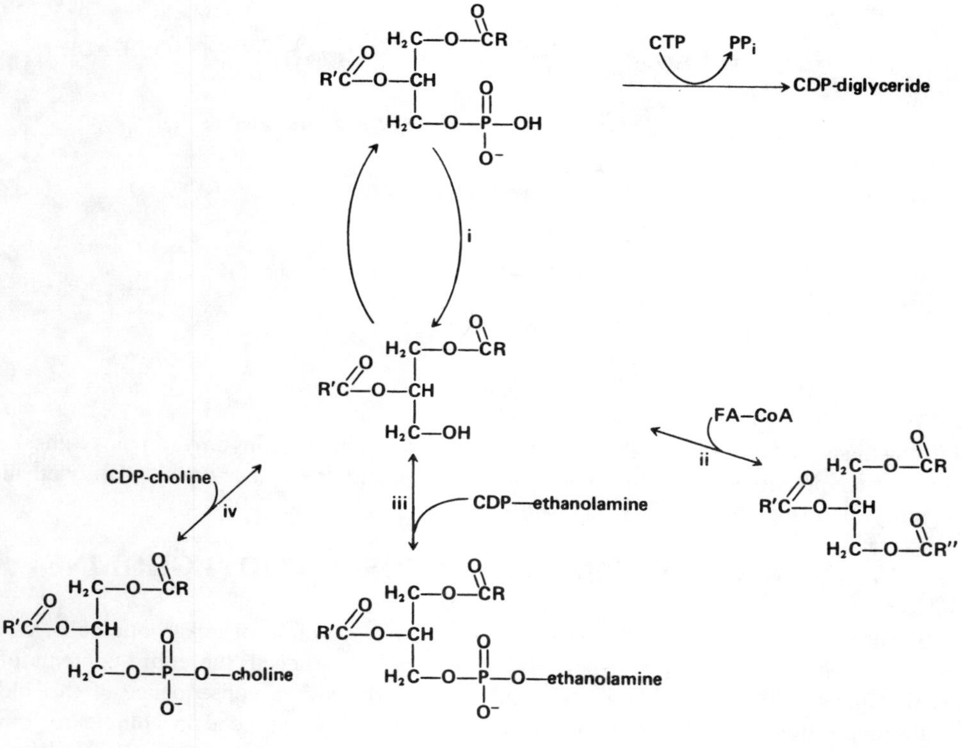

FIGURE 3.

branch point is of obvious importance in determining both the amounts and the fatty acid makeup of all the phospholipids synthesized. The regulation of these two enzymes is discussed in Sections III and VIII (see p. 89).

III. REGULATION OF THE DIACYLGLYCEROL POOL

As outlined in Figure 1, phosphatidic acid molecules may be dephosphorylated to diacylglycerol, which is then utilized for the formation of one of three major products: phosphatidylcholine, phosphatidylethanolamine or triacylglycerol (Figure 3). The enzyme responsible for this dephosphorylation, phosphatidate phosphohydrolase (EC 3.1.3.4), can be reversibly translocated between the soluble and the particulate compartments of cells.[17] The addition of 4 m*M* oleate to hepatocytes increased the proportion of membrane bound enzyme from about 30 to 97% and increased total activity by twofold. It is believed that translocation of phosphatidate phosphohydrolase to the endoplasmic reticulum is influential in achieving the elevated level of triacylglycerol and, to a lesser extent, phosphatidylcholine synthesis observed when tissue levels of free fatty acids rise.

In addition to this acute mechanism of phosphatidate phosphohydroase control, exposure of hepatocytes to high concentrations of cyclic AMP or glucocorticoids for a period of hours raises the activity of phosphatidate phosphohydrolase several fold.[18] This contributes significantly to the cell's capacity for triacylglycerol and very-low-density lipoprotein synthesis. Dietary or hormonal stimulation of phosphatidate phosphohydrolase generally has a much more pronounced effect on the synthesis of triacylglycerol than of phospholipids.

The diacylglycerol pool of the endoplasmic reticulum, supplemented under certain condi-

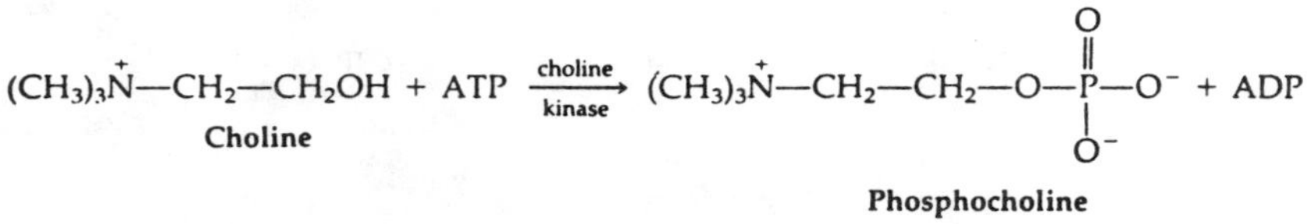

$$(CH_3)_3\overset{+}{N}{-}CH_2{-}CH_2OH + ATP \xrightarrow[\text{kinase}]{\text{choline}} (CH_3)_3\overset{+}{N}{-}CH_2{-}CH_2{-}O{-}\overset{O}{\overset{\|}{P}}(O^-){-}O^- + ADP$$

Choline Phosphocholine

FIGURE 4.

$$\text{phosphocholine} + \text{CTP} \xrightleftharpoons[\text{cytidylyltransferase}]{\text{CTP-phosphocholine}} \text{CDP-choline} + PP_i$$

FIGURE 5.

tions by diacylglycerol formed through other pathways such as sphingomyelin biosynthesis,[19] serves as the precursor for *de novo* synthesis of phosphatidylcholine and phosphatidylethanolamine as described below.

IV. *DE NOVO* SYNTHESIS OF PHOSPHATIDYLCHOLINE

Because it is quantitatively the most significant phospholipid of eukaryotic cells, phosphatidylcholine has undergone especially extensive study; indeed, the entire contents of a recent monograph[20] is devoted to its metabolism. Perhaps as a consequence of this close scrutiny, we are aware of metabolic control processes that transcend in complexity those known to regulate the synthesis of other phospholipids. However, as broader studies are completed, many of the regulatory mechanisms described below may well be recognized to have wider applicability.

Phosphocholine, the polar head group of this major lipid, is formed by the cytosolic enzyme choline kinase (EC 2.7.1.32). The activity (Figure 4), was once suspected to be rate limiting in phosphatidylcholine synthesis. While the enzyme is stimulated by such compounds as polyamines and certain carcinogens,[21] these changes in activity have been determined not to be closely correlated with the rate of phosphatidylcholine formation.

The rate of phosphatidylcholine biosynthesis is now known to be controlled by the enzyme CTP:phosphocholine cytidylyltransferase (EC 2.7.7.15)[22] (Figure 5). Cytidylyltransferase is one of a growing number of enzymes whose activity is enhanced by translocation from the cytosol to membranes — in this instance specifically to the endoplasmic reticulum.[23] The reversible association of cytidylyltransferase with membranes is promoted by free fatty acids, phospholipase C treatment, and many other factors that stimulate phosphatidylcholine synthesis.[24,25] However, no single mechanism responsible for cytidylyltransferase translocation to membranes has been identified. In liver the enzyme is released from membranes when phosphorylated by a cyclic AMP-dependent protein kinase.[26] Phorbol esters stimulate binding of the enzyme to membranes,[27] as does thyrotropin-releasing hormone.[28] Neither of these agents induced translocation of the enzyme in pituitary cells in which protein kinase C had been down-regulated.[28] These and other findings[26] strongly imply that phosphorylation of the protein alters its affinity for membranes. *In vivo* phosphorylation of cytidylyltransferase has recently been detected in HeLa cells,[29] strengthening the arguments for its physiological significance. There still seems a chance that the enzyme's translocation to membranes is an indirect effect of hormone action. The hormone-induced hydrolysis of cellular phospholipids releases diacylglycerols and fatty acids that are themselves capable of mediating enzyme redistribution. It has been shown that the addition of free fatty acids to cells or cell homogenates

diacylglycerol + CDP - choline ⇌ phosphatidylcholine + CMP

CDP - choline: 1, 2 - diacylglycerol
phosphocholinetransferase

FIGURE 6.

promotes both the association with membranes and the activation of cytidylyltransferase,[30] and evidence has also been presented[31] for an induced translocation of the enzyme to membranes by physiological levels of diacylglycerol.

De novo phosphatidylcholine synthesis is completed by the transfer of phosphocholine from CDP-choline to diacylglycerol drawn from a microsomal diacylglycerol pool. The reaction is catalyzed by the intrinsic microsomal enzyme CDP-choline:1,2-diacylglycerol phosphocholinetransferase (EC 2.7.8.2) as illustrated in Figure 6. Past studies often left unresolved the question of whether this enzyme is capable of transferring phosphoethanolamine as well as phosphocholine. The clearest evidence regarding this point comes from genetic studies with *Saccharomyces cerevisiae* mutants defective in phosphocholine or phosphoethanolamine transferase. While the phosphocholine transferase was shown to be specific for CDP-choline,[32] a distinct phosphoethanolamine transferase was capable of utilizing both CDP-choline and -ethanolamine.[33]

In higher plants the cytidylyltransferase exhibits an even broader range of specificity. Phosphocholine, formed by the sequential methylation of phosphoethanolamine, is the predominant base transferred to diacylglycerol in *Lemna paucicostata*, but in soybean (*Glycine max*) phosphomethylethanolamine appears to be preferentially activated by the cytidylyltransferase and utilized to form phosphatidylmonomethylethanolamine.[34] Two additional methylations of the lipid intermediate then complete the synthesis of phosphatidylcholine. On the other hand, carrot (*Daucus carota*) phosphatidylcholine arises via a combination of the two mechanisms, featuring methylation both of the phosphorylated bases and the assembled phospholipids.

Phosphocholine transferase of platelets exerts a preference for diacylglycerols over alkenylacylglycerols, while phosphoethanolamine transferase is less specific.[35] This may explain the predominance of alkenyl ether-linked side chains in ethanolamine phosphoglycerides as compared with choline phosphoglycerides. The two aminoalcohol phosphotransferases of pea and soya bean are also discriminating in their choice of substrates, preferring 1-palmitoy-2-linoleoyldiacylglycerol from a mixture of diacylglycerol molecular species.[36]

Whereas the studies completed to date show that the enzymes contributing to the *de novo* biosynthesis of phosphatidylcholine in different tissues have similar properties, each tissue is at the same time unique in its fine tuning of these reactions. For a discourse on the regulation of choline phospholipid synthesis in one particular tissue, the mammalian heart, the reader is referred to the excellent review by Hatch et al.[37]

V. *DE NOVO* SYNTHESIS OF PHOSPHATIDYLETHANOLAMINE

Phosphatidylethanolamine is formed by a pathway analogous to that responsible for *de novo* phosphatidylcholine synthesis; indeed, some enzymes may participate in the formation of both lipids. For example, it has been shown that ethanolamine kinase, which catalyzes the conversion of ethanolamine to phosphoethanolamine (reaction i, Figure 7) and is distinct from choline kinase, can also phosphorylate choline when tested in purified form.[38] The synthesis of CDP-ethanolamine from phosphoethanolamine and CTP by CTP:phosphoethanolamine cytidylyltransferase (reaction ii, Figure 7) is believed to be the rate-limiting reaction in the pathway.[26] Possible control of this cytidylyltransferase by translocation from cytosol to

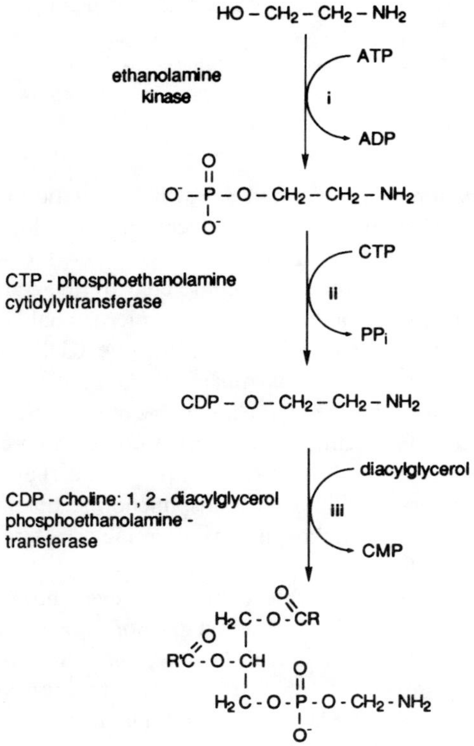

HO – CH_2 – CH_2 – NH_2

ethanolamine kinase (i): ATP → ADP

O^- – P(=O)(O^-) – O – CH_2 – CH_2 – NH_2

CTP - phosphoethanolamine cytidylyltransferase (ii): CTP → PP_i

CDP – O – CH_2 – CH_2 – NH_2

CDP - choline: 1, 2 - diacylglycerol phosphoethanolamine - transferase (iii): diacylglycerol → CMP

H_2C - O - C(=O)R
R'C(=O) - O - CH
H_2C - O - P(=O)(O^-) - O - CH_2 - NH_2

FIGURE 7.

membranes, as occurs with CTP:phosphocholine cytidylyltransferase, has not yet been extensively studied. Phosphoethanolamine transferase (reaction iii, Figure 7) is a microsomal enzyme catalyzing the transfer of phosphoethanolamine from CTP:ethanolamine to either diacylglycerol or 1-alkenyl-2-acyl glycerol.[39]

Questions have been raised regarding the quantitative importance of this *de novo* biosynthetic pathway. Although it is assumed to represent the major source of phosphatidylethanolamine in mammalian cells, Chinese hamster ovary cell mutants defective in either the cytidylyltransferase[40] or the phosphoethanolamine transferase[39] have a virtually normal phosphatidylethanolamine content. The cells are apparently able to draw on one of two alternative pathways: the decarboxylation of phosphatidylserine (see Section VII) or the base exchange reaction transferring free ethanolamine for the serine moiety of phosphatidylserine (see Section VI), to provide their entire complement of phosphatidylethanolamine. Labeling experiments indicated the decarboxylation of phosphatidylserine as the primary mechanism of phosphatidylethanolamine under these circumstances.

VI. PHOSPHOLIPID INTERCONVERSION BY EXCHANGE OF POLAR HEAD GROUPS

The polar head group (base) of certain phospholipids is readily exchanged for a different free base in a Ca^{2+}-stimulated, energy dependent reaction. Much of the phosphatidylserine of mammalian tissues is synthesized by this base exchange reaction catalyzed by the microsomal enzyme phosphatidylcholine:serine O-phosphatidyltransferase. Phosphatidylethanolamine can also serve as the phosphatidyl donor.

Kuge et al.[41] showed the presence in Chinese hamster ovary cells of two different enzymes catalyzing a base exchange of phospholipids with free serine. Mutant cells lacked the ability to utilize phosphatidylcholine as a substrate for serine exchange whereas phosphatidylethanolamine would function in the exchange reaction. The mutant grown in the absence of either phosphatidylserine or phosphatidylethanolamine showed a diminished level of both phospholipids. It was proposed that the exchange of serine for the choline of phosphatidylcholine is the major source of phosphatidylserine (and its product phosphatidylethanolamine) in these cells and that *de novo* biosynthesis of phosphatidylethanolamine from diacylglycerol and CDP-ethanolamine was insufficient to restore a normal lipid composition in the mutant.

VII. FORMATION OF PHOSPHATIDYLETHANOLAMINE AND PHOSPHATIDYLCHOLINE BY ENZYMATIC MODIFICATION OF PHOSPHATIDLYSERINE

Aficionados of lipid metabolism were astonished by the revelation in 1959-60 of a new major pathway for phosphatidylethanolamine and phosphatidylcholine biosynthesis in liver.[42, 43] The overall sequence (reaction V, Figure 1) begins with the decarboxylation of phosphatidylserine to phosphatidylethanolamine, followed by a stepwise methylation of the latter phospholipid, ultimately leading to phosphatidylcholine. Phosphatidylserine decarboxylase, the enzyme catalyzing the first reaction, is associated specifically with the inner mitochondrial membrane.[44] The three methylation steps, on the other hand, are carried out in the endoplasmic reticulum. There has been much conflicting evidence regarding the number of enzymes involved in this methylation sequence.[25] Recent genetic studies of yeast mutants[45] may have resolved the dilemma. Genes for two methyltransferases were found and expressed in deficient cells. One enzyme catalyzed the *in vitro* formation of phosphatidylmonomethylethanolamine only, while a second performed all three methylations. Further genetic analysis[46] suggests that the second methyltransferase normally catalyzes only the last two methylations *in vivo*.

VIII. FORMATION OF CDP-DIACYLGLYCEROLS FROM PHOSPHATIDIC ACID

In competition with the enzyme phosphatidic acid phosphatase (Section III) for the cellular pool of phosphatidic acid in most animals and plants is another enzyme CTP:phosphatidic acid cytidylyltransferase (CDP-DG synthase, EC. 2.7.7.41) (reaction vii, Figure 1). The enzyme from yeast mitochondria has been characterized and found to differ immunologically and in molecular weight from the equivalent enzyme in *E. coli*.[47] Both activities appear to be regulated by growth conditions, and that of yeast is also strongly influenced by the availability of water soluble lipid precursors.[48] The activity is of major metabolic importance in yeast microsomes and mitochondria, but assumes a lesser quantitative significance in higher animals. Little is known regarding preferences of phosphatidic acid molecular species being drawn into the two biosynthetic pathways utilizing this substrate, and any selectivity may have little significance since acyl chain retailoring of the final products is common.

CDP-diglycerides assume a much more important role in bacterial phospholipid synthesis than in animals. They are intermediates in the biosynthesis of all bacterial phospholipids[49] as described below.

IX. *DE NOVO* SYNTHESIS OF PHOSPHATIDYLSERINE, PHOSPHATIDYLINOSITOL, PHOSPHATIDYLGLYCEROL, AND CARDIOLIPIN

These phospholipids are formed from CDP-diacylglycerol (Figure 8), an intermediate arising through the action of CTP:phosphatidate cytidylyltransferase (EC 2.7.7.41) (reaction vii, Figure 1). The enzyme occurs in mitochondria as well as endoplasmic reticulum.[50]

In *Saccharomyces cerevisiae* phosphatidylserine is synthesized by the coupling of serine with CDP-diacylglycerol in a reaction catalyzed by CDP-diacylglycerol:L-serine O-phosphatidyltransferase (phosphatidylserine synthase) (EC 2.7.8.8.) (reaction i, Figure 8). This reaction is of critical importance in yeast, as other major phospholipids, namely phosphatidylcholine and phosphatidylethanolamine, are formed principally from phosphatidylserine.[51]

Phosphatidylserine synthase activity is highly regulated in yeast. The cellular level of this enzyme, as well as that of several others involved in phospholipid synthesis, can be controlled through enzyme repression by inositol, with additional influences by serine, choline, and ethanolamine.[52] A more rapid and direct effect of inositol, when present in higher levels, is to draw CDP-diacylglycerol into the synthesis of phosphatidylinositol (reaction ii, Figure 8) at the expense of phosphatidylserine and its products phosphatidylethanolamine and phosphatidylcholine.[53] Regulation in the yeast system is also exerted through cyclic AMP-dependent phosphorylation of phosphatidylserine synthase, leading to a decrease in its activity.[54] The various influences upon the activity of phosphatidylserine synthase probably determine whether another major phospholipid, phosphatidylcholine, is formed mainly by the stepwise methylation of phosphatidylethanolamine or by the CDP-choline pathway. The methylation pathway appears to predominate when yeast is grown in the absence of inositol and phosphatidylserine synthase is fully derepressed, but when inositol and choline are present, phosphatidylserine synthase formation is repressed, and the CDP-choline pathway assumes a more prominent role.[55]

Phosphatidylserine synthesis in bacteria follows a pattern similar to that in yeast. Gene amplification has been used to cause overproduction of phosphatidylserine synthase in *E. coli*, but this manipulation had little effect on the phosphatidylserine content.[56]

In mammals, phosphatidylserine is thought to be generated mainly through base-exchange reactions (Section VI, p. 88). However, phosphatidylinositol formation has long been considered to involve the coupling of inositol to the diacylglycerol moiety of CDP-diacylglycerol in the endoplasmic reticulum.[50] The responsible enzyme, CDP-diacylglycerol:inositol phosphatidyltransferase (phosphatidylinositol synthase) (EC2.7.8.11) now appears to occur in the plasma membrane of cultured mammalian cells as well as the endoplasmic reticulum. In both membrane locations the enzyme is inhibited by elevated levels of phosphatidylinositol.[57] The discovery that phosphatidylinositol synthesis can be regulated in the plasma membrane is significant in view of the localization in that site of enzymes for polyphosphoinositide formation (see Section X, p. 92).

Phosphatidylglycerol is a lipid that is prevalent in mitochondrial membranes of animal cells, in both mitochondria and chloroplasts of plants and in most bacteria. All of these organelles have a "prokaryotic" character, and the composition of phosphatidylglycerol reflects this. Thus the fatty acid placement is such that oleic acid is preferentially but not exclusively present in the *sn*-l position. This specificity probably reflects that fact that *de novo* "prokaryotic" synthesis of phospholipids is achieved entirely through acylation of glycerol-3-phosphate with fatty acyl-ACP[58] (see Section II, p. 83), using acyltransferases with characteristic preferences for fatty acyl chains.[59] In squash (*Cucurbita moschata*) the initial acylation, at the *sn*-l position of glycerol-3-phosphate, can be carried out by any one of three isomeric

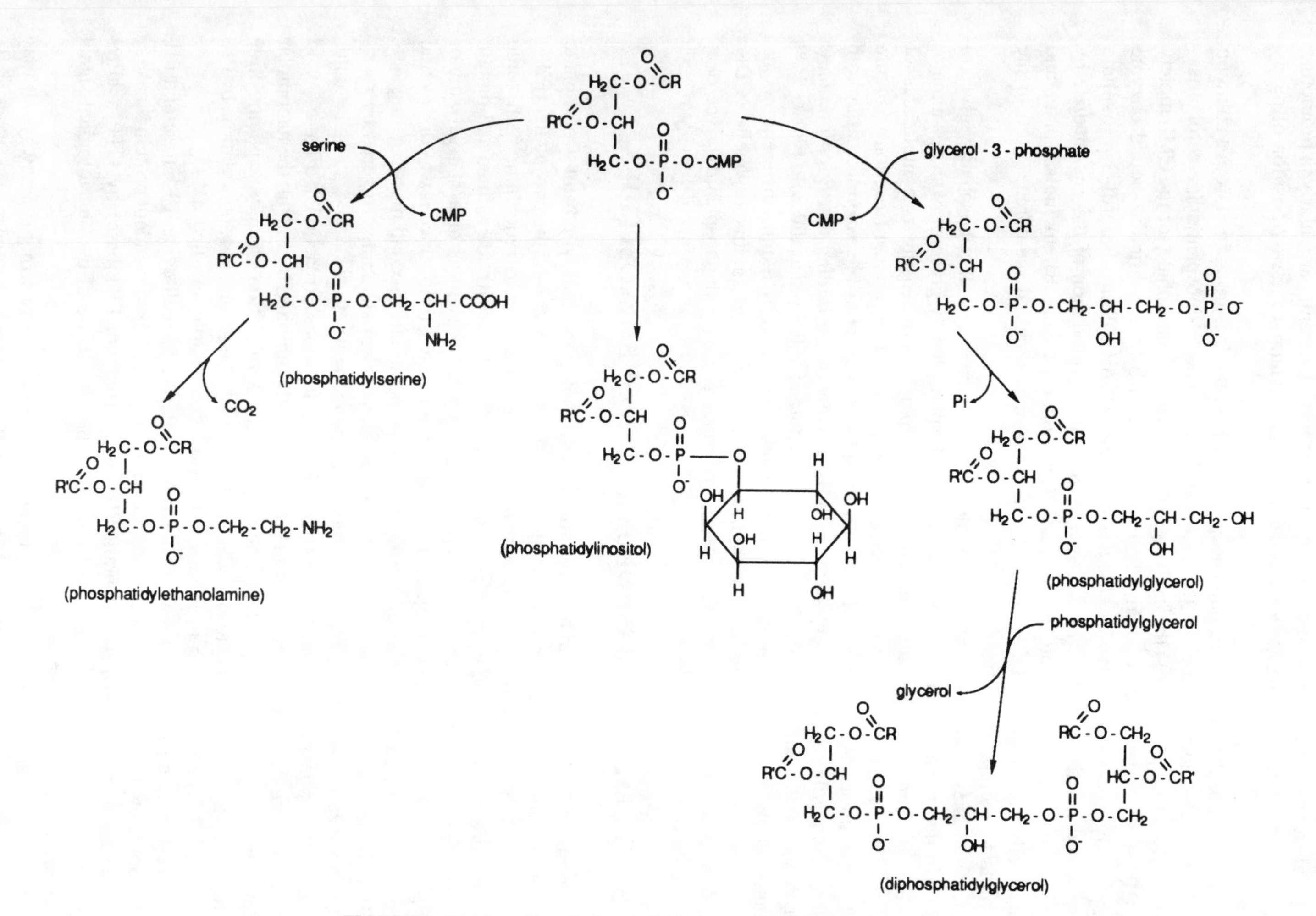

FIGURE 8. The formation of phospholipids from CDP-diacylglycerol in yeast.

forms of acyltransferase.[60] Environmental factors may dictate which isoform is active, and therefore which fatty acid (palmitate or oleate) is installed at the *sn*-l position. The specificity of acyltransferases of plant endoplasmic reticulum is quite different, particularly in promoting the placement of oleate rather than palmitate as the acyl chain at the *sn*-2 position of newly made phosphatidic acid.[61]

All major phospholipids, including phosphatidylethanolamine, are formed from a common precursor, CDP-phosphatidate, in the prokaryotic pathway.[62] Phosphatidylglycerol arises through a two step process: (1) the exchange of *sn*-glycerol 3-phosphate for the CMP moiety of CDP-phosphatidate, (reaction iii, Figure 8) catalyzed by the membrane-bound enzyme phosphatidylglycerol phosphate synthase, CDP-1,2-diacyl-*sn*-glycerol-3-phosphate phosphatidyltransferase, (EC 2.7.8.7), and (2) the dephosphorylation of the intermediate by phosphatidylglycerol phosphate phosphatase (reaction iv, Figure 8). In higher plant and animal cells this synthesis takes place in the inner mitochondrial membrane[63] and in the chloroplast envelope inner membrane.[64,65]

In membranes of bacteria and in mitochondria of eukaryotic cells phosphatidylglycerol may be further metabolized to yield cardiolipin, or bisphosphatidylglycerol, which is a major lipid component there. *E. coli* synthesizes cardiolipin by condensing two molecules of phosphatidylglycerol, producing a molecule of glycerol as the byproduct (reaction v, Figure 8).[66] Little information is available to indicate how the responsible enzyme, cardiolipin synthase, is regulated. Active regulation is likely in view of the finding that *E. coli* mutants overproducing the synthase had at most 1.5 times as much cardiolipin as did wild type cells.[67]

Surprisingly, cardiolipin formation in mitochondria proceeds by a quite different mechanism despite the prokaryotic-like metabolism commonly found in this organelle. CDP-diacylglycerol donates a phosphatidate moiety for condensation with phosphatidylglycerol to yield cardiolipin.[68,69]

X. METABOLISM OF POLYPHOSPHOINOSITIDES

Phosphatidylinositol (PI) can be phosphorylated to form phosphatidylinositol-4-phosphate (PIP) and, after a second phosphorylation, phosphatidylinositol-4,5-bisphosphate (PIP_2) (Figure 9). Whereas PI is synthesized mainly in the endoplasmic reticulum, with some synthesis possibly occurring in the plasma membrane,[57] PIP and PIP_2 are formed and remain located primarily in the plasma membrane, where they are present as quantitatively minor components. PIP_2 has been the subject of intensified research activity since it was discovered to participate in signal transduction across the plasma membrane. Second messengers generated via its hydrolysis are now recognized as key elements in signaling triggered by many hormones, neurotransmitters, growth factors, platelet agonists, etc.[70]

The initial phosphorylation of PI is mediated by PI kinase (reaction i, Figure 9). In chromaffin granules this enzyme is strongly and reversibly inhibited by Ca^{2+} in the nanomolar or low micromolar range, suggesting a negative feedback inhibition of the phosphoinositide cycle (whose action elevates cytosolic Ca^{2+}) at this step.[71] The second phosphorylation, by PIP kinase (EC 2.7.1.68) (reaction ii, Figure 9) is also sensitive to cytosolic Ca^{2+}.[72]

In yeast, where PIP_2-signaling is involved in regulating the cell cycle, PI kinase and PIP kinase appear to be activated through phosphorylation by a cAMP-dependent protein kinase.[73] The activation of a particulate rat brain PIP kinase (but not PI kinase) by GTP analogs implicated GTP-binding proteins in its regulation, but the nature of the physiological signal activating PIP kinase *in vivo* was not identified.[74]

After its formation, PIP is normally metabolized by one of two major pathways. It may undergo dephosphorylation to PIP and thence to PI by phosphomonoesterase attack.[75] By virtue of its location in the plasma membrane, it is also available for participation in PIP_2-

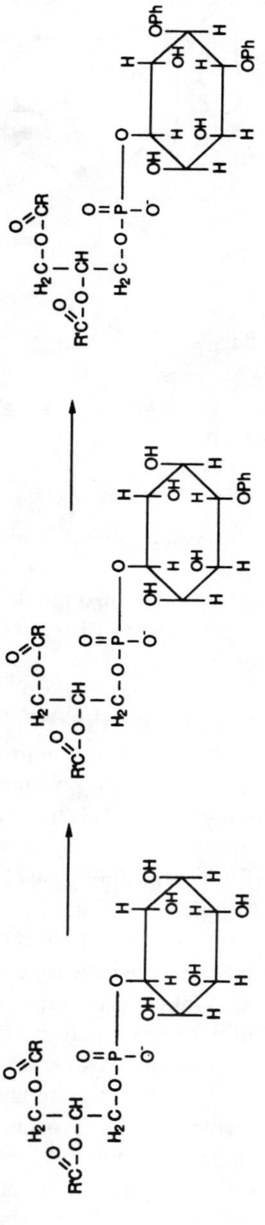

FIGURE 9. The sequential phosphorylation of phosphatidylinositol.

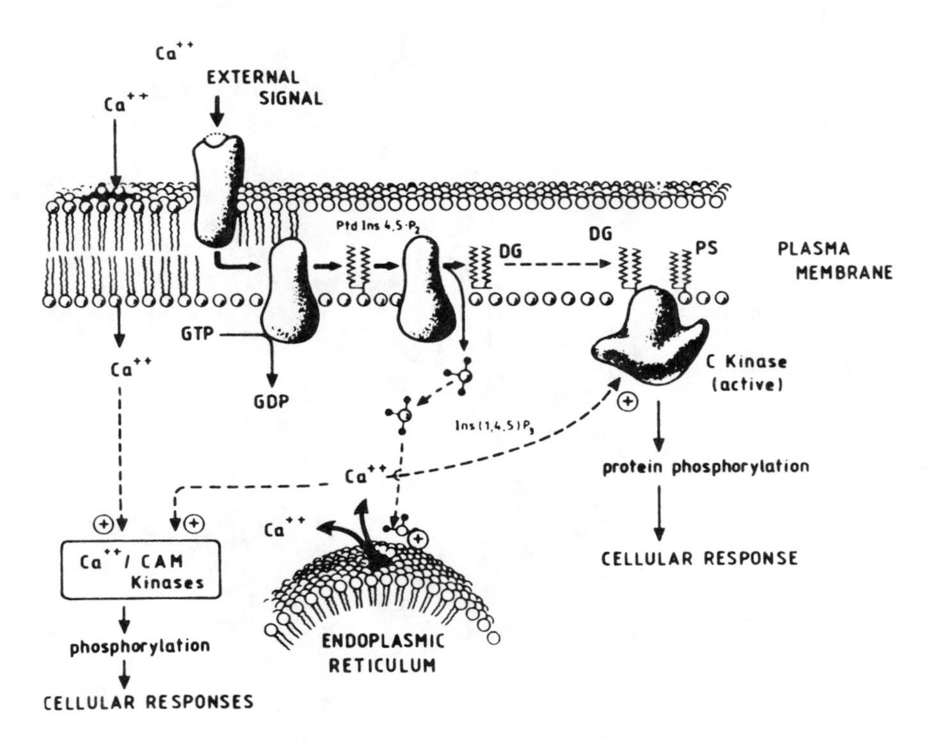

FIGURE 10. Phosphatidylinositol 4,5-bisphosphate-mediated transmembrane signaling pathways.

mediated signal transduction, which occurs widely in animal cells and probably in plant cells as well. Although not all of the plasma membrane PIP_2 is necessarily accessible for use in signaling,[76] the total PIP_2 content can be reduced by as much as 30% within 2 min in stimulated cells. The basic features of the signaling pathway are illustrated in Figure 10.

Transmembrane signaling requires hydrolysis of PIP_2 by phospholipase C. This enzyme is found as a soluble and a membrane-bound form in most animal and plant cells. Its properties have recently been reviewed by Rhee et al.[77] Certain phospholipase C enzymes are specific for PI, PIP, and PIP_2 at low Ca^{2+} concentrations. PIP_2 is the preferred substrate of membrane-associated phospholipase C. When appropriate agonists, whose identities are largely unknown, bind to their receptors, PIP_2-specific phospholipase C is activated through the mediation of a G protein. The mechanism for this activation is poorly understood and appears to differ somewhat from one phospholipase C isozyme to another.[77]

The activity of PIP_2-specific phospholipase C may be regulated by phosphorylation. *In vivo* phosphorylation of the enzyme by protein kinase C has been shown,[78] but it is not known what effect this modification has on phospholipase activity. In BALB/C 3T3 fibroblasts a different phospholipase C isoform is phosphorylated on tyrosine residues following stimulation by epidermal growth factor. This activates the phospholipase and promotes its association with the epidermal growth factor receptor, which is also tyrosine phosphorylated.[79] Conversely, indirect evidence suggests that phospholipase C activity is inhibited in neuroblastoma cells by protein kinase A, presumably through it's phosphorylation of the phospholipase.[80] Evidence has been found for the desensitization of PIP_2-specific phospholipase C activity in vasopressin-treated mammary cells[81] and purinergic receptor agonist-preincubated turkey erythrocytes.[82] The mechanisms responsible for activity loss are not yet clear.

The major products of phospholipase C cleavage of PIP_2 are diacylglycerol and inositol-1,4,5-trisphosphate (1,4,5-IP_3). As indicated in Figure 10, 1,4,5-IP_3 releases Ca^{2+} from intracellular stores, particularly from endoplasmic reticulum cisternae and, in plants,[83] the central vacuole. 1,4,5-IP_3 is also the substrate for a kinase that forms small amounts of 1,3,4,5-IP_4 (reviewed by Berridge and Irvine[84]). Acting in conjunction with IP_3, IP_4 can also affect the cytosolic Ca^{2+} level, perhaps by regulating the transfer of Ca^{2+} between intracellular pools. Other inositol phosphates have been detected including cyclic isomers, 1,3,4,5,6-IP_5, and IP_6.[84] Their function is uncertain but is not thought to be directly associated with signaling.

The bulk of PIP_2-derived IP_3 is dephosphorylated stepwise to 1,4-IP_2, 4-IP, and inositol, with the latter compound being utilized for the resynthesis of PIP_2. The hydrolysis of inositol monophosphate can be inhibited by Li^+, thereby restricting a principal supply of inositol for PIP_2 synthesis. This may explain the puzzling effects of Li^+ upon neural activity.[85]

Diacylglycerol, the other product of PIP_2 cleavage by phospholipase C, also serves as a second messenger. When released in a membrane, it acts in conjunction with phosphatidylserine to stimulate the translocation of inactive protein kinase C from the cytosol to the membrane, where the kinase becomes activated. Phosphatidylserine is specifically required for both the membrane binding and the normal activation of protein kinase C as well as for the binding of phorbol esters, diacylglycerol analogs which induce a persistent activation of the kinase.[86] When protein kinase C remains activated continuously for more than a few moments it appears to become autophosphosphorylated and subsequently down-regulated.[87]

There are multiple isoforms of protein kinase C,[88] and these differ in certain characteristics, such as their capacity for activation by lipids.[89] Protein kinase C is now known to be inhibited by certain lipids, particularly sphingosine.[90]

The operation and regulation of the PIP_2-mediated signal transduction pathway are probably much more complex than pictured in Figure 10. Small amounts of novel polyphosphoinositides, such as the PIP_2 isomer phosphatidylinositol 3,4-bisphosphate and phosphatidylinositol 3,4,5-tris-phosphate, have been reported to arise during neutrophil activation.[91] No physiological function for these compounds is presently known. Sources of diacylglycerol other than from PIP_2 hydrolysis have been proposed and defended with convincing evidence. For example, the hydrolysis of phosphatidylcholine by phospholipase C can in some instances produce ample diacylglycerol to activate protein kinase C.[92] And the stimulation of human neutrophils by chemotactic peptides appears to trigger the formation of diacylglycerol from phosphatidylcholine via the action of phospholipase D, producing phosphatidate, which is then dephosphorylated by phosphatidate phosphohydrolase.[93]

XI. PHOSPHONOLIPIDS

The existence of phosphonolipids in nature has been known since 1963, when ceramide-aminoethylphosphonate was discovered in the sea anemone, *Anthopleura elegantissim*, by Rouser et al.[94] Shortly afterwards, a phosphonate-containing glyceride was isolated from the protozoan *Tetrahymena pyriformis* by Liang and Rosenberg.[95]

The distribution and properties of these lipids have been described by Hilderbrand[96] and by Hori et al.[97] They are prominent membrane components in protozoa, colenterates, and molluscs. After prolonged efforts by many investigators the novel carbon-to-phosphorus linkage in these lipids has recently been confirmed as arising through action by a phosphomutase that catalyzes the reversible rearrangement of phosphoenolpyruvate to phosphonopyruvate[98] (first reaction, Figure 11). Further utilization of the phosphonopyruvate, perhaps by the later reactions shown in Figure 11, leads to the final product, 2-aminoethylphosphonate. Although some early evidence suggested that a lipid intermediate takes part in C-P formation, this has not been supported by recent work.

There is virtually no direct evidence relating to the control of phosphonolipid metabolism. It is known that some phosphonolipid-rich organisms such as *Tetrahymena pyriformis*[99] and

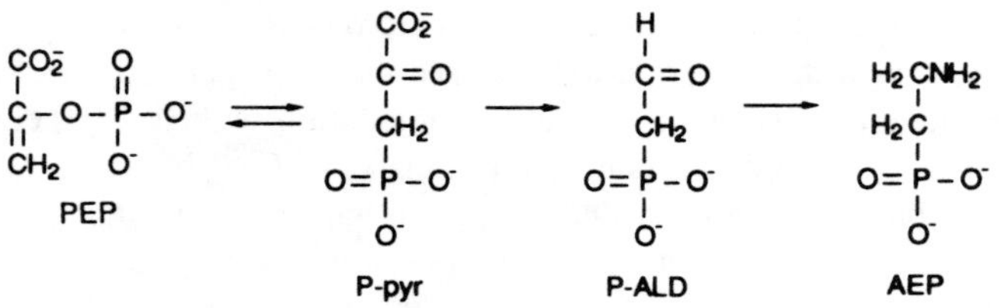

PEP ⇌ P-pyr → P-ALD → AEP

FIGURE 11. (From Barry, R. J., Bowman, E., McQueney, M., and Dunaway-Mariano, D., *Biochem. Biophys. Res. Comm.*, 153, 177–182, 1988. With permission.)

Paramecium aurelia[100] concentrate phosphonoglycerides in extremely high levels in their surface membranes. In *Tetrahymena*, for example, almost 70% of the lipid phosphorus in ciliary membranes is of the phosphonolipid type.[101] In most of these phosphonolipids, an alkyl ether side-chain rather than a fatty acyl group is present at the 1-position of the glyceryl moeity. The enrichment of ether-containing phosphonolipids in a particular membrane has been postulated to stem from their marked resistance to attack by endogenous lipolytic enzymes.[99] This type of stability may also account for the conservation of phosphonolipids in tissues of the oyster during periods when starvation caused depletion of phosphodiester phospholipids.[102]

XII. PHOSPHOLIPIDS CONTAINING ETHER BONDS

Often existing unnoticed among the diacyl-phospholipids of animal tissues are lipids containing an unsaturated or saturated side chain linked to position 1 of the glycerol backbone by an ether bond instead of an acyl bond (Figure 12). These lipids may account for as much as 50 mol% of the total lipid phosphorus in certain mammalian organs, particularly heart and the central nervous system,[103] and they are also prevalent in many other organisms. A comprehensive account of the ether lipids has been assembled by Snyder.[104]

Ether lipid metabolism offers a formidable challenge to the student of membrane lipid regulation. The striking tissue specificity for these compounds leaves little doubt that they serve some unique function, but that function has so far eluded us. Apart from their greater stability to certain chemical and enzymatic reactions,[105,106] we have few clues as to their specific role in cellular activities.

The previously rather esoteric field was suddenly placed in the spotlight by the discovery in 1979 that the very potent platelet activating factor (PAF) is a unique form of ether lipid, 1-alkyl-2-acetyl-*sn*-glycero-3-phosphocholine.[107-109]

Ether lipid biosynthesis begins with the formation of acyl-dihydroxyacetone phosphate and its subsequent conversions to acyl-dihydroxyacetone (Figure 12). The acylation of dihydroxy-acetone phosphate (reaction i, Figure 13) is carried out by an enzyme thought at first to be located in mitochondria and microsomes. However, later studies with guinea pig liver[110] implicated peroxisomes as the intracellular location of dihydroxyacetone-phosphate acyltransferase (EC 2.3.1.42) and of alkyl-dihydroxyacetone-phosphate synthase, and alkyldihydroxyacetone phosphate reductase, the enzymes for steps ii and iii, respectively, of the pathway (Figure 13). Later studies (reviewed by Hajra et al.[111]) indicated that in other animals the latter two enzymes may be distributed between peroxisomes and microsomes. The finding that tissues of Zellweger cerebrohepatorenal syndrome patients, known to lack peroxi-somes, are also deficient in the acyl dihydroxyacetone phosphate pathway enzymes and in ether lipids[112] further underlines the importance of the peroxisome in this metabolic pathway.

Many years were required to establish the mechanism of ether bond formation. The key

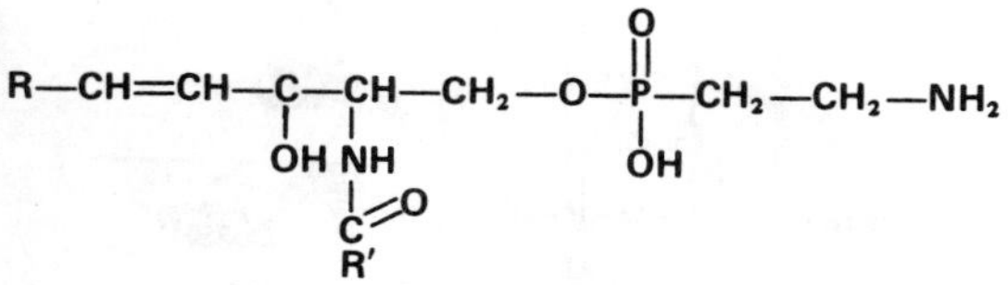

(ceramide-aminoethylphosphonate)

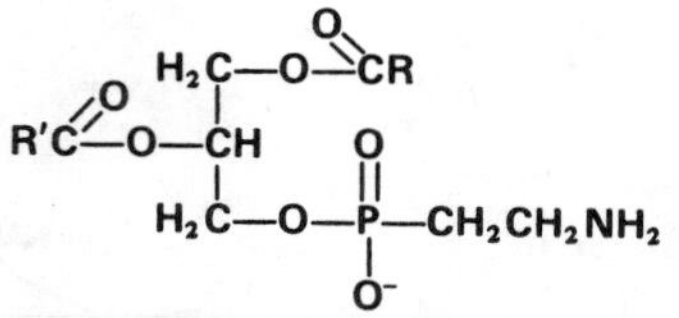

(1, 2-diacylglycero-aminoethylphosphonate)

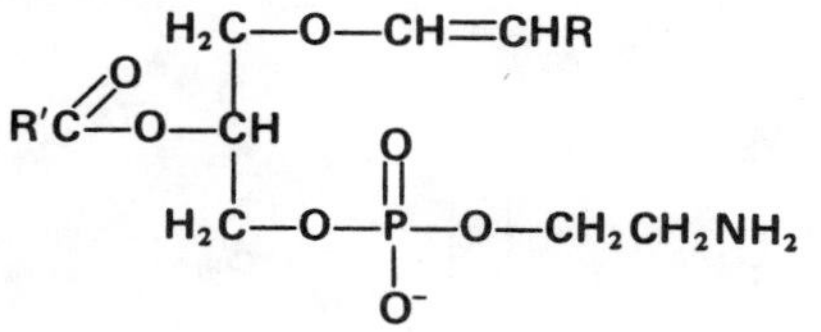

(1-alkenyl-2-acyl-glyceryl-
phosphorylethanolamine)
[plasmenylethanolamine]

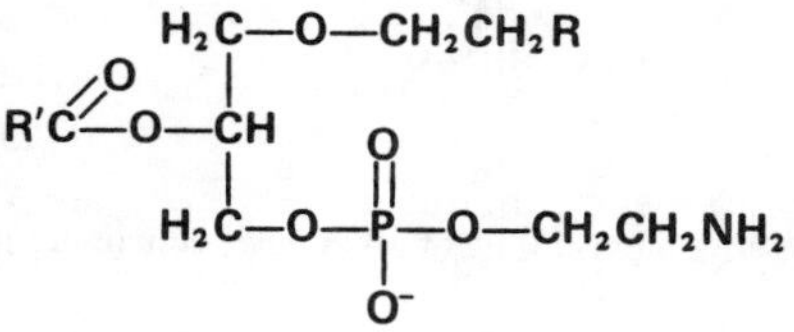

(1-alkyl-2-acyl-glyceryl-
phosphorylethanolamine)
[plasmanylethanolamine]

FIGURE 12.

reaction, catalyzed by alkyl-DHAP synthase, features a coordinated replacement of the acyl chain of acyl-DHAP by an alkyl chain derived from a long chain alcohol (see Snyder[113] for details). Recent studies using rat brain microsomes[114] favor a two step conversion of fatty acid to fatty alcohol, beginning with the transformation of fatty acyl-CoA to fatty aldehyde by fatty acyl-CoA reductase and followed by action of an aldehyde reductase. Brain microsomes are much less efficient in reducing fatty acids containing more than one *cis* -double bond than they are in dealing with saturated and monounsaturated fatty acids.[115] Fatty alcohols may also arise as a byproduct of sphingosine base catabolism (see p. 103).

Following action by alkyl DHAP reductase, the resulting 1-0-alkyl,2-lyso-*sn*-glycero-3-phosphate may be further metabolized by two pathways. In one, acylation of the *sn*-2 position

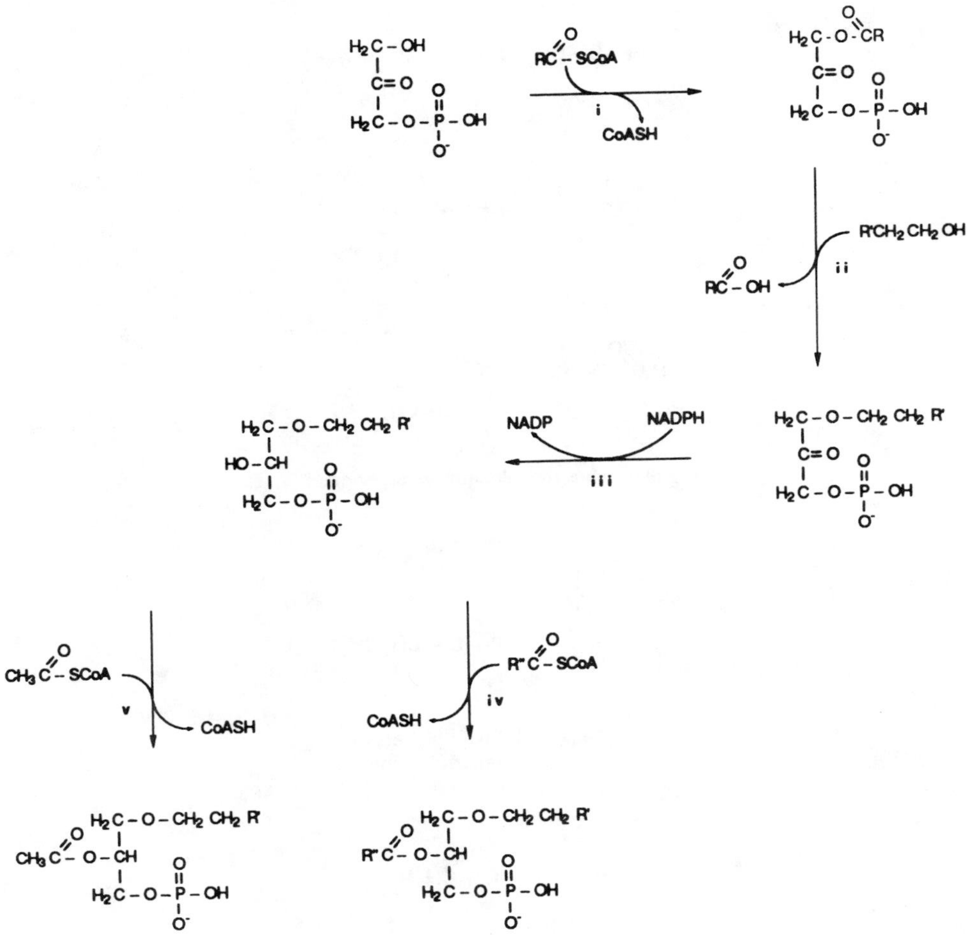

FIGURE 13. The biosynthesis of alkyl ether-containing phospholipids.

by a fatty acyl-CoA (reaction iv, Figure 13) gives rise to the 1-0-alkyl,2-acyl analog of phosphatidic acid, which can be further transformed into other phospholipids as is diacylphosphatidic acid (see Figure 3). Alternatively, 1-0-alkyl,2-lyso-*sn*-glycero-3-phosphate is acetylated (reaction v, Figure 13) to yield the 1-alkyl,2-acetyl analog of phosphatidic acid. This is subsequently dephosphorylated and converted to 1-0-alkyl,2-acetyl-*sn*-glycerophosphoryl-choline (PAF), as indicated in Figure 14, reactions i and ii, by a choline phosphotransferase having properties distinct from those of the enzyme that synthesizes phosphatidylcholine.[116] Recent studies with endothelial cells indicated that agonist-activated PAF synthesis requires an elevation of the cytosolic Ca^{2+} level.[117] Increased activity of protein kinase also raised the cellular content of PAF. Both of these mediators appeared to operate by stimulating the phospholipase A_2 catalyzed hydrolysis of 1-0-alkyl-2-acyl-*sn*-glycero-3-phosphocholine to produce the PAF precursor 1-0-alkyl-2-lyso-*sn*-glycero-3-phosphocholine.

In addition to the *de novo* biosynthetic pathway described above, PAF can be formed by the action of an acetyltransferase on 1-alkyl,2-lyso-*sn*-glycerophosphorylcholine (reaction iii, Figure 14). This second route may be more physiologically significant in the sense that it is activated by the agents known to stimulate cellular production of PAF.[118,119] Little is known

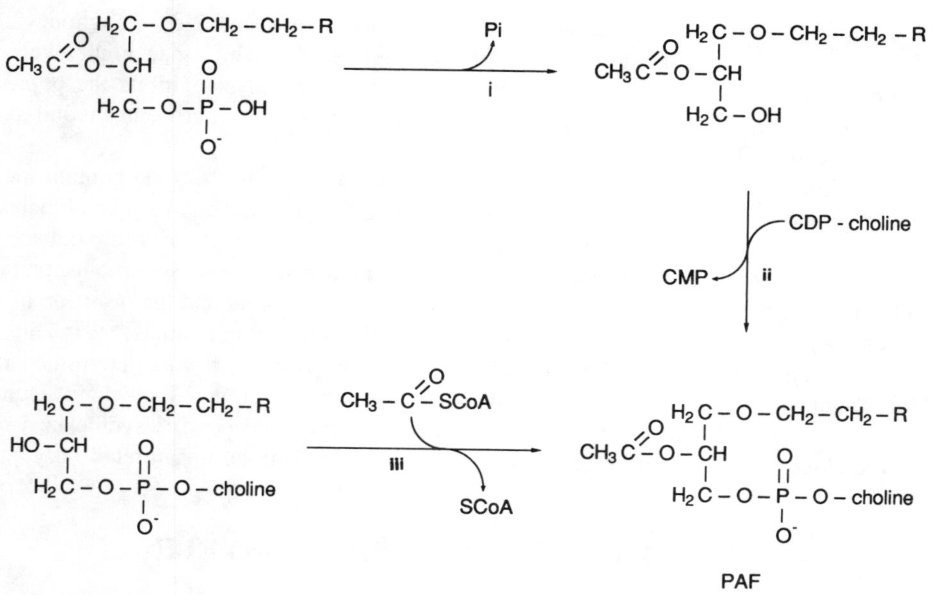

FIGURE 14. Pathways of platelet activating factor formation.

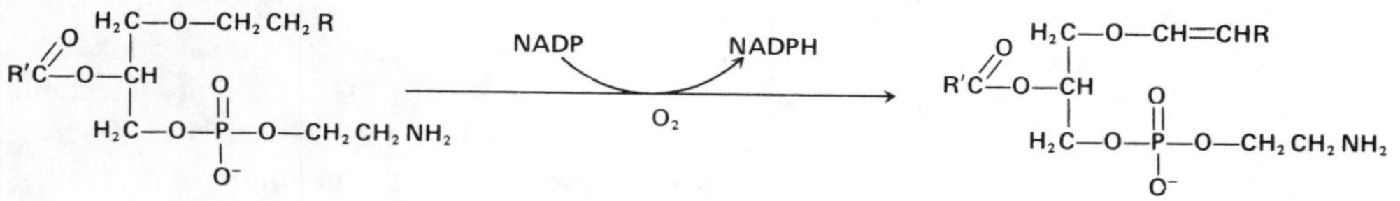

FIGURE 15.

of the factors directly activating the acetyltransferase or the acetylhydrolase responsible for the rapid rise and fall, respectively of tissue PAF.

PAF is degraded by acetylhydrolases that are highly specific for phospholipids containing short acyl chains at the *sn*-2 position (reviewed by Prescott et al.[120]). The enzymes are found in cells as well as plasma, and evidence is accumulating that secretion of the activity into plasma is hormonally regulated.

1-Alkyl-2-acyl-glycerolphospholipids are relatively common in certain animal tissues, e.g., the nervous system and certain tumors, as structural components of the membranes. They also serve as precursors of 1-alk-1-enyl-2-acyl-glycerophospholipids (plasmalogens), which reach high levels in the membranes of select tissues such as heart and brain.

Plasmalogens are synthesized by the insertion of a double bond between the first and second carbon atoms of the alkyl chain found in the alkyl ether analog of phosphatidylethanolamine (Figure 15). The enzyme for this reaction, Δ^1-alkyl desaturase, is located in the endoplasmic reticulum and requires molecular oxygen, a reduced pyridine nucleotide, cytochrome b_5 reductase, and cytochrome b_5 for activity. In many respects the system resembles fatty acid desaturase.[121] Choline-containing plasmalogens, especially prevalent in heart muscle of human and rabbit (but not of rat), may be made from 1-alk-1-enyl-2-acylglycerol released from ethanolamine-containing plasmalogens by the action of phospholipase C.[122]

The catabolism of plasmalogens proceeds via an initial deacylation to lysoplasmalogen followed by microsomal lysoplasmalogen alkenylhydrolase (lysoplasmalogenase) action, yielding a free aldehyde and the glycerylphosphoryl base. Brain, which has high levels of plasmalogens, contains low lysoplasmalogenase activity, while liver exhibits the opposite characteristics. Lysoplasmalogenase purified 200-fold from rat liver microsomes required no cofactors or cations for activity.[123]

Ether lipids are also found in certain types of bacteria.[124] Archaebacteria contain membranes constructed almost entirely of diethers having two C_{20} phytanyl chains. Their biochemistry is discussed in Chapter 6 in conjunction with other isoprenoids. Anaerobic eubacteria contain plasmalogens more closely resembling those of animal cells in possessing nonisoprenoid side chains. However, it appears that the formation of the ether bond and the insertion of the 1,2 double bond proceed by a mechanism quite unlike that utilized in animals.[121,124] Thus by administering glycerol labeled on different carbon atoms with ^3H, it was determined that DHAP is not an intermediate in the formation of the ether bond. Nor were long chain alcohols efficient precursors.[121] It has been speculated that the unsaturated bond is synthesized in a manner analogous to that employed by anaeroic bacteria in forming unsaturated fatty acids (see Chapter 1).

XIII. METABOLISM OF SPHINGOMYELIN

Sphingomyelin is the only phospholipid not containing a glycerol backbone that occurs widely in quantitatively significant amounts. The novel component of sphingomyelin is the long chain or sphingoid base sphingosine (*trans*-4-sphingenine), whose biosynthetic pathway (summarized in Figure 16) has been thoroughly discussed by Kishimoto.[125]

The initial reaction is between palmitoyl-CoA and serine bound to the pyridoxal 5′-phosphate cofactor of serine palmitoyltransferase (EC 2.3.1.50). The lack of detectable intermediates involved in later reactions and the relatively low specific activity of serine palmitoyltransferase, taken together with other properties of the enzyme, indicate that this is the rate limiting reaction of sphingomeylin synthesis.[126]

As indicated in the final reaction of Figure 16, the *trans* double bond at position 4 of the long chain base appears to be inserted subsequent to the acylation of the amino group. This has been demonstrated in rat liver[127] and brain[128] and, more conclusively, in LM cells.[129] Whereas sphingosine is not produced in the free state during sphingolipid biosynthesis, it is released by the rapid metabolic turnover of some of the more complex final products.[130] The amount of sphingosine made available in this way, although low,[131] is comparable in magnitude to the level of diacylglycerol generated during agonist activation of cells, and is now thought to have important regulatory significance in the cell, particularly as a modulator of protein kinase activity.[132,133] In contrast to the pathway of sphingosine biosynthesis in animals, the yeast *Hansenula ciferri* has been shown to produce sphingosine directly by condensation of serine and *trans*-2-hexadecenoyl CoA.[134]

Sphingolipids from a variety of plant sources also contain small amounts of other long chain bases such as phytosphingosine (preferred name 4 D-hydroxysphinganine),[135] bases having an additional hydroxyl group at carbon atom 4 and lacking a double bond between carbons 4 and 5. Although phytosphingosine is generally considered to be a component of plant sphingolipids and is highly enriched in higher plant cell plasma membranes,[136] it can also be formed biosynthetically in animals from dihydrosphingosine.[137]

In addition to the scheme shown in Figure 16, free sphingosine can be acylated by a microsomal enzyme using fatty acyl-CoA. Alternatively, brain tissue appears capable of acylating sphingosine via a reversible reaction involving free fatty acid.[138] The acylation product, ceramide, can be converted to sphingomyelin by a widely occurring enzyme, phosphorylcholine-ceramide transferase, which draws on CDP-choline as a source of phos-

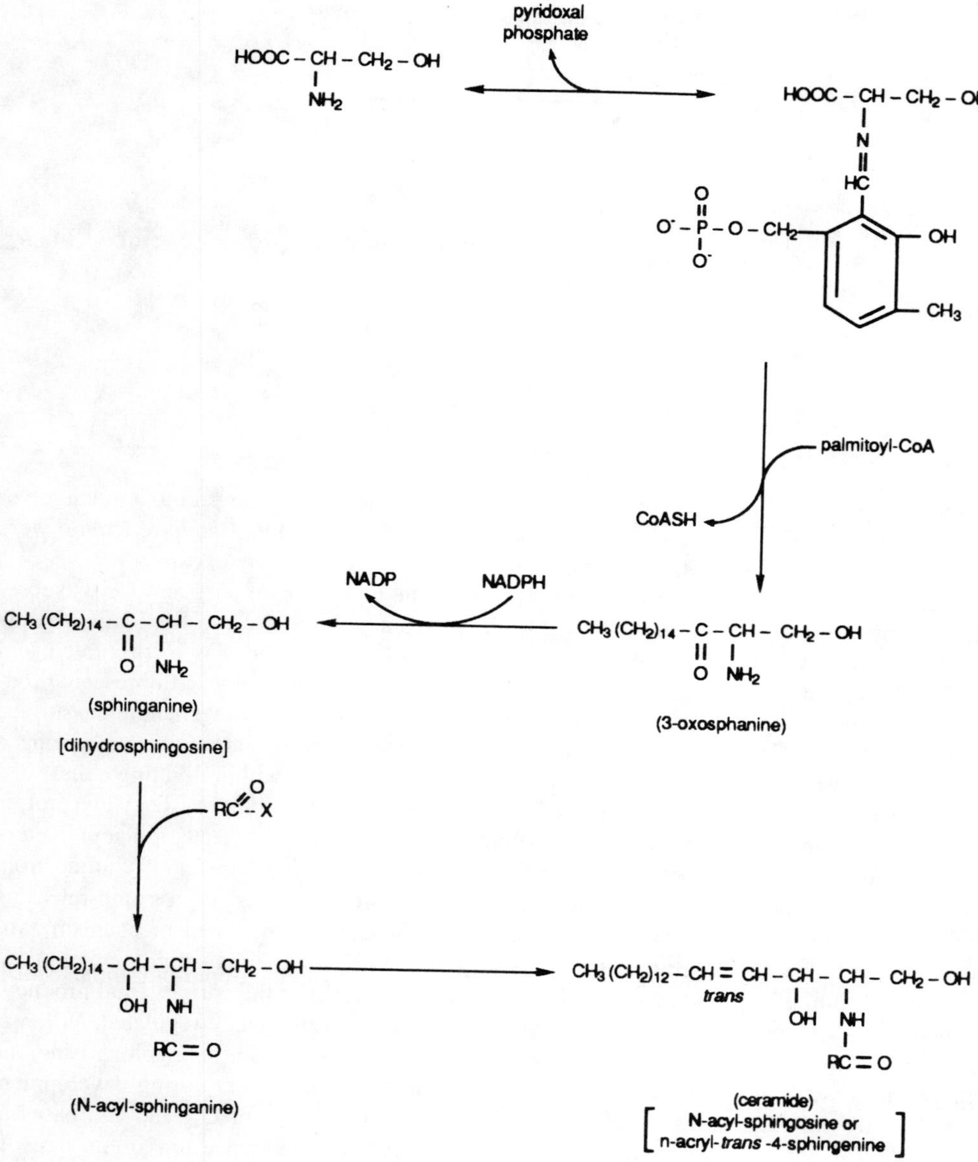

FIGURE 16. Ceramide biosynthesis.

phorylated base. The physiological significance of this reaction has been questioned since it is specific for ceramides having the threo configuration, while natural sphingolipids possess the erythro configuration.

It now seems likely that most tissues form sphingomeylin principally through the transfer of phosphocholine from phosphatidylcholine to ceramide (Figure 17). Although the energy-independent reaction has been reported to take place in the plasma membrane of mouse fibroblasts,[139] more recent studies have indicated that Golgi membranes have a much greater capacity to form sphingomyelin by this reaction. Thus Futterman et al.[140] incubated purified rat liver cell fractions with a radioactive ceramide analog, finding sphingomyelin synthesis to

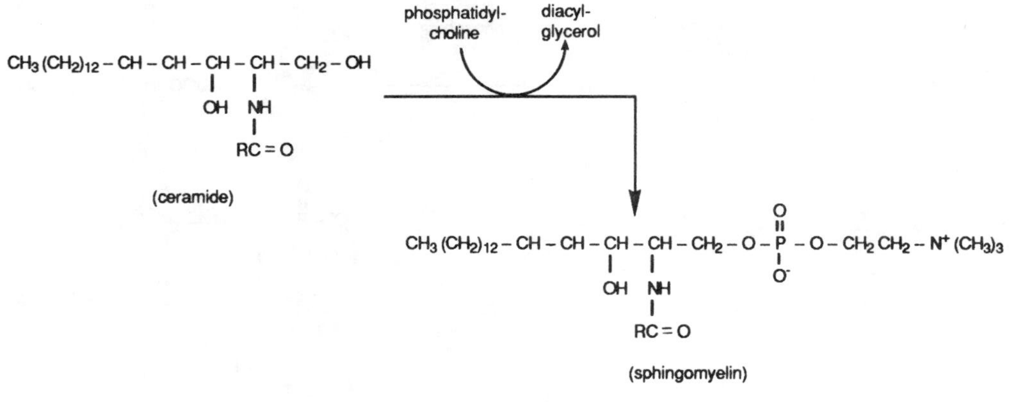

phosphatidylcholine → diacylglycerol

$CH_3(CH_2)_{12}-CH-CH-CH(OH)-CH(NHC(=O)R)-CH_2-OH$

(ceramide)

$CH_3(CH_2)_{12}-CH-CH-CH(OH)-CH(NHC(=O)R)-CH_2-O-P(=O)(O^-)-O-CH_2CH_2-N^+(CH_3)_3$

(sphingomyelin)

FIGURE 17.

occur predominantly in the lumen of the *cis* and *medial* Golgi fractions and to a much lower extent in the *trans* fractions. The plasma membrane and rough endoplasmic reticulum were calculated to account for less than 13% of the total sphingomyelin synthesis.

Further evidence for sphingomyelin synthesis in the Golgi membranes and for its subsequent movement to the plasma membrane via vesicular transport came from microscopic observations of CHO cells.[141] A fluorescent analog of sphingosine was incorporated into sphingomyelin and glucosylceramide in the Golgi apparatus. Both of these fluorescent products were transported to the cell surface by interphase or G_1 cells, but movement to the surface was prevented during mitosis. This behavior is in keeping with earlier work showing a temporary cessation of many other vesicular transport processes within dividing cells.

Sphingomyelin formation is stimulated by added ceramide and inhibited by diacylglycerol.[142] On the other hand, the phorbol ester 12-0-tetradecanoylphorbol-13-acetate, a diacylglycerol analog known to activate protein kinase C, enhanced the *in vivo* transfer of choline from phosphatidylcholine to ceramide by leukemic cells.[143] More recently, thyrotropin-releasing hormone, which activates protein kinase C through a receptor-mediated mechanism, was shown to stimulate the coordinate synthesis of phosphatidylcholine and sphingomyelin.[144]

The selection of ceramide molecular species for conversion to the various final products, principally sphingomyelin, cerebrosides and gangliosides, is obviously regulated. Whereas sphingomyelin always contains almost exclusively the C_{18} long chain base sphingosine, the long-chain base composition of rat brain gangliosides gradually alters during development from having only C_{18} bases in fetal rats to nearly equal amounts of C_{18} and C_{20} bases in adults.[145] It is not yet clear at what stage of biosynthesis the preference for certain bases is expressed.

Likewise, the fatty acids of sphingomyelin are undoubtedly selected by highly regulated pathways. Sphingomyelin of many cells is greatly enriched in very long chain fatty acids (C_{22} and C_{24}). Yet sphingomeylin contains few of the long chain hydroxy fatty acids that characterize the glycosphingolipids (see Chapter 7).

Sphingomyelin is degraded to ceramide and phosphocholine by a phospholipase C type enzyme termed sphingomyelinase (sphingomyelin phosphohydrolase), EC 3.1.4.12 (Figure 18). Sphingomyelinase occurs in a variety of animal tissues. Perhaps most studied is the lysosomal enzyme having a pH optimum of 4.5 to 5.5.[146] The role of this sphingomyelinase is probably to dissemble ingested sphingomeylin, and it does not attack phosphatidylcholine. A lack of the enzyme in the spleen is characteristic of the sphingomyelin storage disorder Niemann-Pick disease.[147]

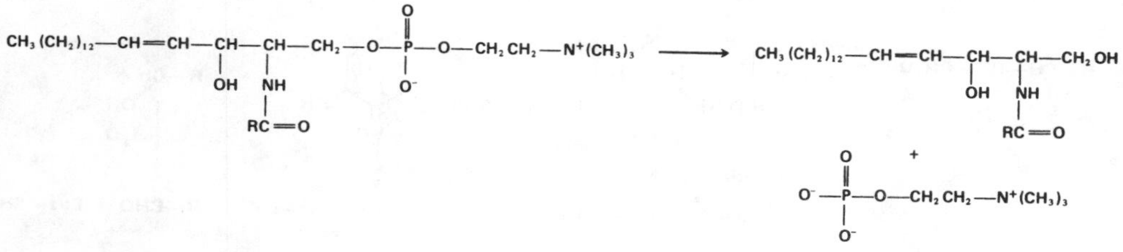

$$CH_3(CH_2)_{12}{-}CH{=}CH{-}CH(OH){-}CH(NH{-}RC{=}O){-}CH_2{-}O{-}P(=O)(O^-){-}O{-}CH_2CH_2{-}N^+(CH_3)_3 \longrightarrow CH_3(CH_2)_{12}{-}CH{=}CH{-}CH(OH){-}CH(NH{-}RC{=}O){-}CH_2OH + O^-{-}P(=O)(O^-){-}O{-}CH_2CH_2{-}N^+(CH_3)_3$$

FIGURE 18.

$$CH_3(CH_2)_{14}{-}CH(OH){-}CH(NH_2){-}CH_2{-}OH \xrightarrow[\text{sphinganine kinase}]{ATP \rightarrow ADP} CH_3(CH_2)_{14}{-}CH(OH){-}CH(NH_2){-}CH_2{-}O{-}P(=O)(O^-){-}O^-$$

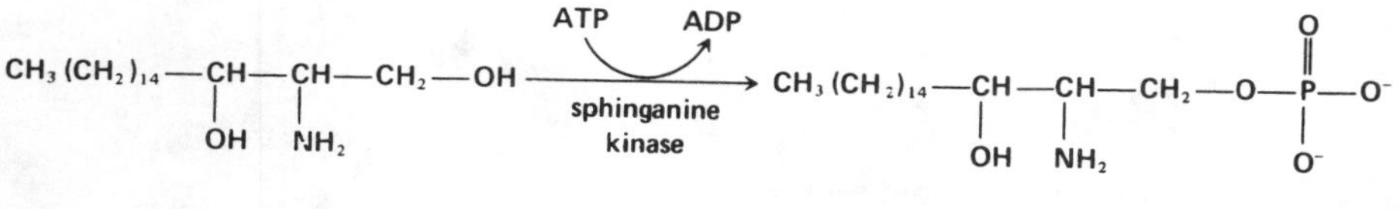

FIGURE 19.

A family of approximately 10 kDa, heat stable glycoproteins, termed saposins, has been found to activate sphingolipid hydrolysis by lysosomal enzymes. One of these, saposin D, stimulates sphingomyelinase activity by 100% when present in less than 1 μM levels.[148]

Neutral sphingomyelinases have been studied in the plasma membrane of rat liver[149] and hen erythrocytes.[150] The localization of the enzyme here positions it to take advantage of the pronounced enrichment of sphingomyelin in this membrane.

The addition of diacylglycerols, but not phorbol esters, has been shown to stimulate an acid sphingomyelinase in pituitary cell homogenates.[151] Enhanced sphingomyelin hydrolysis could be detected even in cells down-regulated for protein kinase C, suggesting that this kinase was not involved. The diacylglycerol effect could arise merely through improving access of the lysosomal sphingomyelinase to its substrate in a different membrane through enhanced fusion. A neutral sphingomyelinase activity also present, presumably in the plasma membrane, was not stimulated.

Ceramide freed as the result of sphingomyelinase action can be cleaved by the acid hydrolase ceramidase to yield a free fatty acid and a long chain base. In cell extracts the enzyme also catalyzes the reverse reaction. This may not be of physiological significance, as the normal substrate for ceramide biosynthesis is fatty acyl-CoA, not free fatty acid. Some mammalian tissues have been reported to contain neutral or alkaline ceramidases that, like the acid ceramidase, catalyze a reversible cleavage of ceramide. The further catabolism of the sphingoid bases is initiated by an ATP-dependent phosphorylation.[152] The reaction, as illustrated with the base sphinganine, is shown in Figure 19.

The phosphorylated base is then cleaved to palmitaldehyde and ethanolamine-phosphate through the action of a pyridoxal phosphate-containing enzyme, sphingosinephosphate lyase.[152,153] A possible mechanism is shown below (Figure 20). The enzyme is found in the particulate fraction (mitochondria and microsomes) or rat liver and also occurs in nonmammalian cells, such as the yeast *Hansenula ciferrii.*

Palmitaldehyde and ethanolamine phosphate, the products of sphingosine degradation, are rapidly recycled into other lipids. Sphingosine arising from dietary sphingolipids is mostly degraded, with the ethanolamine phosphate being efficiently utilized for the synthesis of phosphatidylethanolamine and, to some extent, phosphatidylcholine.[152] The palmitaldehyde is either oxidized to palmitic acid or reduced to hexadecanol by a NADH-dependent alcohol dehydrogenase which is specific for long-chain aldehydes. The hexadecanol thus produced is

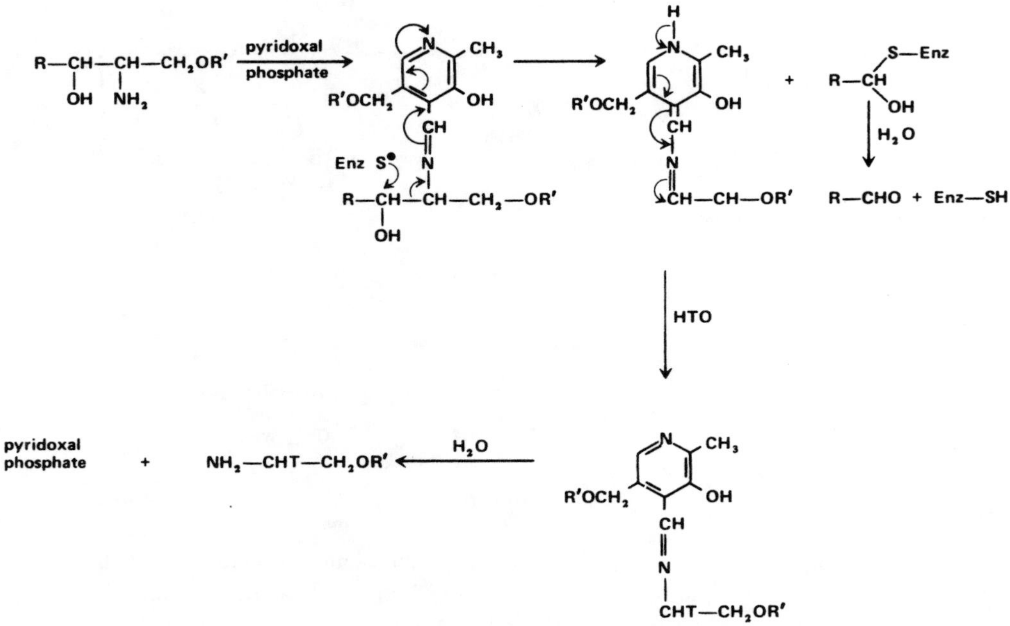

FIGURE 20. A proposed reaction mechanism for the action of sphinganine-1-phosphate lyase. The use of tritiated water (HTO) localized the site of pyridoxal phosphate binding. (Modified from Stoffel, W., *Mol. Cell Biochem.*, 1, 147, 1973; by Shimojo, T., Akino, T., Miura, Y., and Schroepfer, G. J., Jr., *J. Biol. Chem.*, 251, 4448–4457, 1976. With permission.)

utilized in certain tissues for the biosynthesis of phospholipids containing an ether-linked alkyl group at the 1 position of the glyceryl moiety (see Section XI). These may in turn be converted to plasmalogens. In developing brain of young rats, ^{3}H from injected [3-^{3}H]dihydrosphingosine was incorporated exclusively into the ether-linked side chains of plasmalogens or alkyl ether phospholipids, whereas injected [1-^{14}C]palmitate was much less efficiently used in forming the ether lipids.

XIV. REGULATORY FUNCTION OF LIPID-DEGRADING ENZYMES

The previous sections of this chapter have been organized in such a way as to emphasize the biosynthesis of phospholipids. While these biosynthetic pathways will obviously predominate in growing tissues, even rapidly enlarging cells experience a surprisingly active rate of degradation associated with normal metabolic turnover. More recently, it has become clear that many phospholipases have not only basal levels of activity but also higher levels induced by extracellular stimuli. These variable activities are exemplified by the widespread role of certain phospholipases in signal transduction across the cell's plasma membrane.[154]

Enzymes are known which together are capable of hydrolyzing all the ester, ether, and amide linkages of phospholipids. The properties of these enzymes have been thoroughly described in a monograph by Waite.[155] The reactions of principal interest are summarized in Figure 21.

While some phospholipases are nearly ubiquitous, others have a limited distribution. Thus phospholipase D (EC 3.1.4.4) has been reported mainly from certain bacteria and from plants. High activity is found in leaves and stalks of cabbage, seeds of marrow, pea, and soybeans,

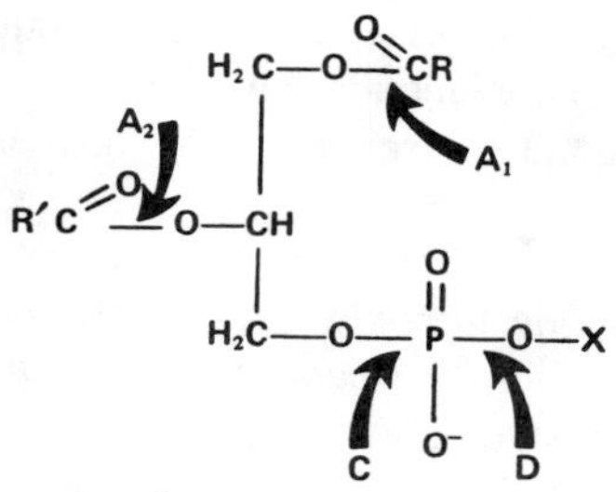

FIGURE 21. The principal sites of hydrolytic attack on phospholipids by enzymes. The letters, A_1, A_2, C, and D represent phospholipases A_1, A_2, C, and D, respectively.

and in the cotyledons of many higher plants. A study of phospholipase D of bean (*Phaseolus vulgaris*) cotyledons[156] indicated an active phospholipase D operating in conjunction with phosphatidic acid phosphatase (EC 3.1.3.4) to reduce phospholipids to diacylglycerols. Both phospholipases were stimulated by physiological levels of Ca^{2+}, and the latter enzyme also showed higher activity in the presence of calmodulin. There is reason to believe that a primary function of plant phospholipase D is to degrade lipid in senescing tissues such as cotyledons.

Phospholipase C, also thought at one time to have a rather restricted occurrence, has recently been the subject of renewed interest. An enzyme of this classification has long been known to occur in the toxins of *Clostridium perfringens, Bacillus cereus, Staphlococcus aureus,* and other prokaryotes.[157] Some have broad substrate specificities, while others, such as an enzyme purified from *Bacillus thuringiensis,*[158] preferentially attacks inositol phospholipids.

A Ca^{2+}-requiring phospholipase C specific for inositol phospholipids, most significantly phosphatidylinositol-4,5-bisphosphate (PIP_2), has been detected in most eukaryotic cells examined, including, more recently, higher plants[159] and algae.[160] Mammalian cells contain at least five distinct forms of inositol phospholipid specific phospholipase C with differing immunological characteristics.[77] The enzyme of greatest interest functions in the plasma membrane to produce diacylglycerol and inositol-1,4,5-trisphosphate from PIP_2 as part of a transmembrane signaling system triggered by hormones, growth factors, chemoattractants, light, and other agonists (see Section X, p. 92, for details of PIP_2 involvement). Activation of the plasma membrane associated phospholipase C appears to be affected by interaction with a GTP-binding protein (G-protein) in a manner thought to resemble that in the more thoroughly studied system for activating adenylate cyclase. At the present time the precise mechanism of G-protein involvement is unclear, but indirect evidence favors activation of phospholipase C by the GTP-linked α subunit of a specific G-protein, sometimes referred to as G_p. Thus in *Xenopus* oocytes phospholipase C activity was inhibited by injecting the G-protein $\beta\gamma$ subunit complex, which is known to associate with (and inactivate) the α subunit.[161] Pertussis toxin, which blocks many G-protein activities, effectively inhibits some forms of phospholipase C but not others.[77]

Most of the phospholipase C forms are associated with both soluble and particulate cell fractions, raising the possibility of their translocation to membranes during activation. Activation may also involve phosphorylation of the protein, since this has been detected *in vivo* as well as *in vitro*. Both protein kinase C and epidermal growth factor receptor have been implicated as the responsible kinases.[77] Existing data suggest that phosphorylation may affect enzyme activity in an indirect way, e.g., by influencing the enzyme's interaction with other regulatory factors.

Apart from the frequently observed hydrolysis of inositol phospholipids, hormone-activated phospholipase C can catalyze the hydrolysis of other phospholipids. Thus phosphatidylcholine hydrolysis follows the binding of a_1-adrenergic agonists to cultured kidney cells.[162] This type of response, perhaps in conjunction with PIP_2 cleavage, would allow flexibility in the proportions of the two second messengers, diacylglycerol and inositol trisphosphate, arising from external stimuli.

Phospholipase A_2 is perhaps the most active of the phospholipases under basal condition, as judged by the normal metabolic turnover rates of phospholipid components. The most important forms of the enzyme from the perspective of our discussion are the membrane associated ones. The extensively studied venom phospholipase A_2s and pancreatic phospholipases are water soluble, function in extracellular spaces, and differ in many other ways from the membrane associated forms. Because they are soluble and therefore easily manipulated, it is the extracellular enzymes which have been used for most mechanistic and kinetic studies.[163,164] The extracellular enzymes may be related to the membrane forms by common features and even common ancestry, as judged by the cross reactivity of cobra (*Naja naja*) venom phospholipase A_2 antibodies with phospholipase A_2 from mammalian sources.[165] On the other hand, extracellular phospholipase A_2 from pancreas or *Crotalus atrox* did not crossreact with several monoclonal antibodies raised to the rat liver mitochondrial phospholipase A_2.[166]

Phospholipase A_2 activity serves many different functions in the cell. Its widespread distribution among most cellular membrane types makes it ideally suited for a key role in the ubiquitous metabolic turnover of phospholipid acyl chains (see Section XV). A tight control must be maintained over this process or else rising levels of the products, fatty acids and lysophospholipids, can severely disrupt the physical properties of the membranes.[167] The reverse interaction, namely, activation of phospholipase A_2 by alterations of membrane fluidity, is also known to occur. The enzyme has been found to attack lipid bilayers much more avidly in the temperature range where a transition from the gel phase to the liquid crystalline phase occurs. Further studies[168] indicate that any kind of bilayer irregularity, such as a small radius of bilayer curvature or a dislocation or other defect of the sort that might be found at the boundary between an expanse of gel phase and liquid crystalline phase, can promote phospholipase action. It is likely that selective phospholipase action of this type routinely occurs in cells under conditions of severe physical or environmental stress.

Many recent findings point to the physiological value of maintaining a free fatty acid pool of variable size, especially in the plasma membrane. The signaling function of phospholipase A_2 in this membrane is thought to operate as follows. The activation of phospholipase A_2 is effected by receptor stimulated G-proteins. For example, the G-protein transducin participates in the activation of phospholipase A_2 by light in mammalian rod outer segments,[169] and use of the GTP analog GTPγS and pertussis toxin indicated that a G-protein controls arachidonate release from phosphatidylinositol, phosphatidylcholine, and phosphatidylethanolamine in neutrophils.[170]

Depending upon the cell type, the agonist involved, and the concentration of Ca^{2+}, different phospholipid classes and different molecular species within each class may be selectively hydrolyzed. In platelets arachidonate-containing molecular species were preferentially hydrolyzed under certain conditions.[171] In neutrophils plasma membrane was the major site of exogenous phosphatidylinositol hydrolysis *in vitro* by phospholipase A_2 while phosphatidylcholine and phosphatidylethanolamine were mainly hydrolyzed in intracellular organelles.[172]

Blood plasma contains a phospholipase A_2 type enzyme that preferentially hydrolyzes the acetyl group from the *sn*-2 position of platelet activating factor. This enzyme, which is mainly associated with low density lipoprotein, also removes the oxidized product of arachidonic acid from the *sn*-2 position of phosphatidylcholine, thereby preventing the accumulation of this potentially toxic compound.[173]

A complex and incompletely resolved relationship exists between G-protein regulation of phospholipase C and A.[174] In some cases, e.g., bradykinin-stimulated aortic endothelial cells, increased phospholipase A_2 activity was observed before increased phospholipase C action.[175] On the other hand, protein kinase C, which normally functions as the result of phospholipase C action, appears to participate in the activation of phospholipase A_2 in neutrophils.[176] The interrelationship is further complicated by the many physiological responses induced by metabolites of the arachidonate produced either directly by phospholipase A_2 or in the two step hydrolysis by phospholipase C/diacylglycerol lipase.[174]

A family of abundant steroid-inducible proteins termed lipocortins has been implicated in the inhibition of tissue phospholipase A_2. Amino acid sequencing studies have shown that the lipocortins are identical to calpactins, cytoskeleton-associated proteins originally studied in a totally different context (reviewed by Davidson et al.[177]). Investigators proposed that these proteins, when phosphorylated by a tyrosine kinase, are separated from an association with phospholipase A_2, thereby activating it to release arachidonic acid from phospholipids for the production of prostaglandins and leukotrienes.

Additional information is needed to confirm the mechanism and physical significance of this scheme. For example, it has not yet been proved that the abundant lipocortins (up to 1% of cellular protein) inhibit phospholipase A_2 activity by binding to the enzyme per se rather than by binding to and reducing the availability of its lipid substrate.[177]

Phospholipase A_1 enzymes, although not as well studied as those specific for the *sn*-2 position, are also widely distributed in nature. They have been characterized mainly in mammalian tissues, particularly liver and heart.[178] Many phospholipases A_1 are soluble enzymes, occurring either in the cytosol, in lysosomes, or as releasable proteins associated with the plasma membrane. Some are active towards triglycerides as well as phospholipids. Cao et al. [179] described a single cytosolic phospholipase A from hamster heart apparently active towards ester bonds at either the *sn*-1 position or the *sn*-2 position of phosphatidylcholine or phosphatidylethanolamine.

Phospholipase A type enzymes are also found in plants. However, since positional specificity often seems to be lacking and the enzymes may even attack other substrates besides phospholipids, the more general term acylhydrolases is commonly used to define them.[180] More detailed studies in recent years have revealed that some of these hydrolases do possess a relatively narrow specificity. Thus microsomal membranes of *Dunaliella salina* have a fatty acyl hydrolase activity exhibiting a preference for phosphatidylglycerol and phosphatidylethanolamine but not phosphatidylcholine or monogalactosyldiacylglycerol.[181]

XV. TURNOVER OF PHOSPHOLIPID ACYL GROUPS

Deacylation of phospholipids at either the *sn*-1 or *sn*-2 position of the glycerol backbone is readily accomplished by phospholipases, as outlined above. When deacylation occurs during normal membrane metabolism, the resulting lysophospholipid is promptly reacylated. The specificity of the acyltransferases involved in this step was determined mainly in the laboratory of Lands.[182] With a few exceptions, e.g., in chloroplasts, the acyltransferase specific for acylation at the *sn*-1 position preferentially transfers saturated fatty acids. In rat liver, the most thoroughly studied tissue, this enzyme seems responsible for replacing palmitic acid, installed at the *sn*-1 position during phosphatidic acid formation, with stearic acid in a sizable proportion of the phospholipid molecules.

A different enzyme esterifies lysophospholipids having a free hydroxyl group at the *sn*-2 position. It preferentially utilizes unsaturated fatty acids in most tissues and is responsible for the gradual insertion of arachidonic acid into liver phospholipids arising from phosphatidic acid acylated in position 2 primarily with oleic or linoleic acid. The initial biosynthesis of certain phospholipid molecular species followed by the programed redistribution of acyl chains into different combinations can readily be detected in isotopic labeling studies.[183,184]

Considerable effort has been devoted to determining the role of these acylation enzymes in establishing the characteristic fatty acid molecular species patterns of various membrane phospholipids. The research has been expedited by the development of gas chromatographic and, more recently, high performance liquid chromatographic techniques[185] capable of resolving and quantifying individual lipid molecular species. Relatively small changes in molecular species composition can lead to significant alterations in membrane fluidity, and this can be important in cells responding to extremes of temperature or other environmental variables (see Chapter 10).

However, apart from the work done with liver and a number of studies utilizing *E. coli*, little is known regarding the specificity of the enzymes with respect to their fatty acid or their lysophospholipid substrates. Aside from the positional specificity, which seems firmly established, it is important to establish whether the fatty acid inserted at a certain position (position 1 or 2) of a phospholipid's glycerol backbone is influenced more strongly by the *in vivo* specificity of the appropriate acyltransferase or by the variety of fatty acyl CoA molecules available at the time of acylation. Okuyama et al.[186] have reported that the critical factor in determining phospholipid molecular species of *E. coli* is the variety of acyl donors available rather than the acyltransferase specificity. Kito et al.,[187] on the other hand, believe that the acyltransferase specificity is of paramount consideration. In all likelihood, both parameters can be critical, depending upon the physiological circumstances. Okuyama et al.[188] observed that the *in vitro* specificity of *E. coli* acyltransferases varied depending upon the concentration of the *sn*-glycerol-3-phosphate acceptor. At low acceptor concentrations, acylation patterns resembling those found *in vivo* were noted, but higher *sn*-glycerol-3-phosphate levels led to a partial loss of acyltransferase specificity.

The specificity of the bacterial acylation reaction described above may be pertinent mainly to the net synthesis of phospholipids, since turnover of fatty acyl chains in bacteria is a slow process at best.[189,190] An exception is the turnover of fatty acids at the *sn*-1 position of phosphatidylethanolamine, caused by deacylation as it functions in the maturation of bacterial lipoproteins. Recycling of the resulting 2-acylglycerophosphorylethanolamine to phosphatidylethanolamine involves an unusual acyltransferase that also functions as an acyl:acyl carrier protein synthetase.[191] This dual function enzyme allows nonesterified fatty acids to be incorporated into the *sn*-1 position of phosphatidylethanolamine while other phospholipid acyl chains must draw on the pool of fatty acyl-ACP (inaccessible to nonesterified fatty acids) for substrates.

Eukaryotic cells maintain a constant metabolic turnover of phospholipid fatty acids, mediated by three enzymatic steps; deacylation (by phospholipase A_1 or A_2), fatty acid activation (by fatty acyl-CoA synthetase), and reacylation (by acyltransferase) (Figure 22). Fatty acyl-CoA can also arise by direct transfer of a fatty acyl chain from phospholipid to coenzyme A.[192,193] The quantitative significance of this latter reaction has not been determined.

The constant recycling of these acyl chains may facilitate the prompt synthesis of new molecular species having different physical properties when the cell's environment changes. It is not clear which one(s) of the participating enzymes is responsible for induced changes in phospholipid molecular species composition. Factors regulating this process are discussed more fully in Chapter 11.

Detailed *in vivo* and *in vitro* studies of phospholipid acyl chain turnover in *Tetrahymena* ciliary membranes[194,195] suggest that it is a tightly coupled process in which intact phospholipids are the only major metabolic pool, since lysophospholipids do not accumulate and free

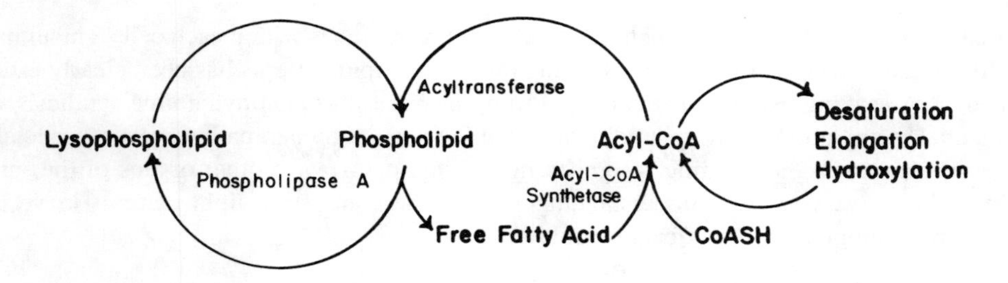

FIGURE 22. Metabolic turnover of phospholipid fatty acids.

fatty acids, especially unsaturated ones, are present in trace amounts throughout the cell.[196] Deacylation of phospholipids via phospholipase A is thought to be the rate-limiting step in the turnover process. This conclusion was supported by the observation that ciliary acyltransferase activity was markedly enhanced by the addition of lysophosphatidylethanolamine.[195]

XVI. OVERVIEW OF PHOSPHOLIPID METABOLISM IN BACTERIA

Typically, bacterial cells have a relatively simple phospholipid composition, featuring phosphatidylethanolamine, phosphatidylglycerol, diphosphatidylglycerol (cardiolipin) and phosphatidylserine as their major components. Phospholipid metabolism, exemplified by reactions studied in *Escherichia coli*, has been reviewed by Cronan and Rock.[197] *De novo* phospholipid synthesis commences with the acylation of glycerol-3-phosphate by acyltransferases specific for fatty acyl-ACP. The key intermediate for all final phospholipid classes is CDP-diacylglycerol. There is no extensive phospholipid degradation and resynthesis during normal growth, although the fatty acid at the *sn*-1 position of phosphatidylethanolamine is selectively utilized for lipoprotein synthesis and replaced from the cellular fatty acid pool. Phospholipid biosynthesis is closely coordinated with bacterial energy metabolism. This control appears to be mediated by the novel nucleotide guanosine 5′-diphosphate-3′-diphosphate acting, perhaps indirectly, on glycerol-3-phosphate acyltransferase.

XVII. OVERVIEW OF PHOSPHOLIPID METABOLISM IN ANIMALS

The pathways available for phospholipid metabolism in animal tissues are considerably more diversified than those in bacteria. Almost all these depicted at the beginning of this chapter in Figure 1 have been observed. Phosphatidic acid, the key precursor for all net phospholipid synthesis, can be formed in several ways. Typically, the scheme coupling diglycerides with CDP-bases provides the bulk of animal membrane phospholipids formed *de novo*. However, the CDP-diglyceride pathway is also active (although it is apparently not involved in phosphatidylserine synthesis as it is in bacteria). The exchange of intact polar head groups is prevalent, as is the conversion of one phospholipid to another via enzymatic head group modification.

The fatty acid composition of phospholipids undergoes a continuous retailoring through deacylation-reacylation reactions and by *in situ* desaturation, thus maintaining the fluidity of the individual membranes at a physiologically optimal level. Ether-linked side chains are frequently present replacing the fatty acid at the 1-position of glycerophospholipids. There is a pronounced specificity of certain lipids for membranes performing specific functions. This creates a remarkable diversity of lipid compositions among the various tissues of a single organism and even among the functionally different membranes within each cell.[5] The impact of different alternate pathways for achieving distinctive lipid patterns has been clearly established.[198] A striking example is the crucial importance of phosphatidylcholine synthesis via sequential methylation of phosphatidylethanolamine in certain mammalian organs, especially liver, in contrast to the seeming insignificance of that pathway to other organs of the same animal. We have very little understanding of how the contrasting lipid patterns serve the special physiological needs of each membrane.

XVIII. OVERVIEW OF PHOSPHOLIPID METABOLISM IN PLANTS

Although phospholipid metabolism in plants has not yet received the intensive study given animal systems, it appears that most of the pathways found in animals also occur here.[199] Cytidine nucleotides are involved in phospholipid biosynthesis through both the CDP diglyceride sequence and through CDP-choline and CDP-ethanolamine. The formation of phosphatidylcholine via the sequential methylation of phosphatidylethanolamine is quantitatively significant. As in animals, there is an active deacylation and reacylation of phospholipids, supporting a retailoring of molecular species when dictated by environmental changes.

The reactions enumerated above take place mainly in the endoplasmic reticulum, as they do in all eukaryotic cells. However, an entirely separate system capable of synthesizing certain phospholipids is also present within the cells' plastids. All plant cells appear to contain plastids, and in photosynthetic cells these can account for more than 70% of the total phospholipid. The major plastid component, phosphatidylglycerol, is made by a plastid-specific pathway resembling in many respects that found in prokaryotic cells. This lipid can be the prevalent phospholipid of the cell. In addition, diacyl moieties needed for the assembly of glycolipids in the plastids (see Chapter 7) are supplied in large amounts from phosphatidylcholine formed in the endoplasmic reticulum. Further complexities in lipid trafficking stem from the fact that all fatty acids supplied to both phospholipid synthesizing systems are made in the plastid (see p. 26).

REFERENCES

1. **White, D. A.,** The phospholipid composition of mammalian tissues, in *Form and Function of Phospholipids,* Ansell, G. B., Dawson, R. M. C., and Hawthorne, J. N., Eds. Elsevier, Amsterdam, 1973, 441–482.
2. **Vance, D. E. and Vance, J. E., Eds.**, *Biochemistry of Lipids and Membranes,* Benjamin/Cummings, Menlo Park, 1985, pp 593.
3. **Thompson, G. A., Jr.,** Phospholipid metabolism in animal tissues, in *Form and Function of Phospholipids,* Ansell, G. B., Dawson, R. M. C., and Hawthorne, J. N., Eds., Elsevier, Amsterdam, 1973, 67–96.
4. **McMurray, W.,** Phospholipids in subcellular organelles and membranes, in *Form and Function of Phospholipids,* Ansell, G. B., Dawson, R. M. C., and Hawthorne, J. N., Eds., Elsevier, Amsterdam, 1973, 205–251.
5. **Cobon, G. S., Crowfoot, P. D., and Linnane, A. W.,** Biogenesis in mitochondria: phospholipid synthesis *in vitro* by yeast mitochondrial and microsomal fractions, *Biochem. J.,* 144, 265–275, 1974.
6. **Hjelmstad, R. H. and Bell, R. M.,** Molecular insights into enzymes of membrane bilayer assembly, *Biochemistry,* 30, 1731–1740, 1991.
7. **Brindley, D. N. and Sturton, R. G.,** Phosphatide metabolism and its relation to triacylglycerol biosynthesis, in *Phospholipids, New Comprehensive Biochemistry Series Vol. 4,* Hawthorne, J. N. and Ansell, G. B., Eds., Elsevier Biomedical, Amsterdam, 1982, 179–213.

8. **Rock, C. O. and Cronan, J. E., Jr.,** Regulation of bacterial membrane lipid synthesis, *Curr. Top. Memb. Transp.*, 17, 207–233, 1982.
9. **Oh-hashi, Y., Inoue, M., Murase, S., Mizuno, M., Kawaguchi, A., and Okuyama, H.,** Enzymatic bases for the fatty acid positioning in phospholipids of *Brevibacterium ammoniagenes, Arch. Biochem. Biophys.*, 244, 413–420, 1986.
10. **Roughan, G. and Slack, R.,** Glycerolipid synthesis in leaves, *Trends Biochem. Sci.*, 9, 283–386, 1984.
11. **Wright, T. M., Rangan, L. A., Shin, H. S., and Raben, D. M.,** Kinetic analysis of 1,2-diacylglycerol mass levels in cultured fibroblasts, *J. Biol. Chem.*, 263, 9374–9380, 1988.
12. **MacDonald, M. L., Mack, K. F., Williams, B. W., King, W. C., and Glomset, J. A.,** A membrane-bound diacylglycerol kinase that selectively phosphorylates arachidonoyl-diacylglycerol, *J. Biol. Chem.*, 263, 1584–1592, 1988.
13. **Maroney, A. C. and Macara, I. G.,** Phorbol ester-induced translocation of diacylglycerol kinase from the cytosol to the membrane in Swiss 3T3 fibroblasts, *J. Biol. Chem.*, 264, 2537–2544, 1989.
14. **Kanoh, H., Yamada, K., Sakane F., and Imaizumi, T.,** Phosphorylation of diacylglycerol kinase *in vitro* by protein kinase C, *Biochem. J.*, 258, 455–462, 1989.
15. **MacDonald, M. L., Mack, K. F., Richardson, C. N., and Glomset, J. A.,** Regulation of diacylglycerol kinase reaction in Swiss 3T3 cells, *J. Biol. Chem.*, 263, 1575–1583, 1988.
16. **Tillman, T. S. and Bell, R. M.,** Mutants of *Saccharomyces cerevisiae* defective in *sn*-glycerol-3-phosphate acyltransferase, *J. Biol. Chem.*, 261, 9144–9149, 1986.
17. **Cascales, C., Mangiapane, E. H., and Brindley, D. N.,** Oleic acid promotes the activation and translocation of phosphatidate phosphohydrolase from the cytosol to particulate fractions of isolated rat hepatocytes, *Biochem. J.*, 219, 911–916, 1984.
18. **Pittner, R. A., Fears, R., and Brindley, D. N.,** Effects of cyclic AMP, glucocorticoids and insulin on the activities of phosphatidate phosphohydrolase, tyrosine aminotransferase and glycerol kinase in isolated rat hepatocytes in relation to the control of triacylglycerol synthesis and gluconeogenesis, *Biochem. J.*, 225, 455–462, 1985.
19. **Pagano, R. E.,** What is the fate of diacylglycerol produced at the Golgi apparatus?, *Trends Biochem. Sci.*, 13, 202–205, 1988.
20. **Vance, D. E., Ed.,** *Phosphatidylcholine Metabolism,* CRC Press, Boca Raton, 1989, pp 249.
21. **Pelech, S. and Vance, D.,** Regulation of phosphatidylcholine biosynthesis, *Biochim. Biophys. Acta.*, 779, 217–251, 1984.
22. **Wright, P., Morand, J., and Kent, C.,** Regulation of phosphatidylcholine biosynthesis in Chinese hamster ovary cells by reversible membrane association of CTP: phosphocholine cytidylyltransferase, *J. Biol. Chem.*, 260, 7919–7926, 1985.
23. **Tercé, F., Record, M., Ribbes, G., Chap, H., and Douste-Blazy, L.,** Intracellular processing of cytidylyltransferase in Krebs II cells during stimulation of phosphatidylcholine synthesis, *J. Biol. Chem.*, 263, 3142–3149, 1988.
24. **Vance, D. E.,** CTP:cholinephosphate cytidylyltransferase, in *Phosphatidylcholine Metabolism,* Vance, D. E., Ed., CRC Press, Boca Raton, 1989, 33–45.
25. **Bishop, W. R. and Bell, R. M.,** Assembly of phospholipids into cellular membranes: biosynthesis, transmembrane movement and intracellular translocation, *Annu. Rev. Cell Biol.*, 4, 579–610, 1988.
26. **Vance, D. E.,** Phospholipid metabolism in eukaryotes, in *Biochemistry of Lipids and Membranes,* Vance, D. E. and Vance, J. E., Eds., Benjamin/Cummings, Menlo Park, 1985, 242–270.
27. **Paddon, H. and Vance D.,** Tetradecanoyl-phorbol acetate stimulates phosphatidylcholine biosynthesis in HeLa cells by an increase in the rate of the reaction catalyzed by CTP:phosphocholine cytidylyltransferase, *Biochim. Biophys. Acta,* 620, 636–640, 1980.
28. **Kolesnick, R.,** Thyrotropin-releasing hormone and phorbol esters induce phosphatidylcholine synthesis in GH3 pituitary cells, *J. Biol. Chem.* 262, 14525–14530, 1987.
29. **Watkins, J. D. and Kent, C.,** Phosphorylation of CTP:phosphocholine cytidylyltransferase *in vivo, J. Biol. Chem.*, 265, 2191–2197, 1990.
30. **Pelech, S., Pritchard, H., Brindley, D., and Vance, D.,** Fatty acids promote translocation of CTP:phosphocholine cytidylyltransferase to the endoplasmic reticulum and stimulate rat hepatic phosphatidylcholine synthesis, *J. Biol. Chem.*, 258, 6782–6788, 1983.
31. **Kolesnick, R. N. and Hemer, M. R.,** Physiologic 1,2-diacylglycerol levels induce protein kinase C-independent translocation of a regulatory enzyme, *J. Biol. Chem.*, 265, 10900–10904, 1990.
32. **Hjelmstad, R. and Bell, R.,** Mutants of *Saccharomyces cerevisiae* defective in *sn*-1,2-diacylglycerol cholinephosphotransferase, *J. Biol. Chem.*, 262, 3909–3917, 1987.
33. **Hjelmstad, R. and Bell, R.,** The *sn*-1,2-diacylglycerol ethanolaminephosphotransferase activity of *Saccharomyces cerevisiae, J Biol. Chem.*, 263, 19748–19757, 1988.

34. **Datko, A. and Mudd, H.,** Phosphatidylcholine synthesis, *Plant Physiology,* 88, 854–861, 1988.
35. **Morikawa, S., Taniguchi, S., Fujii, K., Mori, H., Kumada, K., Fujiwara M., and Fujiwara, M.,** Preferential synthesis of diacyl and alkenylacyl ethanolamine and choline glycerophospholipids in rabbit platelet membranes, *J. Biol. Chem.,* 262, 1213–1217, 1987.
36. **Justin, A., Demandre, C., and Mazliak, P.,** Choline- and ethanolaminephosphotransferases from pea leaf and soya beans discriminate 1-palmitoyl-2-linoleoyldiacylglycerol as a preferred substrate, *Biochim. Biophys. Acta,* 922, 364–371, 1987.
37. **Hatch, G., Choy, K., and Choy, P.,** Regulation of phosphatidylcholine metabolism in mammalian hearts, *Biochem. Cell Biol.,* 67, 67–77, 1989.
38. **Ishidate, K., Nakagomi, K., and Nakazawa, Y.,** Complete purification of choline kinase from rat kidney and preparation of rabbit antibody against rat kidney choline kinase, *J. Biol. Chem.,* 259, 14706–14710, 1984.
39. **Polokoff, M. A., Wing, D. C., and Raetz, C. R. C.,** Isolation of somatic cell mutants defective in the biosynthesis of phosphatidylethanolamine, *J. Biol. Chem.,* 256, 7687–7690, 1981.
40. **Miller, M. A. and Kent, C.,** Characterization of the pathways for phosphatidyl ethanolamine biosynthesis in Chinese hamster ovary mutant and parental cell lines, *J. Biol. Chem.,* 261, 9753–9761, 1986.
41. **Kuge, O., Nishijima, M., and Akamatsu, Y.,** Phosphatidylserine biosynthesis in cultured Chinese hamster ovary cells, *J. Biol. Chem.,* 261, 5795–5798, 1986.
42. **Bremer, J. and Greenberg, D. M.,** Mono- and dimethylethanolamine isolated from rat liver phospholipids, *Biochim. Biophys. Acta,* 35, 287–288, 1959.
43. **Bremer, J., Figard, P. H., and Greenberg, D. M.,** The biosynthesis of choline and its relation to phospholipid metabolism, *Biochim. Biophys. Acta,* 43, 477–488, 1960.
44. **van Golde, L. M. G., Raben, J., Battenberg, J. J., Fleischer, B., Zambrano, F., and Fleischer, S.,** Biosynthesis of lipids in Golgi complex and other subcellular fractions from rat liver, *Biochim. Biophys. Acta,* 360, 179–192, 1974.
45. **Kodaki, T. and Yamashita, S.,** Yeast phosphatidylethanolamine methylation pathway, *J. Biol. Chem.,* 262, 15428–15435, 1987.
46. **McGraw, P. and Henry, S. A.,** Mutations in the *Saccharomyces cerevisiae opi*3 gene: effects on phospholipid methylation, growth, and cross-pathway regulation of inositol synthesis, *Genetics,* 122, 317–330, 1989.
47. **Kelley, M. J. and Carmen, G. M.,** Purification and characterization of CDP-diacylglycerol synthase from *Saccharomyces cerevisiae, J. Biol. Chem.,* 262, 14563–14570, 1987.
48. **Homann, M. J., Bailis, A. M., Henry, S. A., and Carman, G. M.,** Coordinate regulation of phospholipid biosynthesis by serine in *Saccharomyces cerevisiae, J. Bact.,* 169, 3276–3280, 1987.
49. **Finnerty, W. R. and Makula, R. A.,** Microbial lipid metabolism, *CRC Crit. Rev. Microbiol.,* 4, 1–40, 1975.
50. Bell, **R. M. and Coleman, R. A.,** Enzymes of glycerolipid synthesis in eukaryotes, *Annu. Rev. Biochem.,* 49, 459–487, 1980.
51. **Henry, S. A., Klig, L. S., and Loewy, B. S.,** The genetic regulation and coordination of biosynthetic pathways in yeast: amino acid and phospholipid synthesis, *Annu. Rev. Genetics,* 18, 207–231, 1984.
52. **Bailis, A. M., Poole, M. A., Carman, G. M., and Henry, S. A.,** The membrane-associated enzyme phosphatidylserine synthase is regulated at the level of mRNA abundance, *Mol. Cell. Biol.,* 7, 167–176, 1987.
53. **Kelley, M. J., Bailis, A. M., Henry, S. A., and Carman, G. M.,** Regulation of phospholipid biosynthesis in *Saccharomyces cerevisiae* by inositol, *J. Biol. Chem.,* 263, 18078–18085, 1988.
54. **Kinney, A. and Carman, G.,** Phosphorylation of yeast phosphatidylserine synthase *in vivo* and *in vitro* by cyclic AMP-dependent protein kinase, *Proc. Natl. Acad. Sci. U.S.A.,* 85, 7962–7966, 1988.
55. **Kent, C., Carman, G. M., Spence, M. W., and Dowhan, W.,** Regulation of eukaryotic phospholipid metabolism, *FASEB J.,* 5, 2258–2266 1991.
56. **Ohta, A., Waggoner, K., Louie, K., and Dowhan, W.,** Cloning of genes involved in membrane lipid synthesis, *J. Biol. Chem.,* 256, 2219–2225, 1981.
57. **Imai, A. and Gershengorn, M.,** Regulation by phosphatidylinositol of rat pituitary plasma membrane and endoplasmic reticulum phosphatidylinositol synthase activities, *J. Biol. Chem.,* 262, 6457–6459, 1987.
58. **Ray, T. K. and Cronan, J. E., Jr.,** Acylation of glycerol 3-phosphate is the sole pathway of *de novo* phospholipid synthesis in *Escherichia coli, J. Bact.,* 169, 2896–2898, 1987.
59. **Joyard, J. and Douce, R.,** Galactolipid synthesis, in *The Biochemistry of Plants,* Vol. 9., Stumpf, P. K., Ed., Academic Press, New York, 1987, 215–274.
60. **Frentzen, M., Nishida I., and Murata, N.,** Properties of plastidial acyl-(acyl-carrier-protein): glycerol-3-phosphate acyltransferase from the chilling-sensitive plant squash (*Curcubita moschata*), *Plant Cell Physiol.,* 28, 1195–1201, 1987.
61. **Hares, W. and Frentzen, M.,** Properties of the microsomal acyl-CoA:*sn*-1-acyl-glycerol-3-phosphate acyltransferase from spinach (*Spinacia oleracea* L.) leaves, *J. Plant Phys.,* 131, 49–59, 1987.

62. **Raetz, C. R. H.**, Molecular genetics of membrane phospholipid synthesis, *Annu. Rev. Genet.*, 20, 253–295, 1986.
63. **Dennis, E. A. and Kennedy, E. P.**, Intracellular sites of lipid synthesis and the biogenesis of mitochondria, *J. Lipid Res.*, 13, 263–267, 1972.
64. **Andrews, J. and Mudd, J. B.**, Phosphatidylglycerol synthesis in pea chloroplasts, *Plant Physiol.*, 79, 259–265, 1985.
65. **Roughan, P. G.**, Cytidine triphosphate-dependent, acyl-CoA-independent synthesis of phosphatidylglycerol by chloroplasts isolated from spinach and pea, *Biochim. Biophys. Acta*, 835, 527–532, 1985.
66. **Hirschberg, C. B. and Kennedy, E. P.**, Mechanism of the enzymatic synthesis of cardiolipin in *Escherichia coli*, *Proc. Natl. Acad. Sci. U.S.A.*, 69, 648–651, 1972.
67. **Ohta, A., Obara, T., Asami, Y., and Shibuya, I.**, Molecular cloning of the cls gene responsible for cardiolipin synthesis in *Escherichia coli* and phenotypic consequences of its amplification, *J. Bact.*, 163, 506–514, 1985.
68. **Hostetler, K., van den Bosch, H., and van Deenen, L.**, The mechanism of cardiolipin biosynthesis in liver mitochondria, *Biochim. Biophys. Acta*, 260, 507–513, 1972.
69. **Stanacev, N. Z., Davidson, J. B., Stuhne-Sekalec, L., and Domazet, Z.**, The mechanism of the biosynthesis of cardiolipin in mitochondria, *Biochem. Biophys. Res. Commun.*, 47, 1021–1027, 1972.
70. **Michell, R. H., Drummond, A. H., and Downes, C. P., Eds.**, *Inositol Lipids in Cell Signaling*, Academic Press, New York, 1989, pp 534.
71. **Husebye, E. and Flatmark, T.**, Phosphatidylinositol kinase of bovine adrenal chromaffin granules: kinetic properties and inhibition by low concentrations of Ca^{2+}, *Biochim. Biophys. Acta*, 968, 261–265, 1988.
72. **Husebye, E. and Flatmark, T.**, Purification and kinetic properties of a soluble phosphatidylinositol-4-phosphate kinase of the bovine adrenal medulla with emphasis on its inhibition by calcium ions, *Biochim. Biophys. Acta*, 1010, 250–257, 1989.
73. **Kato, H., Uno, I., Ishikawa, T., and Takenawa, T.**, Activation of phosphatidylinositol kinase and phosphatidylinositol-4-phosphate kinase by cAMP in *Saccharomyces cerevisiae*, *J. Biol. Chem.*, 264, 3116–3121, 1989.
74. **Smith, C. D. and Chang, K.-J.**, Regulation of brain phosphatidylinositol-4-phosphate kinase by GTP analogues, *J. Biol. Chem.*, 264, 3206–3210, 1989.
75. **Downes, C. P. and Michell, R. H.**, Inositol phospholipid breakdown as a receptor-controlled generator of second messengers, in *Molecular Mechanisms of Transmembrane Signaling*, Cohen, P. and Houslay, M., Eds., Elsevier, Amsterdam, 1985, 3–56.
76. **Berridge, M. J.**, Inositol trisphosphate and diacylglycerol: two interacting second messengers, *Annu. Rev. Biochem.*, 56, 159–193, 1987.
77. **Rhee, S. G., Suh, P.-G., Ryu, S.-H., and Lee, S. Y.**, Studies of inositol phospholipid-specific phospholipase C., *Science*, 244, 546–550, 1989.
78. **Bennett, C. F. and Crooke, S. T.**, Purification and characterization of a phosphoinositide-specific phospholipase C from guinea pig uterus, *J. Biol. Chem.*, 262, 13789–13797, 1987.
79. **Kumijian, D. A., Wahl, M. I., Rhee, S. G., and Daniel, T. O.**, Platelet-derived growth factor (PDGF) binding promotes physical association of PDGF receptor with phospholipase C, *Proc. Natl. Acad. Sci. U.S.A.*, 86, 8232–8236, 1989.
80. **McAtee, P. and Dawson, G.**, Rapid dephosphorylation of protein kinase C substrates by protein kinase A activators results from inhibition of diacylglycerol release, *J. Biol. Chem.*, 264, 11193–11199, 1989.
81. **Cantau, B., Guillon, G., Alaoui, M. F., Chicot, D., Balestre, M. N., and Devilliers, G.**, Evidence of two steps in the homologous desensitization of vasopressin-sensitive phospholipase C in WRKI cells, *J. Biol. Chem.*, 263, 443–450, 1988.
82. **Martin, M. W. and Harden, T. K.**, Agonist-induced desensitization of a P_{2Y}-purinergic receptor regulated phospholipase C, *J. Biol. Chem.*, 264, 19535–19539, 1989.
83. **Shumaker, K. S. and Sze, H.**, Inositol 1,4,5 triphosphate releases Ca^{2+} from vacuolar membrane vesicles of oak roots, *J. Biol. Chem.*, 262, 3944–3946, 1987.
84. **Berridge, M. J. and Irvine, R. F.**, Inositol phosphates and cell signalling, *Nature*, 341, 197–205, 1989.
85. **Sherman, W. R.**, Inositol homeostasis, lithium, and diabetes, in *Inositol Lipids and Cell Signaling*, Michell, R. H., Drummond, A. H., and Downes, C. P., Eds., Academic Press, New York, 1989, 39–79.
86. **Lee, M. H. and Bell, R. M.**, Phospholipid functional groups involved in protein kinase C activation, phorbol ester binding, and binding to mixed micelles, *J. Biol. Chem.*, 264, 14797–14805, 1989.
87. **Ohno, S., Konno, Y., Akita, Y., Yano, A., and Suzuki, K.**, A point mutation at the putative ATP-binding site of protein kinase C_{α} abolishes the kinase activity and renders it down-regulation-insensitive, *J. Biol. Chem.*, 265, 6296–6300, 1990.

88. **Coussens, L., Parker, P. J., Rhee, L., Yang-Feng, T. L., Chen, E., Waterfield, M. D., Francke, U., and Ullrich, A.,** Multiple, distinct forms of bovine and human protein kinase C suggest diversity in cellular signaling pathways, *Science,* 233, 859–866, 1986.
89. **Walker, J. M., Homan, E. C., and Sando, J. J.,** Differential activation of protein kinase C isozymes by short chain phosphatidylserines and phosphatidylcholines, *J. Biol. Chem.,* 265, 8016–8021, 1990.
90. **Hannun, Y. A., Loomis, C. R., Merrill, A. H., Jr., and Bell, R. A.,** Sphingosine inhibition of protein kinase C activity and of phorbol dibutyrate binding *in vitro* and in human platelets, *J. Biol. Chem.,* 261, 12604–12609, 1986.
91. **Traynor-Kaplan, A. E., Thompson, B. L., Harris, A. L., Taylor, P., Amann, G. M., and Sklar, L. A.,** Transient increase in phosphatidylinositol 3,4- bisphosphate and phosphatidylinositol trisphosphate during activation of human neutrophils, *J. Biol. Chem.,* 264, 15668–15673, 1989.
92. **Exton, J. H.,** Signaling through phosphatidylcholine breakdown, *J. Biol. Chem.,* 265, 1– 4, 1990.
93. **Billah, M. M., Eckel, S., Mullmann, T. J., Egan, R. W., and Siegel, M. I.,** Phosphatidylcholine hydrolysis by phospholipase D determines phosphatidate and diglyceride levels in chemotactic peptide-stimulated human neutrophils, *J. Biol. Chem.,* 264, 17069–17077, 1989.
94. **Rouser, G., Kritchevsky, G., Heller, D., and Lieber, E.,** Lipid composition of beef brain, beef liver, and the sea anemone: two approaches to quantitative fractionation of complex lipid mixtures, *J. Am. Oil Chem. Soc.,* 40, 425–454, 1963.
95. **Liang, C. R. and Rosenberg, H.,** The biosynthesis of the phosphonic analogue of cephalin in *Tetrahymena, Biochim. Biophys. Acta,* 125, 548–562, 1966.
96. **Hilderbrand, R. L., Ed.,** *The Role of Phosphonates in Living Systems,* CRC Press, Boca Raton, 1983.
97. **Hori, T., Horiguchi, M., and Hayashi, A., Eds.,** *Biochemistry of Natural C-P Compounds,* Maruzen, Kyoto, 1984, pp 200.
98. **Barry, R. J., Bowman, E., McQueney, M., and Dunaway-Mariano, D.,** Elucidation of the 2-aminoethylphosphonate biosynthetic pathway in *Tetrahymena pyriformis, Biochem. Biophys. Res. Comm.,* 153, 177–182, 1988.
99. **Kennedy, K. E. and Thompson, G. A., Jr.,** Phosphonolipids: localization in surface membranes of *Tetrahymena, Science,* 168, 989–991, 1970.
100. **Rhoads, D., Beischel, L., Meyer, K., and Kaneshiro, E. S.,** Lipid composition of *Paramecium aurelia* cilia, *J. Cell Biol.,* 75, 223a, 1977.
101. **Nozawa, Y. and Thompson, G. A., Jr.,** Studies of membrane formation in *Tetrahymena pyriformis.* II. Isolation and lipid analysis of cell fractions, *J. Cell Biol.,* 49, 712–721, 1971.
102. **Swift, M. L.,** Phosphono-lipid content of the oyster, *Crassostrea virginica,* in three physiological conditions, *Lipids,* 12, 449–451, 1977.
103. **Horrocks, L. A.,** Content, composition, and metabolism of mammalian and avian lipids that contain ether groups, in *Ether Lipids, Chemistry and Biology,* Snyder, F., Ed., Academic Press, New York, 1972, 177–272.
104. **Snyder, F., Ed.,** *Ether Lipids, Chemistry and Biology,* Academic Press, New York, 1972, pp 433.
105. **Hanahan, D. J.,** Ether-linked lipids: chemistry and methods of measurement, in *Ether Lipids, Chemistry and Biology,* Snyder, F., Ed., Academic Press, New York, 1972, 25–50.
106. **Snyder, F.,** The enzymatic pathways of ether-linked lipids and their precursors, in *Ether Lipids, Chemistry and Biology,* Snyder, F., Ed., Academic Press, New York, 1972, 121–156.
107. **Benveniste, J., Tence, M., Varenne, P., Bidault, J., Boullet, C., and Polonsky, J.,** Semi-synthese et structure purpose du facteur activant les plaquettes (P.A.F.); PAF-acether, un alkyl ether analoque de la lysophosphatidylcholine, *C. R. Acad. Sci. (D) (Paris),* 289, 1037–1040, 1979.
108. **Blank, M. L., Snyder, F., Byers, L. W., Brooks, B., and Muirhead, E. E.,** Antihypertensive activity of an alkyl ether analog of phosphatidylcholine, *Biochem. Biophys. Res. Commun.,* 90, 1194–1200, 1979.
109. **Demopoulos, C. A., Pinckard, R. N., and Hanahan, D. J.,** Platelet activating factor. Evidence for 1-0-alkyl-2-acetyl-*sn*-glyceryl-3-phosphorylcholine as the active component (a new class of lipid chemical mediators), *J. Biol. Chem.,* 254, 9355–9358, 1979.
110. **Hajra, A. K. and Bishop, J. E.,** Glycerolipid biosynthesis in peroxisomes via the acyldihydroxyacetone phosphate pathway, *Ann. N. Y. Acad. Sci.,* 386, 170–182, 1982.
111. **Hajra, A. K., Hori, S., and Webber, K. O.,** The role of peroxisomes in glyceryl ether lipid metabolism, in *Biological Membranes: Aberrations in Membrane Structure and Function,* Karnovsky, M. L., Leaf, A., and Bolis, A. C., Eds., Alan R. Liss, New York, 1988, 99–116.
112. **Datta, N. S., Wilson, G. N., and Hajra, A. K.,** Deficiency of enzymes catalyzing the biosynthesis of glycerol ether lipids in Zellweger syndrome: a new category of metabolic disease, *N. Engl. J. Med.,* 311, 1080–1083, 1984.
113. **Snyder, F.,** Metabolism of platelet activating-factor and related ether lipids: enzymatic pathways, subcellular sites, regulation, and membrane processing, in *Biological Membranes: Aberrations in Membrane Structure and Function,* Karnovsky, M. L., Leaf, A., and Bolis, L. C., Eds., Alan R. Liss, New York, 1988, 57–72.

114. **Takahashi, N., Saito, T., Goda, Y., and Tomita, K.,** Participation of microsomal aldehyde reductase in long-chain fatty alcohol synthesis in the rat brain, *Biochim. Biophys. Acta,* 963, 243–247, 1988.
115. **Natarajan, V. and Schmid, H. H. O.,** Substrate specificity in ether lipid biosynthesis: metabolism of polyunsaturated fatty acids and alcohols by rat brain microsomes, *Biochem. Biophys. Res. Commun.,* 79, 411–416, 1977.
116. **Woodard, D. S., Lee, T., and Snyder, F.,** The final step in the *de novo* biosynthesis of platelet-activating factor, *J. Biol. Chem.,* 262, 2520–2527, 1982.
117. **Whatley, R. E., Nelson, P., Zimmerman, G. A., Stevens, D. L., Parker, C. J., McIntyre, T. M., and Prescott, S. M.,** The regulation of platelet-activating factor production in endothelial cells, *J. Biol. Chem.,* 264, 6325–6333, 1989.
118. **Snyder, F.,** The significance of dual pathways for the biosynthesis of platelet activating factor: 1-alkyl-2-lyso-*sn*-glycero-3-phosphate as a branchpoint, in *New Horizons in Platelet Activating Factor Research,* Winslow, C. M. and Lee, M. L., Eds., John Wiley & Sons, New York, 1987, 13–25.
119. **Ninio, E.,** Regulation of platelet activating factor biosynthesis in various cell types, in *New Horizons in Platelet Activating Factor Research,* Winslow, C. M. and Lee, M. L. Eds., John Wiley & Sons, New York, 1987, 27–35.
120. **Prescott, S. M., Zimmerman, G. A., and McIntyre, T. M.,** Platelet-activating factor, *J. Biol. Chem.,* 265, 17381–17384, 1990.
121. **Paltauf, F.,** Biosynthesis of 1-0-(1′-alkenyl) glycerols (plasmalogens), in *Ether Lipids: Biochemical and Biomedical Aspects,* Mangold, H. K. and Paltauf, F., Eds., Academic Press, New York, 1983, 107–128.
122. **Ford, D. A. and Gross, R. W.,** Identification of endogenous 1-0-alk-1′-enyl-2-acyl-*sn*-glycerol in myocardium and its effective utilization by choline phosphotransferase, *J. Biol. Chem.,* 263, 2644–2650, 1988.
123. **Jurkowitz-Alexander, M., Ebata, H., Mills, J. S., Murphy, E. J., and Horrocks, L. A.,** Solubilization, purification, and characterization of lysoplasmalogen alkenylhydrolase (lysoplasmalogenase) from rat liver microsomes, *Biochim. Biophys. Acta,* 1002, 203–212, 1989.
124. **Goldfine, H. and Langworthy, T. A.,** A growing interest in bacterial ether lipids, *Trends Biochem. Sci.,* 13, 217–221, 1988.
125. **Kishimoto, Y.,** Sphingolipid formation, in *The Enzymes,* 3rd ed., XIV, Boyer, P. D., Ed., Academic Press, New York, 1983, 358–407.
126. **Merrill, A., Wang, E., and Mullins, R.,** Kinetics of long-chain (sphingoid) base biosynthesis in intact LM cells: effects of varying the extracellular concentrations of serine and fatty acid precursors of this pathway, *Biochemistry,* 27, 340–345, 1988.
127. **Stoffel, W.,** Sphingosine metabolism and its link to phospholipid biosynthesis, *Mol. Cell Biochem.,* 1, 147–155, 1973.
128. **Ong, D. E. and Brady, R. N.,** *In vivo* studies on the introduction of the 4-t-double bond of the sphingenine of rat brain ceramides, *J. Biol. Chem.,* 248, 3884–3888, 1973.
129. **Merrill, A. and Wang, E.,** Biosynthesis of long-chain (sphingoid) bases from serine by LM cells, *J. Biol. Chem.,* 261, 3764–3769, 1986.
130. **Medlock, K. and Merrill, A. H.,** Rapid turnover of sphingosine synthesized *de novo* from [^{14}C] serine by Chinese hamster ovary cells, *Biochem. Biophys. Res. Commun.,* 157, 232–237, 1988.
131. **Rafestin-Oblin, M. E., Couette, B., Radanyi, C., Lombes, M., and Baulieu, E. E.,** Mineralocorticosteroid receptor of the chick intestine, *J. Biol. Chem.,* 264, 9304–9309, 1989.
132. **Kolesnick, R. and Clegg, S.,** 1,2-Diacylglycerols, but not phorbol esters activate a potential inhibitory pathway for protein kinase C in GH3 pituitary cells, *J. Biol. Chem.,* 263, 6534–6537, 1988.
133. **Hannun, Y. A. and Bell, R. M.,** Functions of sphingolipids and sphingolipid breakdown products in cellular regulation, *Science,* 243, 500–507, 1989.
134. **DiMori, S., Brady, R., and Snell, E.,** Biosynthesis of sphingolipid bases. IV. The biosynthetic origin of sphingosine in *Hansenula ciferri, Arch. Biochem. Biophys.,* 143, 553–563, 1971.
135. **IUPAC-IUB Commission on Biochemical Nomenclature,** The nomenclature of lipids (recommendations 1976), *Mol. Cell. Biochem.,* 17, 157–171, 1977; also *Lipids,* 12, 455–468, 1977.
136. **Cahoon, E. B. and Lynch, D. V.,** Analysis of glucocerebrosides of rye (*Secale cereale* L. *cv Puma)* leaf and plasma membrane, *Plant Physiol.,* 95, 58–68, 1991.
137. **Crossman, M. and Hirchberg, C.,** Biosynthesis of 4D-hydroxysphinganine by the rat. en bloc incorporation of the sphinganine carbon backbone, *Biochim. Biophys. Acta,* 795, 411–416, 1984.
138. **Gatt, S.,** Enzymatic aspects of sphingolipid degradation, *Chem. Phys. Lipids,* 5, 235–249, 1970.
139. **Marggraf, W.-D., Anderer, A., and Kanfer, J.,** The formation of sphingomyelin from phosphatidylcholine in plasma membrane preparations from mouse fibroblasts, *Biochim. Biophys. Acta,* 664, 61–73, 1981.
140. **Futerman, A. H., Stieger, B., Hubbard, A. L., and Pagano, R. E.,** Sphingomyelin synthesis in rat liver occurs predominantly at the *cis* and *medial* cisternae of the Golgi apparatus, *J. Biol. Chem.,* 265, 8650–8657, 1990.

141. **Kobayashi, T. and Pagano, R. E.,** Lipid transport during mitosis. Alternative pathways for delivery of newly synthesized lipids to the cell surface, *J. Biol. Chem.,* 264, 5966–5973, 1989.
142. **Marggraf, W.-D. and Kanfer, J.,** The phosphorylcholine acceptor in the phosphatidylcholine: ceramide cholinephosphotransferase reaction. Is the enzyme a transferase or a hydrolase?, *Biochim. Biophys. Acta,* 793, 346–353, 1984.
143. **Kiss, A., Deli, E., and Kuo, J. F.,** Phorbol ester stimulation of sphingomyelin synthesis in human leukemic HL60 cells, *Arch. Biochem. Biophys.,* 265, 38–42, 1988.
144. **Kolesnick, R. N.,** Thyrotropin-releasing hormone and phorbol esters stimulate sphingomyelin synthesis in GH_3 pituitary cells, *J. Biol. Chem.,* 264, 11688–11692, 1989.
145. **Rosenberg, A. and Stern, N.,** Changes in sphingosine and fatty acid components of the gangliosides in developing rat and human brain, *J. Lipid Res.,* 7, 122–131, 1966.
146. **Waite, M.,** *The Phospholipases,* Plenum Press, New York, 1987, 148.
147. **Brady, R., Kanfer, J., Mock, M., and Fredrickson, D.,** The metabolism of sphingomyelin, II. Evidence of an enzymatic deficiency in Niemann-Pick disease, *Proc. Natl. Acad. Sci. U.S.A.,* 55, 366–369, 1966.
148. **Morimoto, S., Martin, B., Kishimoto, Y., and O'Brien, J.,** Saposin D: a sphingomyelinase activator, *Biochem. Biophys. Res. Commun.,* 156, 403–410, 1988.
149. **Hostetler, K. Y. and Yazaki, P. J.,** The subcellular localization of neutral sphingomyelinase in rat liver, *J. Lipid Res.,* 20, 456–463, 1987.
150. **Record, M., Loyter, A., and Gatt, S.,** Utilization of membranous lipid substrates by membranous enzymes, *Biochem. J.,* 187, 115–121, 1980.
151. **Kolesnick, R.,** 1,2-Diacylglycerols but not phorbol esters stimulate sphingomyelin hydrolysis in GH_3 pituitary cells, *J. Biol. Chem.,* 262, 16759–16762, 1987.
152. **Stoffel, W.,** Sphingosine metabolism and its link to phospholipid biosynthesis, *Mol. Cell Biochem.,* 1, 147–155, 1973.
153. **Shimojo, T., Akino, T., Miura, Y., and Schroeper, G. J., Jr.,** Sphingolipid base metabolism: stereospecific uptake of proton in the enzymatic conversion of sphinganine 1-phosphate to ethanolamine 1-phosphate, *J. Biol. Chem.,* 251, 4448–4457, 1976.
154. **Dennis, E. A., Rhee, S. G., Billah, M. M., and Hannun, Y. A.,** Role of phospholipases in generating lipid second messengers in signal transduction, *FASEB J.,* 5, 2068–2077, 1991.
155. **Waite, M.,** *The Phospholipases,* Plenum Press, New York, 1987, 332.
156. **Paliyath, G. and Thompson, J. E.,** Calcium- and calmodulin-regulated breakdown of phospholipid by microsomal membranes from bean cotyledons, *Plant Physiol.,* 83, 63–68, 1987.
157. **Waite, M.,** *The Phospholipases,* Plenum Press, New York, 1987, 29–44.
158. **Taguchi, R., Asahi, Y., and Ikezawa, H.,** Purification and properties of phosphatidylinositol-specific phospholipase C of *Bacillus thuringiensis, Biochim. Biophys. Acta,* 619, 48–57, 1980.
159. **McMurray, W. C. and Irvine, R. F.,** Phosphatidylinositol 4,5-biphosphate phosphodiesterase in higher plants, *Biochem. J.,* 249, 877–881, 1988.
160. **Einspahr, K. J., Peeler, T. C., and Thompson, G. A., Jr.,** Phosphatidylinositol 4,5-bisphosphate phospholipase C and phosphomonoesterase in *Dunaliella salina* membranes, *Plant Physiol.,* 90, 1115–1120, 1989.
161. **Moriary, T., Gillo, B., Carty, D., Premont, R., Landau, E., and Iyengar, R.,** $\beta\gamma$ subunit of GTP-binding proteins inhibits muscarinic receptor stimulation of phospholipase C, *Proc. Natl. Acad. Sci. U.S.A.,* 85, 8865–8869, 1988.
162. **Slivka, S., Meier, K., and Insel, P.,** α_1-adrenergic receptors promote phosphatidylcholine hydrolysis in MDCK-D1 cells, *J. Biol. Chem.,* 263, 12242–12246, 1988.
163. **Scott, D. L., White, S. P., Otwinowski, Z., Yuan, W., Gelb, M. H., and Sigler, P. B.,** Interfacial catalysis: the mechanism of phospholipase A_2, *Science,* 250, 1541–1546, 1990.
164. **Jain, M. K. and Berg, O. G.,** The kinetics of interfacial catalysis by phospholipase A_2 and regulation of interfacial activation: hopping versus scooting, *Biochim. Biophys. Acta,* 1002, 127–156, 1989.
165. **Masliah, J., Kadiri, C., Pepin, D., Rybkine, T., Etienne, J., Chambaz, J., and Bereziat, G.,** Antigenic relatedness between phospholipases A_2 from *Naja naja* venom and from mammalian cells, *FEBS Lett.,* 222, 11–16, 1987.
166. **deJong, J. G. N., Amesz, H., Aarsman, A. J., Lenting, H. B. M., and van den Bosch, H.,** Monoclonal antibodies against an intracellular phospholipase A_2 from rat liver and their cross-reactivity with other phospholipases A_2, *Eur. J. Biochem.,* 164, 129–135, 1987.
167. **Golan, D. E., Furlong, S. T., Brown, C. S., and Caulfield, J. P.,** Monopalmitoylphosphatidylcholine incorporation into human erythrocyte ghost membranes causes protein and lipid immobilization and cholesterol depletion, *Biochemistry,* 27, 2661–2667, 1988.
168. **Wilschut, J. C., Regts, J., Westenberg, H., and Scherphof, G.,** Action of phospholipases A_2 on phosphatidylcholine bilayers: effects of the phase transition, bilayer curvature, and structural defects, *Biochim. Biophys. Acta,* 508, 185–196, 1978.

169. **Jelsema, C.,** Light activation of phospholipase A_2 in rod outer segments of bovine retina and its modulation by GTP-binding proteins, *J. Biol. Chem.,* 262, 163–168, 1987.
170. **Nakashima, S., Nagota, K., Ueeda, K., and Nozawa, Y.,** Stimulation of arachidonic acid release by guanine nucleotide in saponin-permeabilized neutrophils: evidence for involvement of GTP-binding protein in phospholipase A_2 activation, *Arch. Biochem. Biophys.,* 261, 375–383, 1988.
171. **Takamura, H., Narita, H., Dork, H., Tanuka, K., Matsuura, T., and Kito, M.,** Differential hydrolysis of phospholipid molecular species during activation of human platelets with thrombin and collagen, *J. Biol. Chem.,* 262, 2262–2269, 1987.
172. **Bulsinde, J., Diez, E., Schüller, A., and Mollinedo, F.,** Phospholipase A_2 activity in resting and activated human neutrophils, *J. Biol. Chem.,* 263, 1929–1936, 1988.
173. **Seilmamer, J. J., Pruzanski, W., Vadas, P., Plant, S., Miller, J. A., Kloss, J., and Johnson, L. K.,** Cloning and recombinant expression of phospholipase A_2 present in rheumatoid arthritic synovial fluid, *J. Biol. Chem.,* 264, 5335–5338, 1989.
174. **Burgoyne, R. D., Cheek, T. R., and O'Sullivan, A. J.,** Receptor-activation of phospholipase A_2 in cellular signalling, *Trends Biochem. Sci.,* 12, 332–333, 1987.
175. **Kaya, H., Patton, G. M., and Hong, S. L.,** Bradykinin-induced activation of phospholipase A_2 is independent of the activation of polyphosphoinositide-hydrolyzing phospholipase C., *J. Biol. Chem.,* 264, 4972–4977, 1989.
176. **McIntyre, T., Reinhold, S., Prescott, S., and Zimmerman, G.,** Protein kinase C activity appears to be required for the synthesis of platelet-activating factor and leukotriene B4 by human neutrophils, *J. Biol. Chem.,* 262, 15370–15376, 1987.
177. **Davidson, F., Dennis, E., Powell, M., and Glenney, J.,** Inhibition of phospholipase A_2 by "lipocortins" and calpactins, *J. Biol. Chem.,* 262, 1698–1705, 1987.
178. **Waite, M.,** *The Phospholipases,* Plenum Press, New York, 1987, 79–110.
179. **Cao, Y., Tam, S., Arthur, G., Chen, H., and Choy, P.,** The purification and characterization of a phospholipase A in hamster heart cytosol for the hydrolysis of phosphatidylcholine, *J. Biol. Chem.,* 262, 16927–16935, 1987.
180. **Galliard, T.,** Degradation of acyl lipids: hydrolytic and oxidative enzymes, in *The Biochemistry of Plants,* Vol. 4, Stumpf, P. K., Ed., Academic Press, New York, 1980, 85–116.
181. **Norman, H. and Thompson, G. A., Jr.,** Activation of a specific phospholipid fatty acyl hydrolase in *Dunaliella salina* microsomes during acclimation to low growth temperature, *Biochim. Biophys. Acta,* 875, 262–269, 1986.
182. **Lands, W. E. M. and Crawford, C. G.,** Enzymes of membrane phospholipid metabolism in animals, in *The Enzymes of Biological Membranes,* Vol. 2, Martonosi, A., Ed., Plenum Press, New York, 1976, 3–85.
183. **Yoshioka, S., Nakashima, S., Okano, Y., and Nozawa, Y.,** Arachidonic acid mobilization among phospholipids in murine mastocytoma P-815 cells: role of ether-linked phospholipids, *J. Lipid Res.,* 27, 939–944, 1986.
184. **Lwie, K., Wiegand, R., and Anderson, R.,** Docosahexaenoate-containing molecular species of glycerophospholipids from frog retinal rod outer segments show different rates of biosynthesis and turnover, *Biochemistry,* 27, 9014–9020, 1988.
185. **Smith, L. A. and Thompson, G. A., Jr.,** Analysis of lipids by high performance liquid chromatography, *Modern Methods of Plant Analysis,* 5, 149–169, 1987.
186. **Okuyama, H., Yamada, K., Kameyama, Y., Ikezawa, H., Akamatsu, Y., and Nojima, S.,** Regulation of membrane lipid synthesis in *Escherichia coli* after shifts in temperature, *Biochemistry,* 16, 2668–2673, 1977.
187. **Kito, M., Ishinaga, M., Nishihara, M., Kato, M., Sawada, S., and Hata, T.,** Metabolism of the phosphatidylglycerol molecular species in *Escherichia coli, Eur. J. Biochem.,* 54, 55–63, 1975.
188. **Okuyama, H., Yamada, K., Ikezawa, H., and Wakil, S.,** Factors affecting the acyl selectivities of acyltransferases in *Escherichia coli, J. Biol. Chem.,* 251, 2487–2492, 1976.
189. **Joseleau-Petit, D. and Kepes, A.,** Degraded and stable phosphatidylglycerol in *Escherichia coli* inner and outer membranes, and recycling of fatty acyl residues, *Biochim. Biophys. Acta,* 711, 1–9, 1982.
190. **Rock, C.,** Turnover of fatty acids in the 1-position of phosphatidylethanolamine in *Escherichia coli, J. Biol. Chem.,* 259, 6188–6194, 1984.
191. **Cooper, C., Hsu, L., Jackowski, S., and Rock, C.,** 2-acylglycerolphosphoethanolamine acyltransferase/acyl-acyl carrier protein synthetase is a membrane-associated acyl carrier protein binding protein, *J. Biol. Chem.,* 264, 7384–7389, 1989.
192. **Irvine, R. and Dawson, R. M.,** Transfer of arachidonic acid between phospholipids in rat liver microsomes, *Biochem. Biophys. Res. Commun.,* 91, 1399–1405, 1979.
193. **Trotter, J. and Ferber, E.,** CoA-dependent cleavage of arachidonic acid from phosphatidylcholine and transfer to phosphatidylethanolamine in homogenates of murine thymocytes, *FEBS Lett.,* 128, 237–241, 1981.

194. **Ramesha, C. S. and Thompson, G. A., Jr.,** Cold stress induces *in situ* phospholipid molecular species changes in cell surface membranes, *Biochim. Biophys. Acta,* 731, 251–260, 1983.
195. **Ramesha, C. S. and Thompson, G. A., Jr.,** The mechanism of membrane response to chilling, *J. Biol. Chem.,* 259, 8706–8712, 1984.
196. **Ryals, P. and Thompson, G. A., Jr.,** Alteration of the composition and size of the free fatty acid pool of *Tetrahymena* responding to low-temperature stress, *Biochim. Biophys. Acta,* 919, 122–131, 1987.
197. **Cronan, J. E., Jr. and Rock, C. O.,** Biosynthesis of membrane lipids, in *Escherichia coli and Salmonella typhimurium,* Neidhardt, F. C., Ed., American Society for Microbiol., Washington, 1987.
198. **Snyder, F., Ed.,** *Mammalian Lipid Metabolism,* Vol. 1, Plenum Press, New York, 1977.
199. **Moore, T. S., Jr.,** Phospholipid biosynthesis, *Annu. Rev. Plant Physiol.,* 33, 235–259, 1982.

Chapter 5

STEROL METABOLISM

I. INTRODUCTION

Sterols are major lipid constituents of membranes in most organisms, the principal exception being bacteria. Many animal species, e.g., the members of Insecta and Crustacea, require sterols but cannot synthesize them. In these cases, dietary sterols are utilized, frequently after some enzymatic modification.

Representative of the sterols found as membrane components are cholesterol, the principal sterol of mammals; ergosterol, the sterol of yeast and a number of other lower forms; and ß-sitosterol, a common sterol of higher plants (Figure 1). Most of the membrane sterols are, like ergosterol and ß-sitosterol, variants of the basic cholesterol structure with additional alkylation on the hydrocarbon side chain.

Relatively minor structural differences can cause large changes in the way sterols are dealt with in animals. For example, ß-sitosterol is absorbed by rat intestine only one tenth to one fifth as well as is cholesterol.

Cells that do contain sterols cannot continue to grow if the sterol supply is interrupted. This is most easily shown in organisms having a dietary sterol requirement. However, the discovery that certain sterol derivatives, such as 25-hydroxycholesterol and 7-ketocholesterol, are extremely potent inhibitors of cholesterol biosynthesis[1,2] has permitted workers to demonstrate a growth requirement of cultured mammalian cells for cholesterol or an analogous sterol.

Many findings suggest that cells exert strict control of the sterol level in membranes. Under normal conditions, every cell type maintains a characteristic sterol concentration, and, as described in Chapter 1, analysis of functionally different organelles isolated from a variety of cell types generally revealed a significantly higher sterol to phospholipid molar ratio in plasma membranes than in other membranes. The stabilizing effect of sterols on membrane fluidity (see p. 11) must be of great physiological importance, since depletion of the normal content in erythrocytes[3] and *Acholeplasma*[4] led to increased cell fragility and abnormal permeability properties.

While most cells are capable of regulating the rate of sterol synthesis, they usually have no capacity for sterol degradation. Apart from a quantitatively insignificant utilization of cholesterol for steroid hormone synthesis, a mammal's disposal of excess cellular cholesterol is totally dependent upon the unique ability of the liver to convert cholesterol to bile salts. Thus, a complex system has evolved for the exchange of cholesterol between circulating plasma and cells of the various tissues. The inability of the circulating plasma lipoproteins to channel large excesses of dietary cholesterol away from tissue deposition sites and into the liver may be a causative factor in cardiovascular disease.[5]

Certain unicellular organisms, lacking the benefits of a liver, are totally incapable of degrading sterols. When the protozoan *Tetrahymena pyriformis* is maintained in a nutrient-free medium, it can sustain itself for several days by scavenging its own carbohydrates, proteins, and phospholipids for energy. As the cell shrinks to 20% of its normal size, all membrane components but the sterol-like compound tetrahymanol are depleted, ultimately resulting in a membrane tetrahymanol to phospholipid ratio three times the usual value.[6] The organisms, while physiologically abnormal, can survive these altered lipid proportions if resupplied with nutrients.

In the following sections of this chapter, the known mechanisms for maintaining optimal membrane sterol levels will be considered.

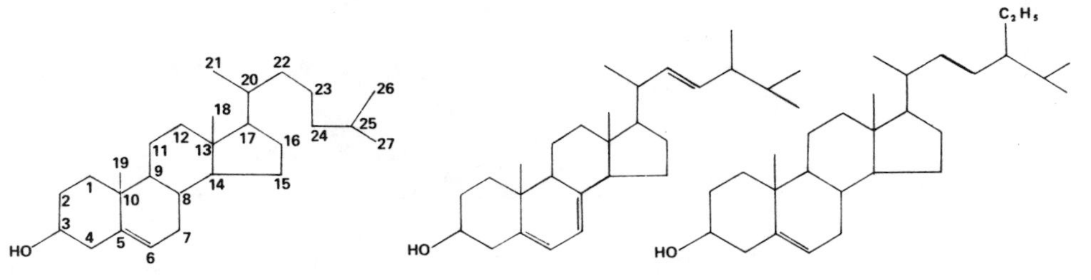

Cholesterol ($C_{27}H_{46}O$) Ergosterol ($C_{28}H_{44}O$) β-Sitosterol ($C_{29}H_{50}O$)

FIGURE 1.

II. THE MECHANISMS OF STEROL BIOSYNTHESIS

The pathway for the formation of cholesterol has been studied extensively over the years. It will be presented here in outline form only. A thorough treatment of historical aspects and details of the individual reactions is available in a recent monograph.[7]

The initial reactions beginning with acetate, the source of all sterol carbon atoms, and leading to the key sterol intermediate squalene are depicted in Figure 2. Two molecules of acetyl-CoA are condensed by the enzyme acetoacetyl-CoA thiolase, yielding acetoacetyl-CoA. A third acetyl-CoA is then joined at the keto group of acetoacetyl-CoA, giving rise to hydroxymethylglutaryl-CoA (HMG-CoA) and free CoA. The enzyme responsible for this reaction is HMG-CoA synthase.

HMG-CoA has been thought to stand at an important metabolic branch point. It may be utilized for the synthesis of isoprenoid compounds, or under different physiological conditions, it may be degraded to acetoacetate and thence to ketone bodies (Figure 3). HMG-CoA exists in mammalian cells in two metabolic pools, one in mitochondria and one in the extramitochondrial compartment. The mitochondrial HMG-CoA alone is convertible to ketone bodies, and it is not available for cholesterol synthesis, which takes place outside the mitochondria.[8] Thus, the branch point in the pathway may realistically be considered to occur prior to HMG-CoA formation.

Cytoplasmic HMG-CoA is next converted to mevalonic acid. This step is catalyzed by the key regulatory enzyme, HMG-CoA reductase, which will be described in more detail later. The reaction itself is a rather complex and still poorly understood two-stage reduction involving first the NADPH-requiring conversion of the HMG-CoA thioester group to an aldehyde, producing the apparently enzyme-bound intermediate mevaldic acid (not shown in Figure 2). The second NADPH-specific reduction gives rise to mevalonic acid. HMG-CoA reductase and all subsequent enzymes involved in cholesterol biosynthesis are associated with the endoplasmic reticulum and have also been reported to occur in rat liver peroxisomal membranes.[9]

Activation of mevalonic acid through a series of phosphorylations culminates in the formation of Δ^3-isopentenyl pyrophosphate, a compound which may be considered equivalent to the long-sought isoprene precursor of all natural isoprenoid substances. The versatility of the enzymes utilizing isopentenyl phyrophosphate is exemplified by the isomerization and condensation steps leading to the well-established squalene intermediate, farnesyl pyrophosphate.

The terminal reactions of squalene biosynthesis are catalyzed by the microsomal enzyme squalene synthetase. Two molecules of farnesyl pyrophosphate are bound stepwise to the enzyme in a proposed "ping-pong" mechanism requiring the participation of NADPH.[10] The

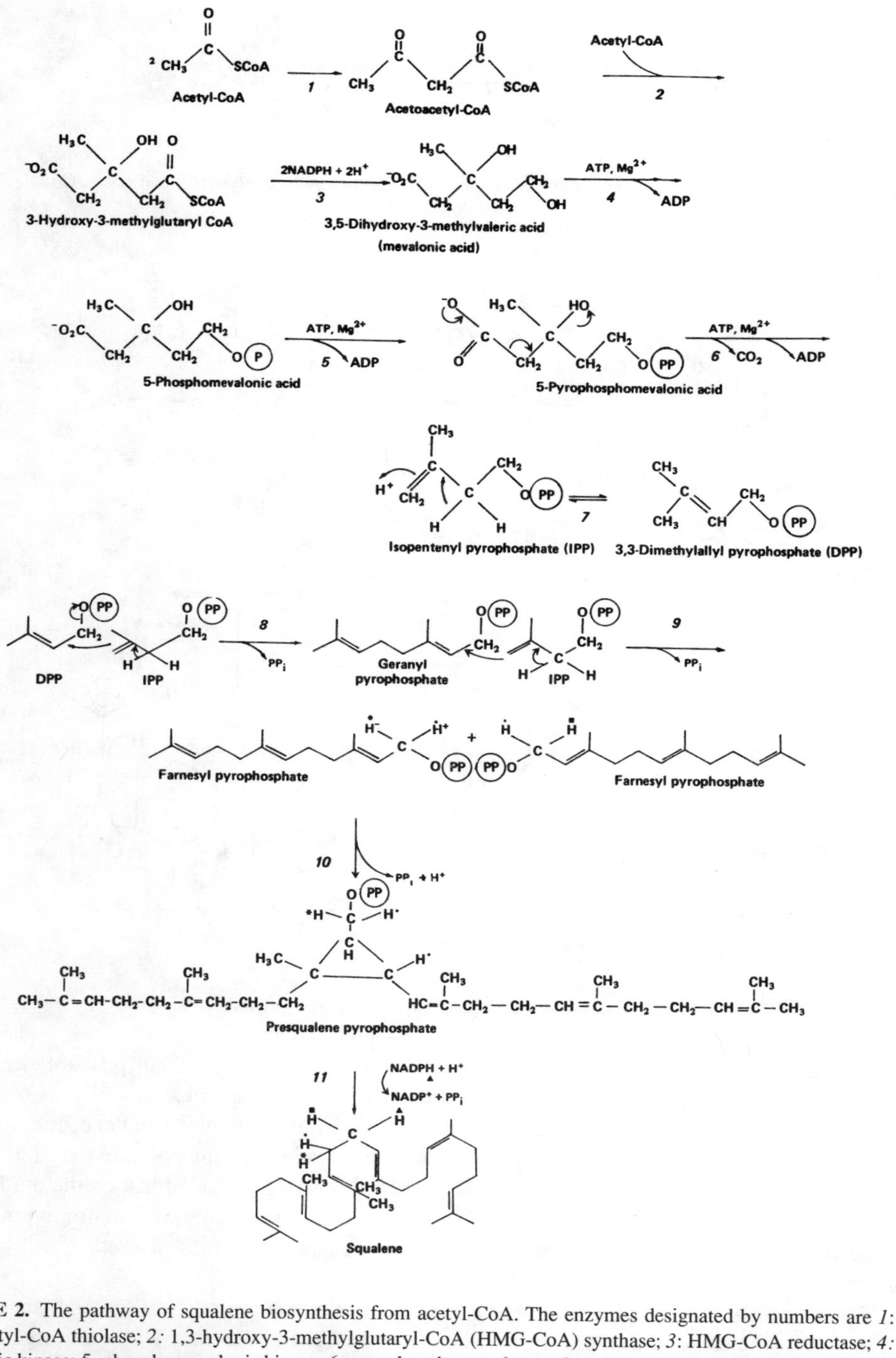

FIGURE 2. The pathway of squalene biosynthesis from acetyl-CoA. The enzymes designated by numbers are *1:* acetoacetyl-CoA thiolase; *2:* 1,3-hydroxy-3-methylglutaryl-CoA (HMG-CoA) synthase; *3:* HMG-CoA reductase; *4:* mevalonic kinase; *5:* phosphomevalonic kinase; *6:* pyrophosphomevalonate decarboxylase; *7:* isopentenyl pyrophosphate isomerase; *8:* geranyl pyrophosphate synthetase; *9:* farnesyl pyrophosphate synthetase, *10:* presqualene synthetase; *11:* squalene synthetase, P = phosphate; PP = pyrophosphate.

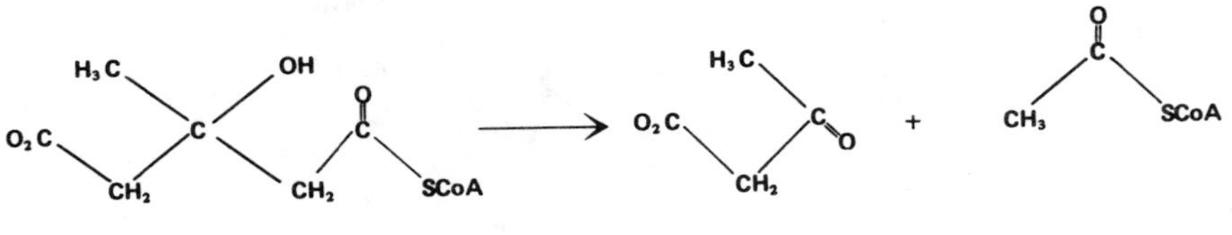

FIGURE 3. The degradation of hydroxymethylglutaryl-CoA to acetoacetate and acetyl-CoA.

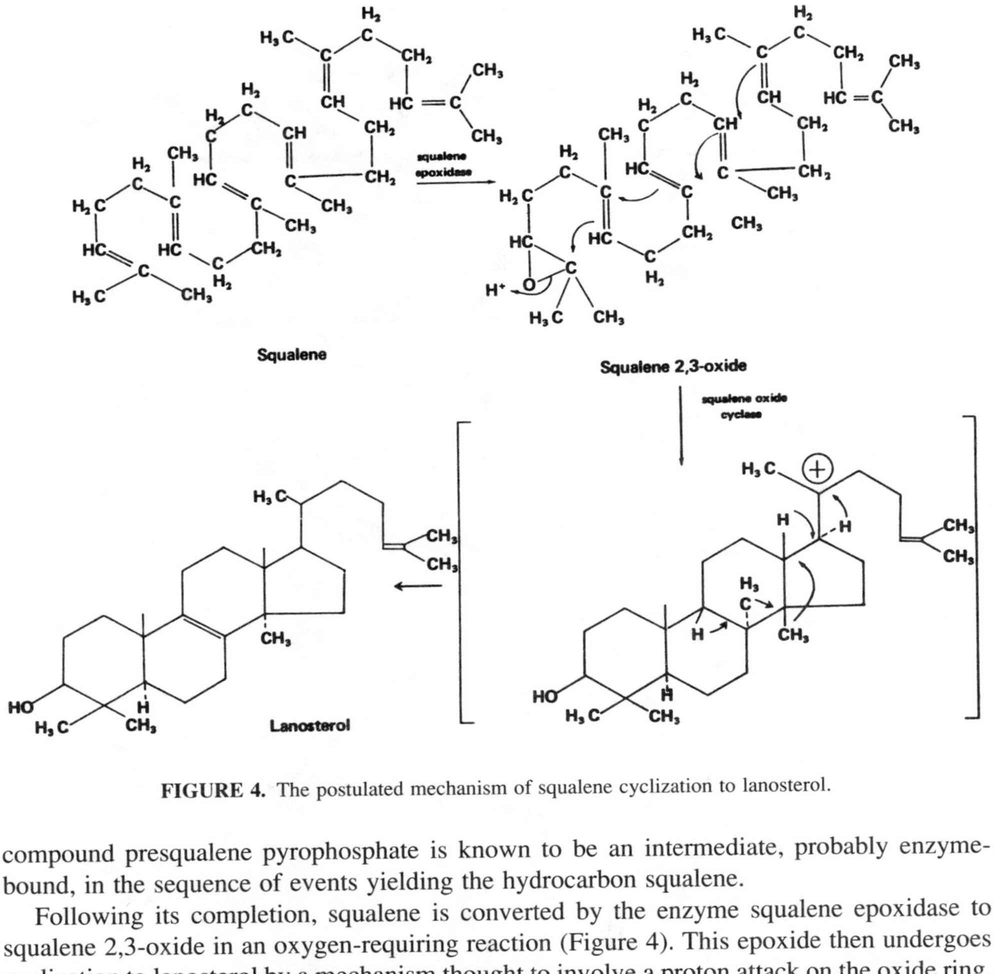

FIGURE 4. The postulated mechanism of squalene cyclization to lanosterol.

compound presqualene pyrophosphate is known to be an intermediate, probably enzyme-bound, in the sequence of events yielding the hydrocarbon squalene.

Following its completion, squalene is converted by the enzyme squalene epoxidase to squalene 2,3-oxide in an oxygen-requiring reaction (Figure 4). This epoxide then undergoes cyclization to lanosterol by a mechanism thought to involve a proton attack on the oxide ring, followed by a series of electron shifts and ring closures. The unstable carbonium ion intermediate is transformed into lanosterol by the intramolecular migration of methyl and hydride groups as shown in Figure 4.

Lanosterol is capable of undergoing a variety of enzymatic modifications that sometimes give rise to the same compound by different routes. This is the case with its conversion to cholesterol (Figure 5). The pathway shown on the left is considered to be the principal one operative in mammals. Reduction of the double bond at the 24-25 position of the side chain, rearrangement of the double bond in ring B, and sequential removal of the two methyl groups at position 4 are the steps involved. The methyl groups are both lost in a sequence of events initiated by oxidation of the 3-hydroxy group to a ketone. The 4 α-methyl group is lost first

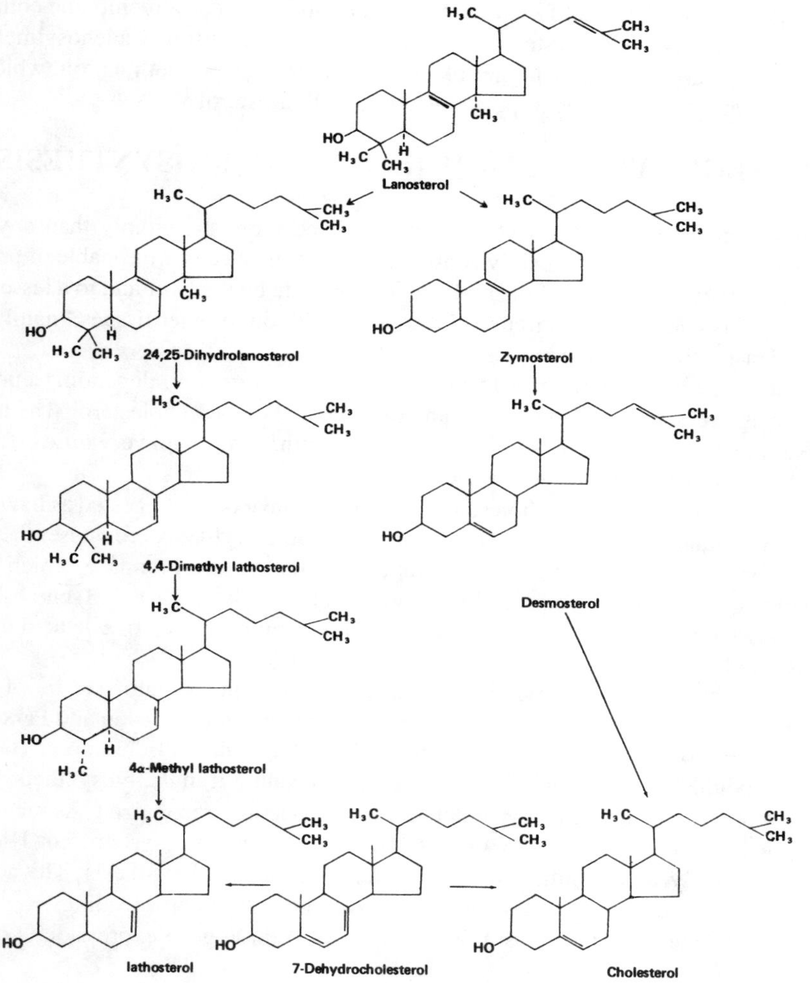

FIGURE 5. The conversion of lanosterol to cholesterol in mammals.

by an oxidative reaction producing CO_2. The keto group is then reduced back to a 3 β-ol with the concomitant rearrangement of the methyl group originally present at the 4β position to the 4α position. The 3β-ol is again oxidized to a keto group prior to oxidative loss of the final methyl group. The enzyme, 4α-methyl sterol oxidase, which eliminates the 4α-methyl group of 4α-methyllathosterol, is thought by some investigators to be of significance in the regulation of cholesterol biosynthesis. Its regulatory capabilities are discussed in Section III.

The latter stages of cholesterol biosynthesis, in which microsomal enzymes act upon squalene and subsequent insoluble intermediates, require the involvement of a cytosolic protein complex composed of low molecular weight (approximately 16,000) protomers known as the sterol carrier protein (SCP).[11] The function of this complex is somewhat analogous to that of the acyl carrier protein (Chapter 2, Section II) in the sense that it presumably helps to position intermediates on the active sites of synthetic enzymes. However, it is not covalently linked to the sterol intermediate and not tightly bound to the catalytic protein. It appears that there are 2 SCPs, each acting at a different part of the biosynthetic pathway.

Some organisms further modify cholesterol before utilizing it as a membrane component. Yeast enzymes alkylate the cholesterol side chain at position 24, using *S*-adenosylmethionine to attach a methyl group.[12] Many higher plants carry out a second methylation, which gives an ethyl group at position 24, as in poriferasterol and β-sitosterol.[13]

III. REGULATION OF CHOLESTEROL BIOSYNTHESIS

In mammalian tissues, which have been investigated more thoroughly than any others, cholesterol synthesis is very stringently controlled. While most cells are capable of producing cholesterol, approximately 90% of the bodily needs is made by the liver and to a lesser extent, the intestine. This cholesterol is circulated through the blood to other tissues, mainly as part of the low density lipoprotein complex.

The actual amount of cholesterol formed *de novo* varies greatly, depending upon many factors, such as stress, diurnal light cycles, and the intake of dietary cholesterol. The influence of these parameters was found to be most pronounced in the liver, with very little effect being noted in slices from 15 extrahepatic tissues.[14]

A number of steps involved in cholesterol metabolism have been suggested as having some regulatory importance.[15,16] These are (1) hydroxymethylglutaryl-CoA reductase, the enzyme responsible for the synthesis of mevalonic acid; (2) 4α-methyl sterol oxidase, which removes the 4α-methyl group during the conversion of lanosterol to cholesterol; and (3) cholesterol 7α-hydroxylase, which catalyzes the rate limiting and first committed step of bile acid formation from cholesterol.

The first of these three enzymes is considered by most investigators to be of primary importance. Its properties have been comprehensively reviewed by Gibson and Parker,[17] and are briefly summarized below. However, before proceeding with a discussion of HMG-CoA reductase, it should be pointed out that the enzymes preceding it in the biosynthetic pathway may play a more important regulatory function than is generally recognized. As mentioned in Section II, it is the acetoacetyl-CoA thiolase-HMG-CoA synthase system, not HMG-CoA reductase, that catalyzes the initial committed step of cholesterol synthesis. The activity of these enzymes, particularly the synthetase, is depressed by cholesterol feeding,[15,18,19] but not by added mevalonate. The mechanism of HMG-CoA synthase gene expression is discussed below.

Thus HMG-CoA reductase is still widely accepted as the enzyme most influential in regulating cholesterogenesis. It has long been recognized that *de novo* cholesterol synthesis can be rapidly and drastically reduced by the inclusion of cholesterol in the diet. Figure 6 illustrates how decreased activity of HMG-CoA reductase was pinpointed as the factor responsible for the inhibition. Characterization of this response has been complicated by the superimposition of fluctuations in reductase activity due to other, often less obvious causes. There is, for example, a very reproducible diurnal rhythm of reductase activity in the liver and intestine of rats. In animals exposed to a normal schedule of 12 hr light and 12 hr dark and fed *ad libitum*, HMG-CoA reductase activity (and cholesterol synthesis) rises to a sharp peak near the middle of the dark phase and declines during the light phase. Surprisingly, the rhythm in reductase activity persists in complete darkness or light, and it is not dependent upon the presence or absence of food.

During the investigation of this curious phenomenon, it was discovered that the diurnal rise in reductase activity could be blocked by inhibitors of protein synthesis. Later studies confirmed that a cyclic variation in the synthesis of reductase leads to an increase in the quantity of the enzyme beginning just before the dark period (Figure 7).

The underlying cause of the cycling HMG-COA reductase activity appears to be hormonal.[21] Insulin and thyroid hormone can, under certain conditions, stimulate a marked rise

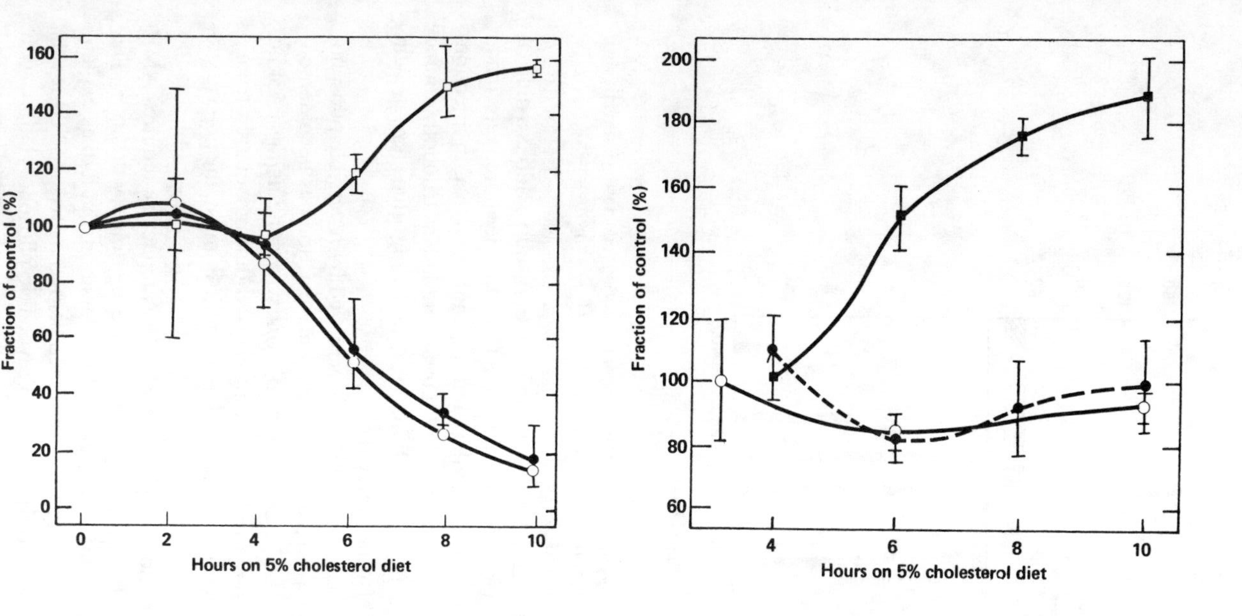

FIGURE 6. Effect of dietary cholesterol on enzymes of hepatic cholesterol biosynthesis. Rats were fed diets containing 5% cholesterol starting at 8 AM, and their livers were analyzed during the following 10 hr. Activity is expressed as the percent of that observed in livers of rats fed normal diets. Left: cholesterol synthesis from acetate (●), HMG-CoA reductase activity (○), and liver cholesterol content (□). Right: incorporation of radioactivity from [2-^{14}C]mevalonate into cholesterol (●), incorporation of radioactivity from [1-^{14}C]acetyl-CoA into HMG-CoA (○) and liver cholesterol content (■). (From Shapiro, D. J. and Rodwell, V. W., *J. Biol. Chem.* 246, 3210, 1971. With permission.)

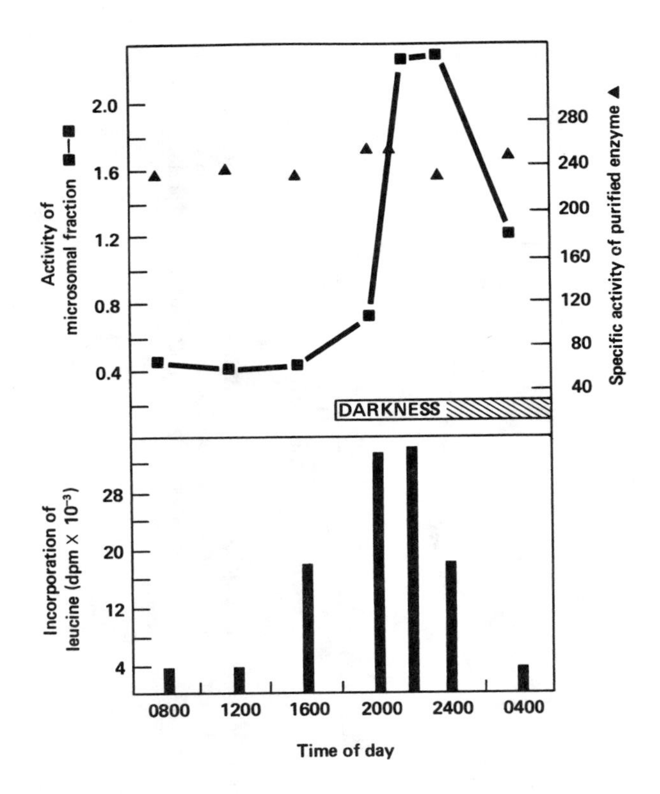

FIGURE 7. HMG-CoA reductase activity and incorporation of [^{3}H]leucine into purified reductase. (From Higgins, M., Kawachi, T., and Rudney, H., *Biochem. Biophys. Res. Commun.*, 45, 138, 1971. With permission.)

in reductase synthesis (for example, see Figure 8). The serum level of insulin exhibits a diurnal rhythm that coincides with the rhythm of HMG-CoA reductase. Furthermore, rats made diabetic by treatment with drugs lose the diurnal rhythm, but regain it upon insulin injection. Diabetic-hypophysectomized rats do not respond to insulin treatment unless triiodothyronine has been previously administered. Effects of other hormones, including glucagon, growth hormone, and adrenal hormones have also been noted.[21]

Short term regulation of HMG-CoA reductase activity is achieved by reversible phosphorylation of the enzyme. Inactivation of the enzyme by protein kinase action has been demonstrated in many species and cell types.[17] The early *in vitro* work was done using a 53kDa enzyme that was later discovered to be a cytosolic fragment cleaved during isolation from a 97kDa integral membrane protein.[17] Apart from being on the cytosolic domain, the precise site of phosphorylation is not known.

The reductase kinase responsible for phosphorylating HMG-CoA reductase requires AMP for full activity and is termed the AMP-activated protein kinase by some workers.[23] The physiological significance of the AMP modulation is not fully understood, but it does explain the high levels of post-mortem reductase phosphorylation (see below) as tissue ATP levels drop precipitously. The reductase kinase is itself subject to reversible phosphorylation, with the phosphorylated form being the active one. Regulation of reductase kinase kinase and reductase phosphatase figure in the control of HMG-CoA reductase activity, as described below.

When dealing with an enzyme such as HMG-CoA reductase, which experiences fluctuations in absolute quantity as well as in phosphorylation regulated activity of existing mol-

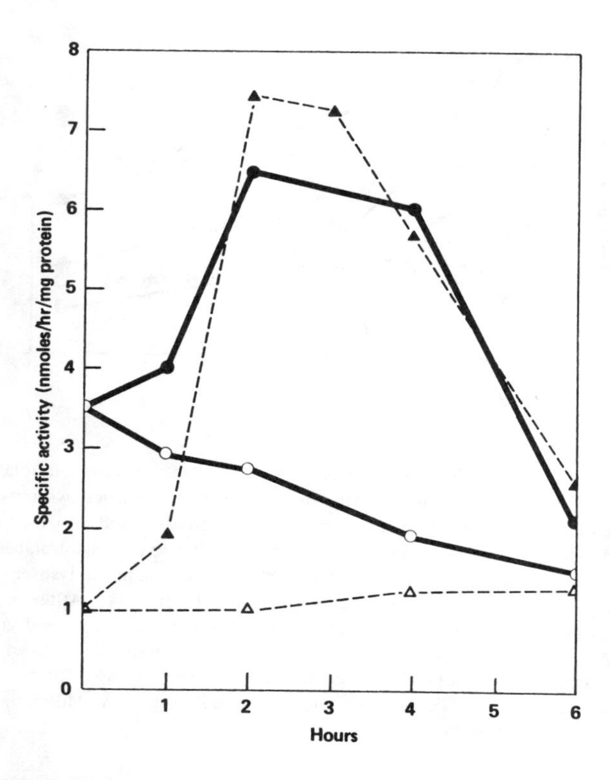

FIGURE 8. Effect of insulin on HMG-CoA reductase activity as a function of time. At 0800 hr insulin was administered to normal and 2-day diabetic rats in doses of 3 units and 6 units per 100 g body weight, respectively. The animals were killed at the specified time intervals and their hepatic HMG-CoA reductase activities were determined. Each point represents the average of values obtained from assaying two to four rats individually. ○—○ normal control rats; ●—● normal rats treated with insulin; △- - -△, diabetic rats; ▲- - -▲, diabetic rats treated with insulin. The groups of normal and diabetic rats treated with insulin for 2 hr were significantly different ($p < 0.05$) from their corresponding control groups by student's test. (From Lakshmanan, M. R., Nepokroeff, C. M., Ness, G. C., Dugan, R, E., and Porter, J. W., *Biochem. Biophys. Res. Commun.*, 50, 704, 1973. With permission.)

ecules, special care must be taken to ascertain which one of these factors is responsible for observed changes. By acting to prevent phosphorylation or dephosphorylation during reductase preparation and assay, Nordstom et al.[24] measured the "expressed" activity and compared it with total activity generated by the addition of protein phosphatase. This should yield data on both the amount and relative activation of HMG-CoA reductase present. However, it was not recognized for some years that the highly phosphorylated and inactive state of the enzyme as isolated from tissues was actually due to active phosphorylation during purification. Only by cold-clamping, cooling the tissue very rapidly prior to homogenization, could the true *in vivo* phosphorylation state be preserved.[25] By taking these precautions, it is now clear that the expressed activity of HMG-CoA reductase in rat liver varies from 28 to 80% of total activity during the diurnal cycle.

Many studies of HMG-CoA reductase regulation by phosphorylation have been described in recent reviews.[17,26] These can be summarized as falling into two major classes: feedback control and hormonal control. In the former category, mevalonic acid and cholesterol have been shown to inactivate the reductase by depressing protein phosphatase activity and in some cases stimulating both reductase kinase and reductase kinase kinase activities.

Hormones can affect reductase activity indirectly through feedback control. Thus an increase in insulin will, by generally stimulating growth, create a greater demand for choles-

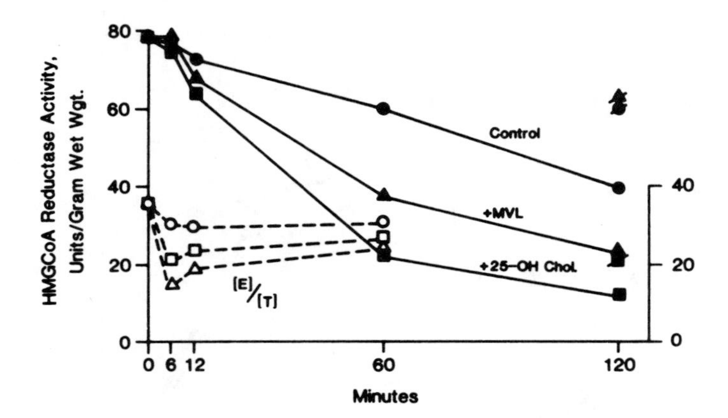

FIGURE 9. A decrease in the ratio (E/T) of HMG-CoA reductase expressed activity to total activity precedes the enhanced rate of decline of total reductase activity in rat hepatocytes treated with mevalonic acid or 25-hydroxycholesterol. Hepatocytes from cholestyramine-fed rats were prepared and incubated under standard conditions as described in reference 28, with the following additions made at zero time: control, basal conditions (●, ○); mevalonate, 1.0 m*M* (▲, △); 25-hydroxycholesterol, 25 μ*M* (■, □); propylamine (a lysosomotropic agent), 15 m*M* (●); mevalonate + propylamine (▲); 25-hydroxycholesterol + propylamine (■). At the indicated times, aliquots of cells were removed, microsomes were isolated, and HMG-CoA reductase expressed and total (after protein phosphatase treatment) activities were determined. Total HMG-CoA reductase activity (*closed symbols*) is given in units of microsomal reductase activity/g of cell wet weight; the expressed reductase activity (*open symbols*) is given as the ratio of expressed/total activity (E/T), shown as percent. (From Parker, R. A., Miller, S. J., and Gibson, D. M., *J. Biol. Chem.*, 264, 4877–4887, 1989. With permission)

terol and decrease its availability for feedback inhibition. However, hormones such as insulin and glucagon have a more direct effect. Raising the serum insulin level (by glucose injection) or lowering it (by anti-insulin injection) led to an increase and a decrease, respectively, in the *in vivo* expressed activity of HMG-CoA reductase of rat liver within 30 min.[27] Glucagon injection did not affect the phosphorylation state of the enzyme unless insulin secretion was blocked at the same time. In the latter case reductase phosphorylation increased and expressed activity decreased. It therefore appears that insulin plays a dominant role in regulating the phosphorylation of hepatic HMG-CoA reductase. In this same study total HMG-CoA reductase activity was found to change little over the experimental period and over a 4-h period of starvation. On the other hand, 24 hours of starvation caused an 80% decrease in total HMG-CoA reductase activity.

Phosphorylation (and inactivation) of HMG-CoA reductase appears to predispose the enzyme to degradation by proteases. Using hepatocytes in which a rapid phosphorylation of the reductase was induced by adding either mevalonolactone or 25-hydroxycholesterol, Parker et al.[28] observed an enhanced loss of total reductase that was preceded by a decline in expressed activity (reflecting the increased phosphorylation) (Figure 9). Although not yet conclusively demonstrated, the pathway for enhanced HMG-CoA reductase degradation would seem to involve (1) phosphorylation of the intact 97kDa membrane-bound enzyme, (2) proteolytic cleavage of a 53kDa phosphorylated fragment, and (3) further proteolysis of this nonmembrane-associated fragment, possibly by lysosomal action. A direct involvement of the HMG-CoA reductase membrane domain in the initial stages of degradation was indicated by the mevalonate-regulated loss of a fusion protein constructed to include only that hydrophobic portion of the enzyme coupled to a reporter protein.[29]

In addition to the short term regulation of HMG-CoA reductase activity by phosphorylation-dephosphorylation, there is a longer term control through end product repression of the enzyme's synthesis. Recent substitution mutagenesis studies with hamster fibroblasts have

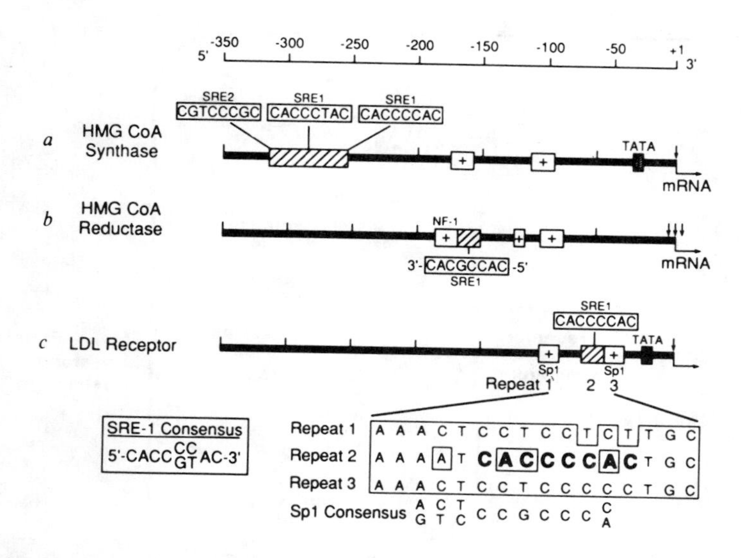

FIGURE 10. Sterol regulatory elements (SRE) in the 5′ flanking regions of the genes for (a) HMG-CoA synthase, (b) HMG-CoA reductase, and (c) the LDL receptor. +, sequences that are necessary for transcription in the absence and presence of sterols. The nucleotide sequences of repeats 1, 2, and 3 in the LDL receptor gene promoter are shown in c. Identical nucleotides are boxed. The core eight nucleotides comprising the SRE-1 in repeat 2 are denoted by bold type. The consensus sequence for binding of the general transcription factor Sp1, also necessary along with SRE-1 for high level transcription in the absence of sterols, is shown below the repeat 3. sequence. (From Goldstein, J. L. and Brown, M. S., *Nature,* 343, 425–430, 1990. With permission.)

located a *cis*-acting sequence of 20 base pairs in the 5′ flanking region of the HMG-CoA reductase gene that is responsible for sterol-dependent repression of transcription.[30] Based upon the behavior of mutant promoters, the active sequence was proposed to be an octanucleotide showing a 7 base out of 8 match with a previously identified regulatory element in the low density lipoprotein receptor promoter, and one in the promoter region of the HMG-CoA synthase gene. Both of these latter genes are also repressible by sterol.

Recent findings regarding these closely related base sequences, designated sterol regulatory elements (SRE) have been reviewed by Goldstein and Brown.[31] All three of the genes have at least one copy of the principal regulatory element (SRE-1). Binding of sterols to these elements is influenced to varying degrees by nearby base sequences (Figure 10). The function of SRE-1 in the low density lipoprotein receptor promoter and the HMG-CoA synthase promoter regions is a positive one, i.e., it enhances transcription in the absence of sterols. There is suggestive but as yet incomplete evidence that the SRE-1 of the HMG-CoA reductase gene may actively repress transcription in the presence of sterols.

HMG-CoA reductase levels are also regulated at the translational level. Synthesis of the reductase m-RNA continues at approximately one-eighth of the maximal rate even when repression by sterols is in full effect. This insures that HMG-CoA reductase activity remains available for the synthesis of nonsterol isoprenoids if needed (see Chapter 6). If excess sterols are present and the nonsterol isoprenoid needs are also met by supplying mevalonic acid, translation of HMG-CoA reductase m-RNA is further reduced by fivefold.[32]

Other factors may also influence HMG-CoA reductase gene expression. Phorbol esters have been reported to induce a rapid rise in m-RNA for both HMG-CoA reductase and low density lipoprotein receptor.[33] Further studies are needed to confirm the specificity and mechanism of this apparent derepression.

While HMG-CoA reductase activity is widely considered to be modulated through feedback control by cholesterol and mevalonate, it has been difficult to rule out a key regulatory

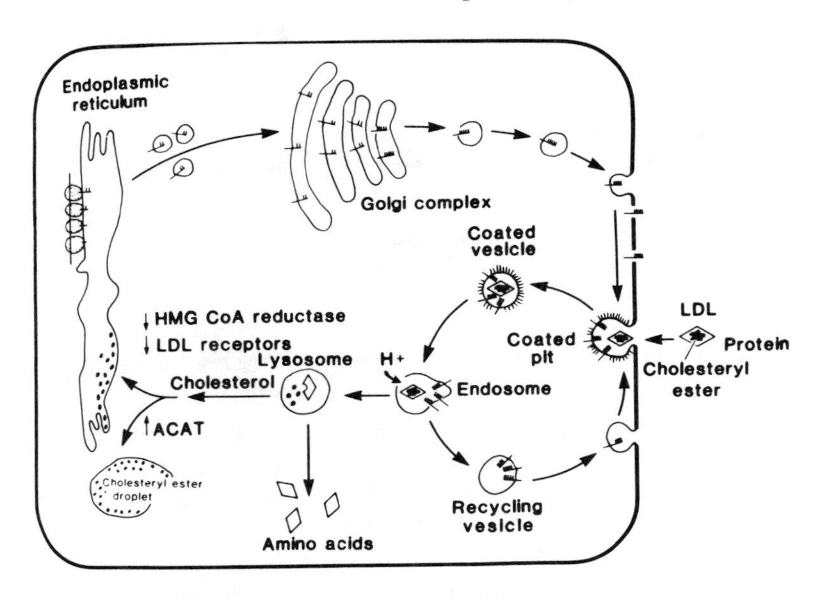

FIGURE 11. The LDL receptor pathway in mammalian cells. The receptor undergoes a circuitous itinerary, beginning life as a newly synthesized protein on the endoplasmic reticulum. The remainder of the itinerary is discussed in detail by Brown and Goldstein.[39] HMG-CoA reductase denotes 3-hydroxy-3-methylglutaryl-CoA reductase; ACAT denotes acyl-CoA;cholesterol acyltransferase. Vertical arrows indicate regulatory effects. (From Brown, M. S. and Goldstein, J. L., *Science*, 232, 34–47, 1986. With permission.)

role for partly oxidized sterols. Smith[34] has reviewed the nature and tissue distribution of oxysterols. The six principle oxysterols of mouse liver, present in free and esterified form, were shown to account for less than 0.1% of the cholesterol and cholesterol esters present.[35] However, the levels of certain oxysterols, particularly 24-hydroxycholesterol, and 25-hydroxycholesterol, rose after cholesterol was fed to mice previously maintained on a cholesterol-free diet, and the increase was correlated with repression of HMG-CoA activity.

Nonhepatic cells of mammals obtain cholesterol principally from low density lipoproteins (LDL) circulating in the plasma. The typical LDL particle is an aggregate of one 400 kDa protein, 800 molecules of phospholipid, 500 molecules of cholesterol, and about 1500 molecules of cholesterol ester.[36] The cholesterol esters become associated with the LDL in the liver,[37] although much of the esterification itself is effected in the blood plasma by the enzyme lecithin:cholesterol acyltransferase (EC 2.3.1.43).[38] The circulating LDL particles are taken up by cells through the mediation of the LDL receptor, a glycoprotein having a protein molecular mass of 93 kDa and associated with coated pits on the plasma membrane. The receptor with its bound LDL is internalized by endocytosis. Following lysosomal fusion with the endocytotic vesicle, the protein moiety of the LDL is degraded and the cholesterol esters are hydrolysed (Figure 11). Free cholesterol thus released or an oxygenated derivative formed from it suppresses endogenous cholesterol synthesis in the host cell by inhibiting HMG-CoA reductase activity through accelerated degradation of the phosphorylated enzyme protein or through reduced transcription of the HMG-CoA reductase gene as described above.

The LDL-derived cholesterol further influences the cell's free cholesterol balance by (1) activating the intracellular cholesterol esterifying enzyme acyl CoA:cholesterol acyltransferase (ACAT) and suppressing synthesis of LDL receptors by lowering the concentration of receptor mRNA. Controlling the number of plasma membrane LDL receptors is of vital importance in the metabolism of LDL in both liver and other cell types.[36]

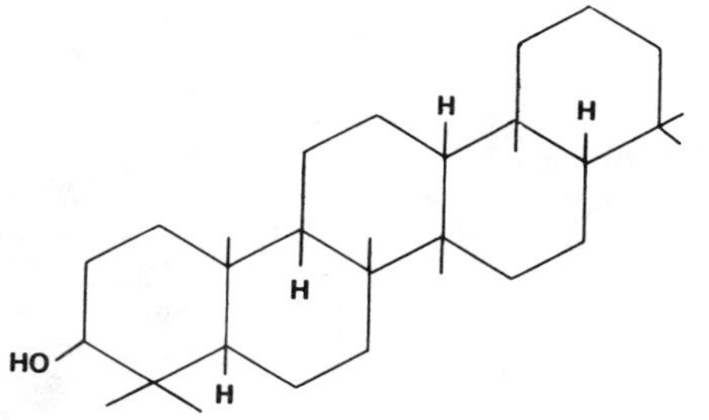

FIGURE 12. Tetrahymanol.

IV. REGULATION OF STEROL ANALOG METABOLISM

Some organisms contain no sterols proper but instead have structurally similar molecules. The best known example is *Tetrahymena pyriformis*, which synthesizes the pentacyclic triterpenoid, tetrahymanol (Figure 12). The rate of tetrahymanol biosynthesis is obviously under close regulation, since the ratio of tetrahymanol to phospholipid is maintained at a characteristic value in whole cells and in isolated organelles,[40] but the type of control is not clear. Interestingly, growth of *Tetrahymena* in a medium containing ergosterol results in inhibition of tetrahymanol synthesis and eventual replacement of tetrahymanol by sterols in the growing cells.[41,42] There appear to be several metabolic blocks imposed by the added ergosterol. The initial inhibition, occurring between acetate and mevalonate in the synthetic pathway, could not be shown to result from HMG-CoA reductase inactivation, since the *in vitro* activity of this key enzyme was unaffected in ergosterol-grown cells.[43] There were two other apparent sites of inhibition; one between mevalonate and squalene, and another during the nonoxidative cyclization of squalene to tetrahymanol. The pronounced alteration of the composition of other membrane lipids and the physical properties of various membranes due to ergosterol replacement of tetrahymanol[42,44] may create a membrane environment not conducive for normal action of the biosynthetic enzymes. The inhibitory effects of ergosterol are for the moment of uncertain physiological significance, since dietary supplementation with tetrahymanol fails to inhibit *de novo* tetrahymanol synthesis.[43]

V. REGULATION OF STEROL CATABOLISM

Our understanding of this process is confined almost exclusively to studies of cholesterol biochemistry. The sole pathway having any quantitative impact upon mammalian membrane composition is the conversion of cholesterol to bile acids in the liver. The reactions involved are illustrated in Figure 13.

The transformation of cholesterol to 7α-hydroxycholesterol is the key regulatory step,[45] although further regulation at the branch point following 7α-hydroxy-4-cholesten-3-one has not been ruled out.

Cholesterol 7α-hydroxylase (cholesterol 7α-monooxygenase, EC 1.14.13.17) is a mixed function oxidase found in the smooth endoplasmic reticulum of liver. It acts primarily on cholesterol, although some other sterols, e.g., cholestanol, can also serve as substrates. The presence of an ethyl group at C-24, as in β-sitosterol, almost completely destroys the sterol's susceptibility to 7α-hydroxylase action.[46]

Substrate for the 7α-hydroxylase is derived from the liver microsomal pool of cholesterol.[47] Isotope studies indicate that there is a preferential utilization of newly synthesized

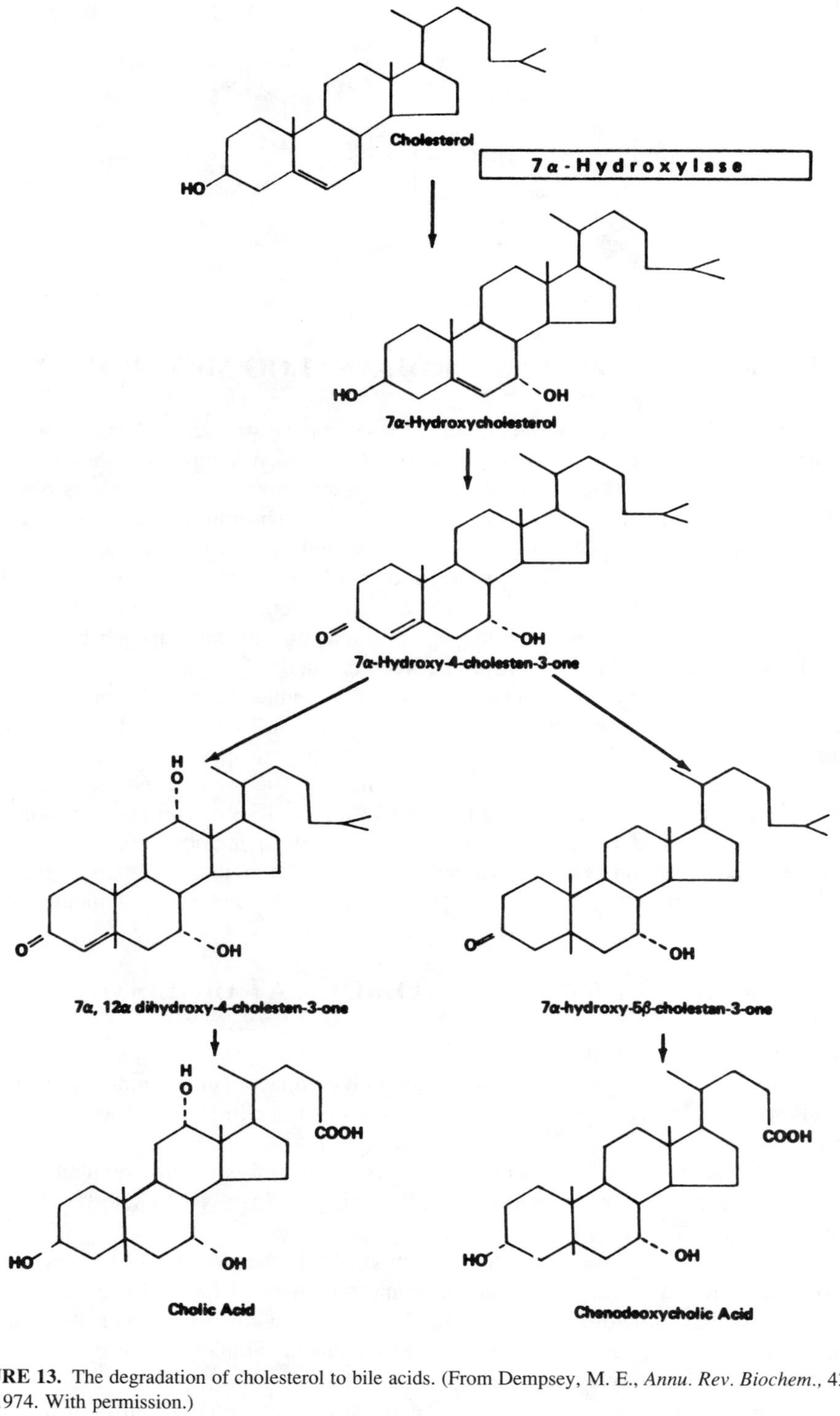
Cholesterol
7α-Hydroxylase
HO
7α-Hydroxycholesterol
OH
7α-Hydroxy-4-cholesten-3-one
7α, 12α dihydroxy-4-cholesten-3-one
7α-hydroxy-5β-cholestan-3-one
COOH
Cholic Acid
Chenodeoxycholic Acid

FIGURE 13. The degradation of cholesterol to bile acids. (From Dempsey, M. E., *Annu. Rev. Biochem.*, 43, 967–990, 1974. With permission.)

liver cholesterol, although cholesterol brought into the liver from other tissues can also be metabolized.

The enzyme is inhibited in rat liver by bile acids (except taurochenodeoxycholate) administered for 1 week.[48] Conversely, bile fistulas cause an increase in bile acid synthesis of about tenfold.[49,50] The increase appeared due to enhanced synthesis of the enzyme or some closely related protein since the increased activity was blocked by actinomycin D.[51] These experiments and related studies[52] led to an estimate of 2 to 3 hours for the 7α-hydroxylase half-life in the rat.

Regulation of 7α-hydroxylase activity has recently been confirmed to occur at the transcriptional level. Chronic biliary diverted rats exhibited a 4.5-fold increase in 7α-hydroxylase mass and a l0-fold increase in steady state m-RNA levels as compared to nonoperated controls.[53] Intraduodenal infusion of taurocholate caused a 72% decrease in 7α-hydroxylase mass and a 74% decrease in m-RNA levels as compared to bilary diverted controls. It is not yet clear whether bile salts repress enzyme transcription by binding directly to a control element of the 7α-hydroxylase gene promoter region or function in some more indirect way. Interestingly, in the same study[53] cholesterol feeding to nonoperated rats led to a 291% increase in 7α-hydroxylase poly(A) RNA and a doubling of its transcriptional activity. The promoter for cholesterol 7α-hydroxylase may have both bile salt- and sterol-responsive elements.

REFERENCES

1. **Chen, H. W., Kandutsch, A. A., and Waymouth, C.,** Inhibition of cell growth by oxygenated derivatives of cholesterol, *Nature (London),* 251, 419–421, 1974.
2. **Brown, M. S. and Goldstein, J. L.,** Suppression of 3-hydroxy-3-methylglutaryl coenzyme A reductase activity and inhibition of growth of human fibroblasts by 7-ketocholesterol, *J. Biol. Chem.,* 249, 7306–7314, 1974.
3. **Bruckdorfer, K. R., Demel, R. A., Deglier, J., and van Deenen, L. L. M.,** The effect of partial replacement of membrane cholesterol by other steroids on the osmotic fragility and glycerol permeability of erythrocytes, *Biochim. Biophys. Acta,* 183, 334–345, 1969.
4. **Rottem, S., Yashouv, J., Ne'eman, Z., and Razin, S.,** Cholesterol in mycoplasma membranes. Composition, ultrastructure, and biological properties of membranes from *Mycoplasma mycoides* var. *capri* cells adapted to grow with low cholesterol concentrations, *Biochim. Biophys. Acta,* 323, 495–508, 1973.
5. **Porter, R. and Knight, J., Eds.,** *Atherogenesis: Initiation Factors,* Ciba Symposium 12, Elsevier, Amsterdam, 1973.
6. **Thompson, G. A., Jr., Bambery, R. J., and Nozawa, Y.,** Environmentally produced alterations of the tetrahymanol phospholipid ratios in *Tetrahymena pyriformis* membranes, *Biochim. Biophys. Acta,* 260, 630–638, 1972.
7. **Danielsson, H. and Sjövall, J., Eds.** *Sterols and Bile Acids,* Elsevier, Amsterdam, 1985, pp 447.
8. **Dempsey, M. E.,** Regulation of steroid biosynthesis, *Annu. Rev. Biochem.,* 43, 967–990, 1974.
9. **Appelkvist, E.-L., Reinhart, M., Fischer, R., Billheimer, J., and Dallner, G.,** Presence of individual enzymes of cholesterol biosynthesis in rat liver peroxisomes, *Arch. Biochem. Biophys.,* 282, 318–325, 1990.
10. **Beytia, E., Quereshi, A. A., and Porter, J. W.,** Squalene synthetase. III Mechanism of the reaction, *J. Biol. Chem.,* 248, 1856–1867, 1973.
11. **Vahouny, G. V., Chanderbhan, K., Harroubi, A., Noland, B. J., Pastuszyn, A., and Scallen, T. J.,** Sterol carrier and lipid transfer proteins, *Adv. Lipid Res.,* 22, 83–113, 1987.
12. **Saat, Y. A. and Block, K. E.,** Effect of a supernatant protein on microsomal squalene epoxidase and 2,3-oxidosqualene-lanosterol cyclase, *J. Biol. Chem.,* 251, 5155–5160, 1976.
13. **Benviniste, P.,** Sterol biosynthesis, *Annu. Rev. Plant. Physiol.,* 37, 275–308, 1986.
14. **Goodwin, T. W.,** Membrane-bound enzymes in plant sterol biosynthesis, in *The Enzymes of Biological Membranes,* Vol. 2, Martonosi, A., Ed., Plenum Press, New York, 1976, 207–223.

15. **Dempsey, M. E.,** Regulation of steroid biosynthesis, *Annu. Rev. Biochem.*, 43, 967–990, 1974.
16. **Popjak, G.,** Specificity of enzymes of sterol biosynthesis, *Harvey Lect.*, 65, 127–156, 1969–1970.
17. **Gibson, D. M. and Parker, R. A.,** Hydroxymethylglutaryl coenzyme A reductase, in *The Enzymes XVIII*, Boyer, P. D. and Krebs, E. G., Eds., Academic Press, New York, 1987.
18. **Sugiyama, T., Clinkenbeard, K., Moss, J., and Lane, M. D.,** Multiple cytosolic forms of hepatic β-hydroxyl-β-methylglutaryl CoA synthase: possible regulatory role in cholesterol synthesis, *Biochem. Biophys. Res. Commun.*, 48, 255–261, 1972.
19. **Panini, S. R., Schnitzer-Polokoff, R., Spencer, T. A., and Sinensky, M.,** Sterol-independent regulation of 3-hydroxy-3-methylglutaryl-CoA reductase by mevalonate in chinese hamster ovary cells, *J. Biol. Chem.*, 264, 11044–11052, 1989.
20. **Shapiro, D. J. and Rodwell, V. W.,** Regulation of hepatic 3-hydroxy-3-methylglutaryl coenzyme A reductase and cholesterol synthesis, *J. Biol. Chem.*, 246, 3210–3216, 1971.
21. **Dugan, R. E. and Porter, J. W.,** Hormonal regulation of cholesterol synthesis, in *Biochemical Actions of Hormones*, Vol. 4, Litwack, G., Ed., Academic Press, New York, 1977, 197–247.
22. **Higgins, M., Kawachi, R., and Rudney, H.,** The mechanism of diurnal variation of hepatic HMG-CoA reductase activity in the rat, *Biochem. Biophys. Res. Commun.*, 45, 138–144, 1971.
23. **Hardie, D. G., Carling, D., and Sim, A. T. R.,** The AMP-activated protein kinase: a multisubstrate regulator of lipid metabolism, *Trends Biochem. Sci.*, 14, 20–23, 1989.
24. **Nordstrom, J. L., Rodwell, V. W., and Mitschelen, J. J.,** Interconversion of active and inactive forms of rat liver hydroxymethylglutaryl-CoA reductase, *J. Biol. Chem.*, 252, 8929–8934, 1977.
25. **Eason, R. A. and Zammit, V. A.,** Diurnal changes in the fraction of 3-hydroxy-3-methylglutaryl-CoA reductase in the active form in rat liver microsomal fractions, *Biochem. J.*, 220, 739–745, 1984.
26. **Preiss, B., Ed.,** *Regulation of HMG-CoA Reductase*, Academic Press, New York, 1985, pp 330.
27. **Easom, R. A. and Zammit, V. A.,** Acute effects of starvation and treatment of rats with anti-insulin serum, glucagon, and catecholamines on the state of phosphorylation of hepatic 3-hydroxy-3-methylglutaryl-CoA reductase *in vivo*, *Biochem. J.*, 241, 183–188, 1987.
28. **Parker, R. A., Miller, S. J., and Gibson, D. M.,** Phosphorylation of native 97-kDa 3-hydroxy-3-methylglutaryl-coenzyme A reductase from rat liver, *J. Biol. Chem.*, 264, 4877–4887, 1989.
29. **Chun, K. T., Bar-Nun, S., and Simoni, R. D.,** The regulated degradation of 3-hydroxy-3-methylglutaryl-CoA reductase requires a short lived-protein and occurs in the endoplasmic reticulum, *J. Biol. Chem.* 265, 22004–22010, 1990.
30. **Osborne, T. F., Gil, G., Goldstein, J. L., and Brown, M. S.,** Operator constitutive mutation of 3-hydroxy-3-methylglutaryl coenzyme A reductase promoter abolishes protein binding to sterol regulatory element, *J. Biol. Chem.*, 263, 3380–3387, 1988.
31. **Goldstein, J. L. and Brown, M. S.,** Regulation of the mevalonate pathway, *Nature*, 343, 425–430, 1990.
32. **Nakanishi, M., Goldstein, J. L., and Brown, M. S.,** Multivalent control of 3-hydroxy-3-methylglutaryl coenzyme A reductase, *J. Biol. Chem.*, 263, 8929–8937, 1988.
33. **Auwerx, J. H., Chait, A., and Deeb, S. S.,** Regulation of the low density lipoprotein receptor and hydroxymethylglutaryl coenzyme A reductase genes by protein kinase C and a putative negative regulatory protein, *Proc. Natl. Acad. Sci. U.S.A.*, 86, 1133–1137, 1989.
34. **Smith, L. L.,** Cholesterol autooxidation 1981–1986, *Chem. Phys. Lipids*, 44, 87–125, 1987.
35. **Saveier, S. E., Kandutsch, A. A., Gayen, A. K., Swahn, D. K., and Spencer, T. A.,** Oxysterol regulators of 3-hydroxy-3-methylglutaryl-CoA reductase in liver, *J. Biol. Chem.*, 264, 6863–6869, 1989.
36. **Brown, M. S. and Goldstein, J. L.,** A receptor mediated pathway for cholesterol homeostasis, *Science*, 232, 34–47, 1986.
37. **Nestel, P. J.,** The regulation of lipoprotein metabolism, in *Plasma Lipoproteins*, Gotto, A. M., Jr., Ed., Elsevier, Amsterdam, 1987, 153–182.
38. **Jonas, A.,** Lecithin cholesterol acyltransferase in *Plasma Lipoproteins*, Gotto, A. M., Jr., Ed., Elsevier, Amsterdam, 1987, 299–333.
39. **Brown, M. S. and Goldstein, J. L.,** The LDL receptor and HMG-CoA reductase — two membrane molecules that regulate cholesterol homeostasis, *Curr. Topics Cell Regul.*, 26, 3–15, 1985.
40. **Thompson, G. A., Jr., Bambery, R. J., and Nozawa, Y.,** Further studies of the lipid composition and biochemical properties of *Tetrahymena pyriformis* membrane systems, *Biochemistry*, 10, 441–447, 1971.
41. **Conner, R. L., Mallory, F. B., Landrey, J. R., Ferguson, K. A., Kaneshiro, E. S., and Ray, E.,** Ergosterol replacement of tetrahymanol in *Tetrahymena* membranes, *Biochem. Biophys. Res. Commun.*, 44, 995–1000, 1971.
42. **Nozawa, Y., Fukushima, H., and Iida, H.,** Studies of *Tetrahymena* membranes. Modification of surface membrane lipids by replacement of tetrahymanol by exogenous ergosterol in *Tetrahymena pyriformis*, *Biochim. Biophys. Acta*, 406, 248–263, 1975.

43. **Beedle, A. A., Munday, K. A., and Wilton, D. C.,** Studies on the biosynthesis of tetrahymanol in *Tetrahymena pyriformis.* The mechanism of inhibition by cholesterol, *Biochem. J.,* 142, 57, 1974.
44. **Kasai, R., Sekiya, T., Okano, Y., Nagao, S., Ohki, K., Ohnishi, S., and Nozawa, Y.,** Adaption of membrane lipids to temperature changes in *Tetrahymena.* Regulation of acyl chain composition of membrane phospholipids in ergosterol-replaced cells, *Maku (Membrane),* 2, 301–312, 1977.
45. **Bjorkhem, I.,** Mechanism of bile acid biosynthesis in mammalian liver, in *New Comprehensive Biochemistry,* Vol. 12, Danielsson, H. and Sjovan, J., Eds., Elsevier, Amsterdam, 1985, 231–278.
46. **Boyd, G. S., Brown, M. J. G., Hattersley, N. G., Lawson, M. E., and Suckling, K. E.,** Rat liver cholesterol 7α-hydroxylase, in *Advances in Bile Acid Research. III. Bile Acid Meeting, Freiburg i. Br.,* Mantern, S., Hackenschmidt, J., Back, P., and Gerok, W., Eds., Fik. Schattaur Verlag, Stuttgart, 1974, 45–51.
47. **Myant, N. B. and Mitropoulos, K. A.,** Cholesterol 7α-hydroxylase, *J. Lipid Res.,* 18, 135–153, 1977.
48. **Shefer, S., Hauser, S., Lapar, V., and Mosbach, E. H.,** Regulatory effects of sterols and bile acids on hepatic 3-hydroxy-3-methylglutaryl CoA reductase and cholesterol 7α–hydroxylase in the rat, *J. Lipid Res.,*14, 573–580, 1973.
49. **Thompson, J. C. and Vars, H. M.,** Biliary excretion of cholic acid and cholesterol in hyper-, hypo-, and euthyroid rats, *Proc. Soc. Exp. Bio. Med.,* 83, 246–248, 1953.
50. **Eriksson, S.,** Biliary excretion of bile acids and cholesterol in bile fistula rats. Bile acids and steroids, *Proc. Soc. Exp. Biol. Med.,* 94, 578–582, 1957.
51. **Einarsson, K. and Johansson, G.,** Effect of actinomycin D and puromycin on the conversion of cholesterol into bile acids in bile fistula rats. Bile acids and steroids 206, *FEBS Lett.,* 1, 219–222, 1968.
52. **Renson, J. J., Van Cantfort, J., Robaye, B., and Gielen, J.,** Mesures de la demi-vie la cholésterol 7α-hydroxylase, *Arch. Int. Physiol.,* 77, 972–973, 1969.
53. **Pandak, W. M., Li, Y. C., Chiang. J. Y. L., Studer, E. J., Gurley, E. C., Heuman, D. M., Vlahcevic, Z. R., and Hyleman, P. B.,** Regulation of cholesterol 7α-hydroxylase mRNA and transcriptional activity by taurocholate and cholesterol in the chronic bilary diverted rat, *J. Biol. Chem.,* 266, 3416–3421, 1991.

Chapter 6

THE REGULATION OF NONSTEROID ISOPRENOID METABOLISM

I. INTRODUCTION

In addition to the nearly ubiquitous sterols, many other isoprenoid compounds are found as structural elements of membranes. A large percentage of these serve some type of highly specialized function and occur in such low concentrations that they have little direct effect upon the physical properties of the membrane. Included in this category are vitamin A, ubiquinone, the tocopherols, etc.

I have rather arbitrarily chosen to bypass the above compounds and to discuss instead only certain isoprenoids and isoprenoid derivatives that are relatively important as bulk components of membranes. Of these, three will be considered — chlorophyll and carotenoids of plants, and the phytanyl ether lipids of certain bacteria. The placements of these compounds in the general isoprenoid biosynthetic pathway is illustrated in Figure 1.

Regulation of isoprenoid biosynthesis at the HMG-CoA reductase step might be the first possibility to consider, in view of this enzyme's importance in controlling animal isoprenoid metabolism. However, very little is known regarding HMG-CoA reductase in bacteria or in chlorophyll and carotenoid producing plants. There is even no general agreement as to where the plant enzyme is localized, with evidence for its presence in microsomes, plastids and mitochondria all being more or less equivocal.[1] In plants more so than in animals there can be massive and sudden diversions of isoprenoid precursors into specific products, e.g., chlorophyll during leaf greening or carotenoids during fruit ripening, while maintaining the synthesis of other products at a steady rate. Consequently, much attention has been directed towards understanding regulatory steps at later stages of biosynthesis, near some of the many branch points. The examples described below illustrate the current status of this research.

II. CHLOROPHYLL

Chlorophyll occurs most commonly in two forms, chlorophyll a and chlorophyll b, differing in the substituents of their porphyrin ring (Figure 2). In the chloroplast grana of most green plants, they are found together in a molar ratio of 2:1 (a:b).

In a typical higher plant, such as spinach, chlorophyll a + b amount to 30 to 40 mol% of the total chloroplast lipids.[2] Electrophoresis of disrupted chloroplasts suggests that most chlorophyll a and b molecules may be quite firmly associated with integral chloroplast proteins *in vivo*.[3] Chlorophyll biosynthesis has been thoroughly discussed by many authors (e.g., Rudiger and Schoch[4]) and is briefly summarized in Figure 3.

The first specific precursor for the porphyrin ring of chlorophyll is 5-aminolaevulinic acid, which can be formed in animals and plants by two distinct pathways, the so-called Shemin pathway (A) and the more recently discovered C_5 pathway, shown here with alternative routes (B)[4] (Figure 4). Condensation reactions lead to the formation of protoporphyrin IX, which marks the branch point of the haem and chlorophyll biosynthetic pathways. Further modification leads to the key intermediate protochlorophyllide, whose reduction to chlorophyllide a is dependent upon light. Finally, chlorophyll synthetase transfers the phytol group from phytol diphosphate to chlorophyllide a to yield the final product, chlorophyll a. The enzymes for all these biosynthetic steps are found in various compartments within the chloroplast and probably in other plastid types as well.

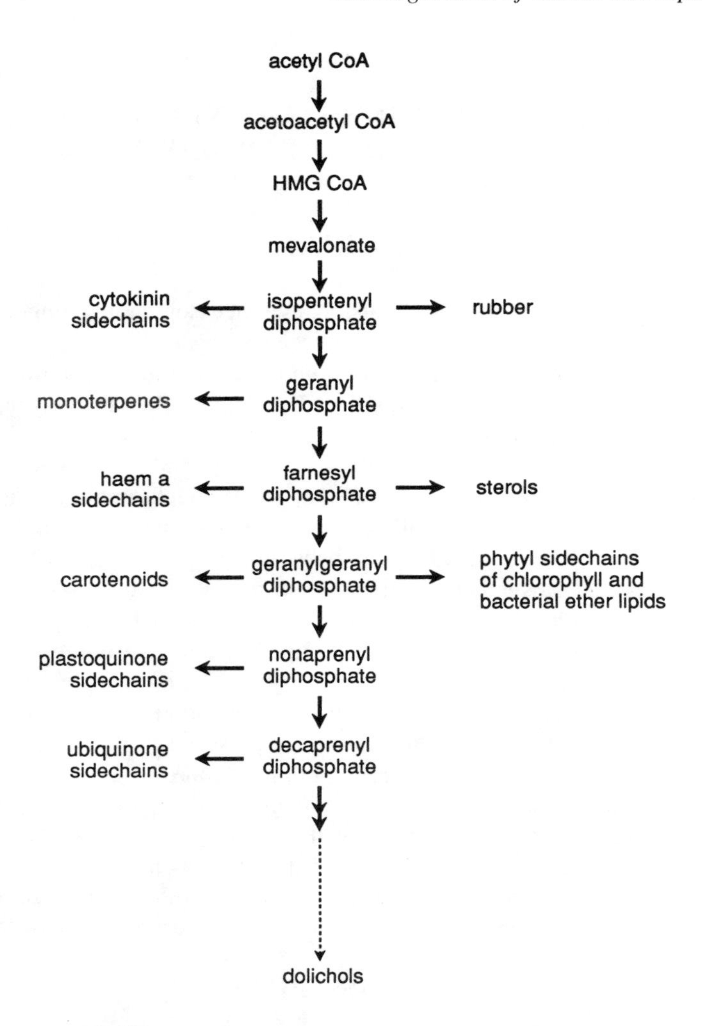

FIGURE 1. Biosynthetic interrelationships of isoprenoid classes.

Although several enzymes in the chlorophyll biosynthetic pathway appear subject to metabolic regulation,[4] the principal regulation is imposed at the protochlorophyllide reduction step. Transfer of a seedling from darkness to light permits this reduction to proceed, resulting in massive chlorophyll synthesis.[1] However, under normal conditions, protochlorophyllide does not accumulate in the dark because 5-aminolaevulinic acid synthesis is inhibited, due to feedback inhibition by either protochlorophyllide itself or by haem.[4] By mechanisms not clearly understood, the supply of phytol diphosphate is regulated to match the demand for chlorophyll synthesis. Some regulation of HMG CoA reductase, mevalonate kinase, and mevalonate 5-phosphate kinase has been detected in greening leaves,[1] but it is not clear that these changes are sufficient to account for the sharp fluctuations in phytol diphosphate synthesis.

The regulation of chlorophyll synthesis has been the subject of intensive research for many years. This interest is well deserved since chlorophyll formation is itself the rate limiting reaction in the development of the entire chloroplast. The role of chlorophyll in chloroplast development is best observed during the "greening" of dark-grown angiosperm seedlings (gymnosperms, most algae, and some ferns can form chlorophyll in the dark). When removed

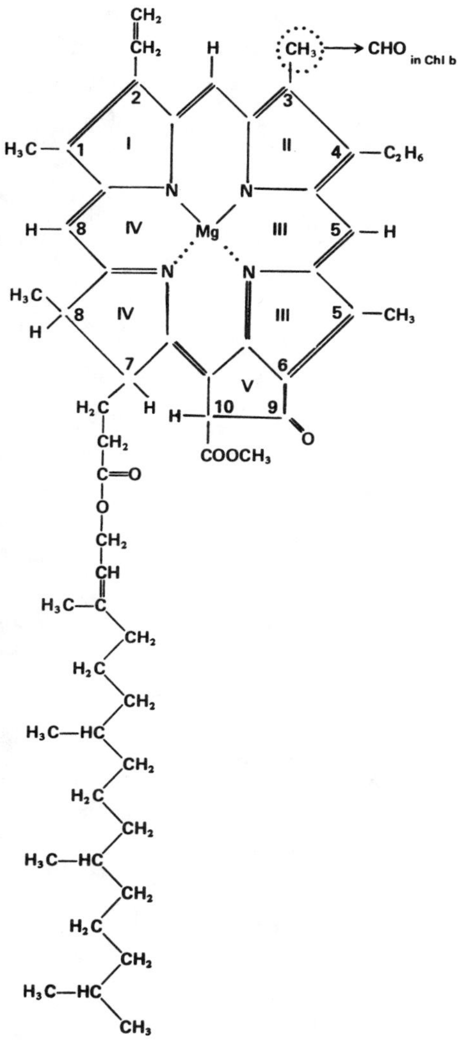

FIGURE 2. Chlorophyll a. In chlorophyll b, the methyl group at position 3 is replaced by an aldehyde group.

from darkness, these seedlings contain no functional chloroplasts, but do have numerous rudimentary chloroplasts, known as proplastids. The light requirement for chloroplast development is outlined in Figure 5.

Etiolated leaves kept in darkness accumulate protochlorophyllide a. Upon illumination, this is converted to chlorophlyllide a, which is then esterified with phytol, yielding the final product, chlorophyll a. During the initial period of light, chlorophyll formation takes place rather slowly, but it soon increases to a constant, rapid rate.

In etiolated plants or greening plants returned to darkness, the only chlorophyll precursor known to accumulate is protochlorophyllide a. There is considerable evidence (summarized by Bogorad[5]) that protochlorophyllide a exerts control on the biosynthetic pathway at the level of δ-aminolevulinic acid synthesis. For example, a tenfold increase in protochlorophyllide a was achieved by administering δ-aminolevulinic acid to etiolated leaves in darkness.[7] Various observations indicate that this inhibition may involve the action of protochlorophyll (and

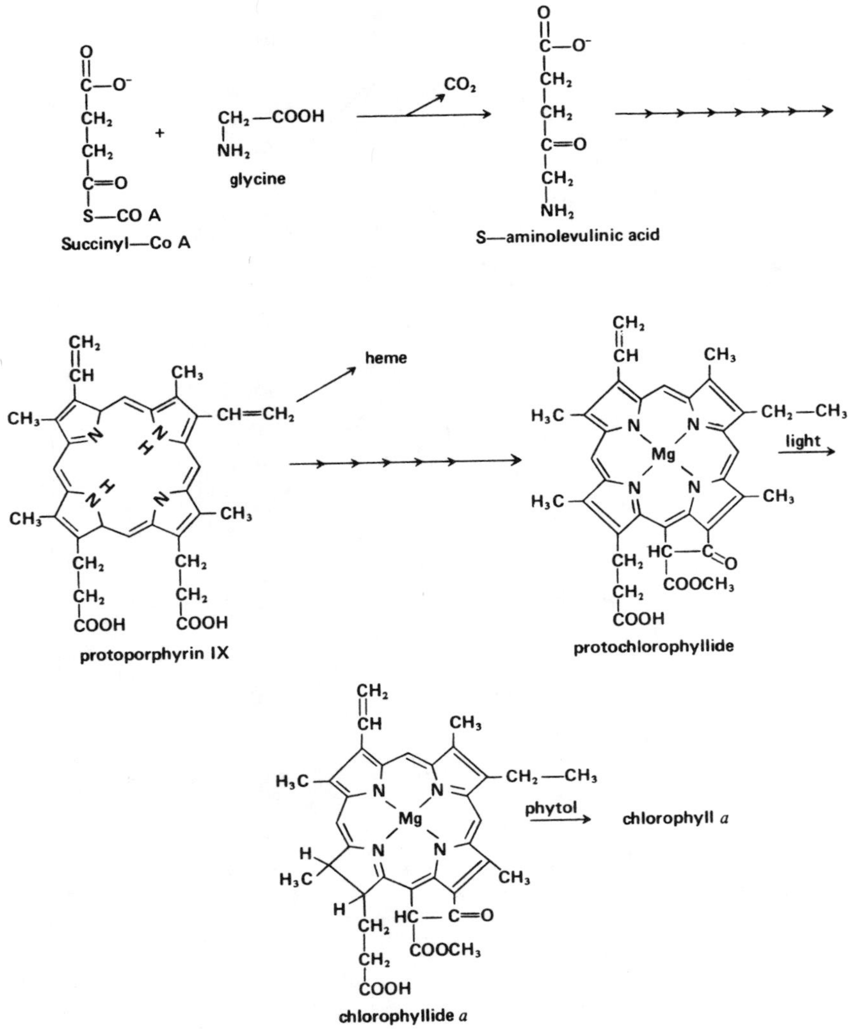

FIGURE 3. Summary of the chlorophyll a biosynthetic pathway.

possibly chlorophyllide) on a regulatory protein controlling δ-aminolevulinic acids synthase, as opposed to direct action on the enzyme itself. Some of the evidence has been interpreted as supporting a role for the regulatory protein in blocking continued replacement of the rapidly turning over δ-aminolevulinic acid synthase.[8]

Thus, two points of inhibition are apparent. First, the absence of light prevents the conversion of protochlorophyllide to chlorophyll. Second, the increased level of protochlorophyllide blocks, by feedback inhibition, further flux through the pathway. The lifting of inhibition at the first site depletes the backlog of protochlorophyllide, releasing the second block. Therefore, it seems that light is directly and indirectly the prime factor regulating chlorophyll synthesis. The fact that illumination with red light decreases the lag in rapid chlorophyll synthesis while far red light reverses the decrease implicate phytochrome.[9,10] Some functional product of phytochrome action may be a translocatable compound since irradiation of the embryonic axis of dark-grown bean seedlings speeds chlorophyll synthesis in the leaves.[11]

CHLOROPHYLLS

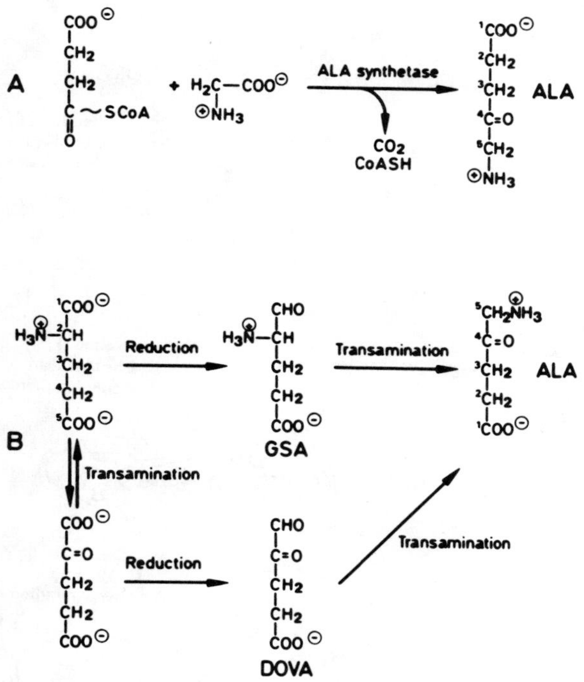

FIGURE 4. Formation of 5-aminolevulinic acid (ALA) by (A) the Shemin pathway and (B) the C_5 pathway with alternative routes. (From Rüdiger, W. and Schock, S., *Plant Pigments*, Goodwin, T. W., Ed., Academic Press, New York, 1988, 1–59. With permission.)

Any scheme to explain the metabolic regulation of chlorophyll synthesis must take into account the fact that δ-aminolevulinic acid and other compounds in the porphyrin biosynthetic pathway, as far as protoporphyrin IX, are common precursors of both chlorophyll and heme. Since heme production does not fluctuate in etiolated and greening plants to the same marked extent noted for chlorophyll, there must be a distinct regulatory mechanism permitting independent control. Various possibilities have been considered.[12,13] The rapid incorporation of added ^{14}C-δ-aminolevulinic acid into both chlorophyll and heme rules out the possibility that one of the products is formed in a metabolically isolated compartment. A more likely avenue of control is the inhibition of δ-aminolevulinic acid synthesis by protoheme. Light overcomes this protoheme inhibition, perhaps acting through the protochlorophyllide conversion reaction. In this manner, both the needs for heme-containing proteins and chlorophyll can be regulated.

During the senescence of fescue leaves, chlorophyll is degraded to a polar green pigment by dephytylation.[14] Much of the phytol that is released is esterified, probably to long chain fatty acids such as linolenic acid.[15] An increase in the α-tocopherol content of leaves during the period of senescence characterized by a decline in the chlorophyll content has led to the suggestion that phytol may be a precursor of α-tocopherol.[16]

III. CAROTENOIDS

β-Carotene is the principal carotene of most plant tissues and is an essential component of the light transferring machinery of chloroplasts. In spinach, it comprises 2 mol% of the

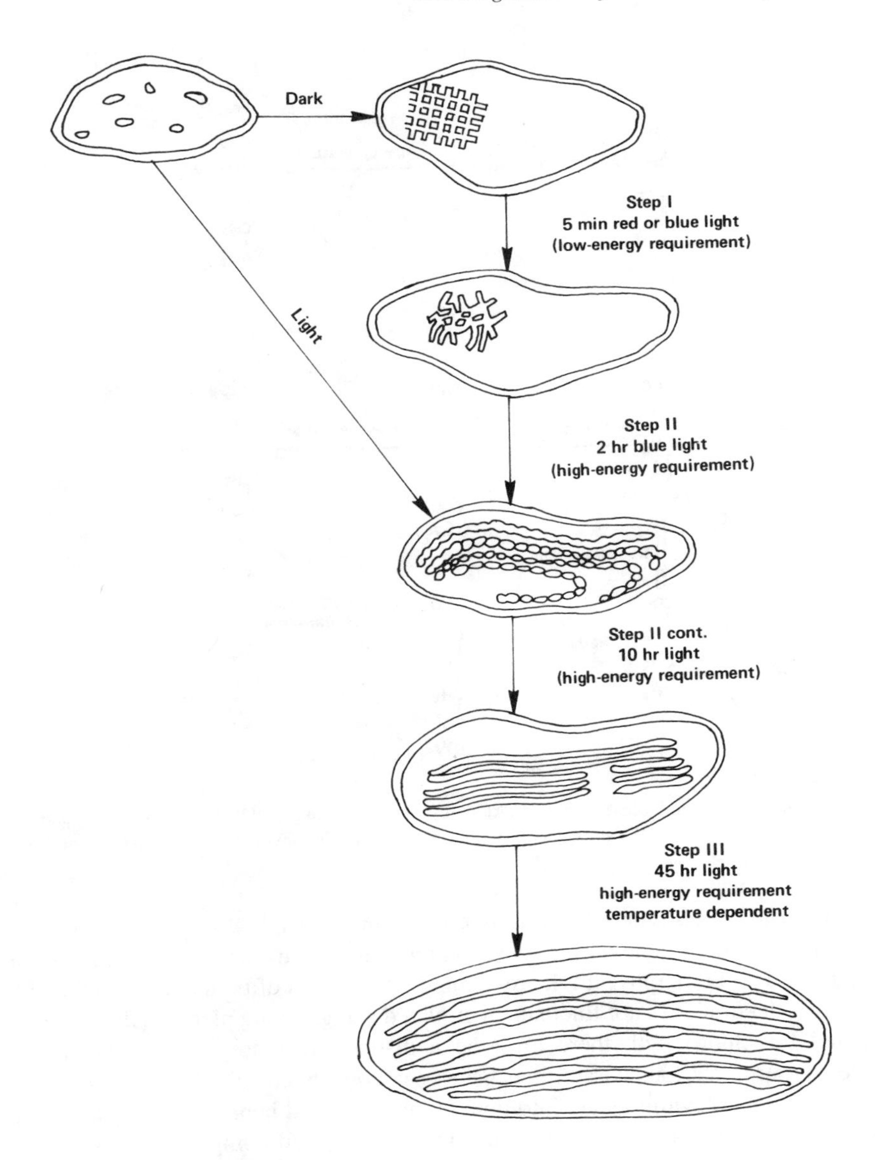

FIGURE 5. Membrane changes during light-induced chloroplast development. For a discussion of energy requirements, see reference (From Devlin, R. M. and Barker, A. V., *Photosynthesis,* Van Nostrand Reinhold, New York, 1971, 304. With permission.)

chloroplast lipids and is usually accompanied by its oxygenated counterparts, the xanthophylls, which may account for another 6 to 7 mol% of the total lipids.[17]

Carotenes make up a much higher proportion of the lipids in certain pigmented plant tissues, such as the tomato or red pepper fruits. Here the major pigment, lycopene, is eventually shed from degenerating membranes of the ripening fruit and transformed into intracellular globules or crystals.[18]

Carotenes are synthesized only by plants. As is the case with all isoprenoids, they are formed from the isoprene unit, isopentenyl pyrophosphate, whose biosynthesis was discussed in the previous chapter (see p. 120). The reactions governing its condensation to yield the initial C_{40} isoprenoid phytoene have been reviewed by Britton [19] and by Nes and McKean.[20]

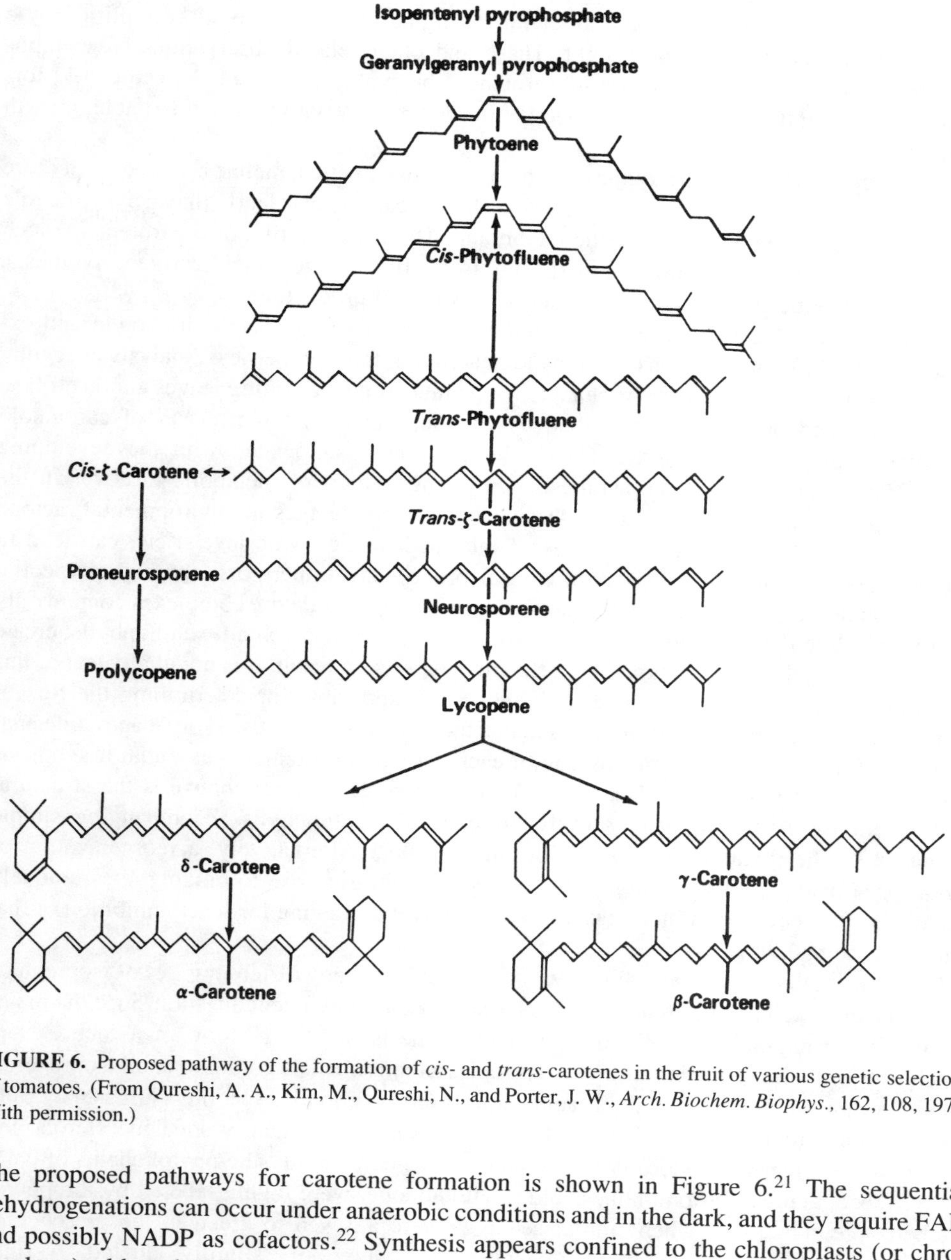

FIGURE 6. Proposed pathway of the formation of *cis-* and *trans*-carotenes in the fruit of various genetic selections of tomatoes. (From Qureshi, A. A., Kim, M., Qureshi, N., and Porter, J. W., *Arch. Biochem. Biophys.*, 162, 108, 1974. With permission.)

The proposed pathways for carotene formation is shown in Figure 6.[21] The sequential dehydrogenations can occur under anaerobic conditions and in the dark, and they require FAD and possibly NADP as cofactors.[22] Synthesis appears confined to the chloroplasts (or chromoplasts), although at least some of the necessary mevalonic acid may be imported from the cytosol.[19]

The rate of carotene biosynthesis is subject to sudden and very dramatic fluctuations. Whereas an unripe tomato fruit contains none of the red lycopene pigment, the onset of ripening can lead to the synthesis of more than 1 mg of lycopene per day. The genetics of carotene biosynthesis in fungi and higher plants is fairly well understood due to the availability of a variety of mutants lacking particular pigments. However, very little is yet known concerning the biochemical mechanism controlling their synthesis. The lag in carotenoid synthesis following the illumination of etiolated maize seedlings could be abolished by a brief pretreatment of the seedlings with red light. This effect could be nullified by a subsequent treatment with far-red light. These and other related observations have implicated phytochrome in the regulation of carotenoid biosynthesis.[23] In algae, carotenoid formation is frequently stimulated by a low nitrogen supply or other unfavorable growth conditions.[19]

The synthesis of carotenoids of the chloroplast photosynthetic lamellae is a process of more relevance to the subject of this monograph since here the pigments clearly play a dynamic role in the light-harvesting apparatus of the membrane. The concentration of β-carotene is much reduced in dark-grown seedlings, but exposure to the light triggers rapid carotene synthesis, displaying a kinetic pattern similar to that of chlorophyll production.

Because most of the enzymes of carotenoid biosynthesis are membrane-bound and extremely labile, little progress has been made in studying their properties. Analysis of regulation by charting *in vivo* patterns of change in ripening fruit or greening leaves can itself be a very complicated process. As described by Britton,[19] at least three phases of carotenoid biosynthesis may be encountered: (1) the bulk synthesis associated with the developing chloroplast membranes, (2) the synthesis continuing as part of metabolic turnover in the mature organelle, and (3) renewed synthesis triggered by changes in environmental factors, e.g., light intensity. Differing pool sizes of intermediates, enzyme levels, etc. can lead to contradictory interpretations regarding control processes. Furthermore, the highly specific localization of the carotenoids, such that the photosystem I and II core complexes contain only β-carotene while the associated light harvesting complexes also contain xanthophylls, create difficulties in establishing metabolic interrelationships. As a result, it is not at all clear at this time which step(s) of the biosynthetic sequence is responsible for determining the flux of material through the pathway. It would seem that the availability of satisfactory macromolecular acceptor sites for the specific disposition of each carotenoid product is essential if synthesis of that product is to continue. The most obvious example of strict control is the structural requirement for fixed proportions of β-carotene and xanthophylls accompanying simple expansion of the thylakoid membrane during greening. Alternatively, a reorganization of thylakoid structure caused by high light intensity might give rise to binding sites uniquely suited to sequester other carotenoids, thereby selectively relieving feedback inhibition of that branch of the pathway.

Goodwin[24] attempted to explain how formation of the many different types of isoprenoids found in a plant cell can be controlled. This is an especially pertinent question since the major known site for regulation of animal isoprenoid synthesis is at the HMG-CoA reductase step, which is prior to the formation of mevalonic acid, an intermediate common to all isoprenoids. Goodwin showed that ^{14}C-mevalonic acid administered to etiolated seedlings just as they were first exposed to light labeled only the extraplastidic isoprenoids, such as squalene, sterols, and ubiquinone. Conversely, chloroplast isoprenoids, i.e., carotenoids, the phytol chains of chlorophyll, tocopherols, plastoquinones, and phylloquinone, were highly labeled by the photosynthetic products made when similar seedings were allowed to green in the presence of $^{14}CO_2$. In this case, the cytoplasmic isoprenoids were only slightly labeled. A tight compartmentation from the synthesis of mevalonic acid onwards was suggested as a mechanism for achieving control. It is pertinent to recall at this point that two separate pools of HMG-CoA have been indicated in animal cells, one in the cytosol and one in mitochondria.[25]

β-carotene, a major end product of carotenoid biosynthesis in plants, is of extreme importance to the animals that ingest it in their diet. It is symmetrically cleaved to yield two molecules of retinol (vitamin A), known through the pioneering work of Wald[26] to form retinal, the light absorbing moiety of rhodopsin. More recently, it has been recognized that vitamin A supports a wide variety of nonvisual functions, particularly those involved in cellular differentiation. To carry out these systemic functions, vitamin A is converted to all *trans*-retinoic acid, a product that is present at low levels in a variety of cell types.[27] Retinoic

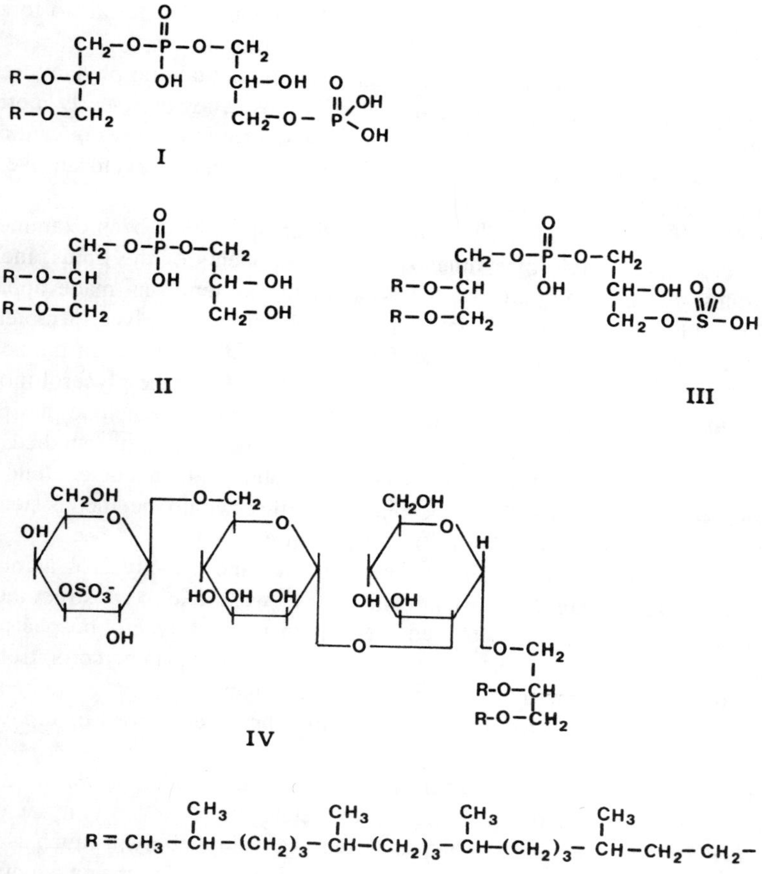

FIGURE 7. Structures of the major polar lipids in *Halobacterium cutirubrum:* I. Phosphatidylglycerophosphate (PGP), II. Phosphatidylglycerol (PG), III. Phosphatidylglycerosulfate (PGS), IV. Glycolipidsulfate (GLS). In all compounds, the alkyl group, R, is 3R,7R,11R, 15-tetramethylhexadecyl and is linked to glycerol by ether bonds.

acid arises through sequential action by retinol and retinal dehydrogenase. These retinoids are unstable in their free form and are protected *in vivo* either by existing as fatty acid esters, the storage form of retinol, or as complexes with specific proteins. Thus cellular retinol binding proteins and retinoic acid binding proteins have been characterized, and the latter is probably involved in the expression of retinoic acid's biological activities.[28]

IV. LIPIDS OF ARCHAEBACTERIA

Some years ago, an unusual class of phospholipids and glycolipids was discovered in the membranes of the bacterium, *Halobacterium cutirubrum*. This organism is normally found in extremely saline environments and grows optimally in water containing 25 to 30% salt.[29] Characterization of the lipids of this and, more recently, the related species, *H. halobium*,[29] revealed that essentially all hydrocarbon side chains of the membrane structural lipids were phytanyl groups linked to the glycerol backbone by ether linkages. The basic nature of these lipids is shown in Figure 7, Structure I.

A wide variety of additional glycerol isopranyl ether lipids have been found in other archaebacteria, including thermoacidophiles and methanogens.[31] Among the unusual lipids represented in these organisms are the tetraethers (Figure 7, Structure 2), in which two diether moieties of the type shown in Structure 1 are coupled together tail to tail to form a single tetraether molecule capable of spanning the membrane lipid bilayer.

Since moderately halophilic and nonhalophalic bacteria contain no phytanyl ether lipids, one can only speculate about the evolution of these lipids. They are clearly more stable to chemical and enzymatic hydrolysis and to peroxidation than are the more conventional lipids. This stability undoubtedly contributes to the ability of the archaebacteria to survive such great extremes of temperatures, pH, and salt.

Biosynthetic pathways for the *H. cutirubrum* diether lipids have been examined in some depth.[29] The cells are capable of synthesizing all components of the lipids, including the phytanyl groups. A novel property of the biosynthetic system was made apparent from incorporation studies involving specifically labeled glycerol. The glycerol moiety retained 100% of its ^{3}H when 1(3)-^{3}H-glycerol was fed, but none of the ^{3}H when the administered glycerol was labeled at the C-2 position.[29] It was concluded that the glycerol moiety of the ether lipids must have been generated entirely through a dihydroxyacetone phosphate intermediate. However, in this instance, unlike the case of the thoroughly studied monoalkyl glyceryl ether synthesis (see p. 96), a dihydroxyacetone phosphate precursor could not be free to equilibrate with glyceraldehyde 3-phosphate, for that would be inconsistent with the observed complete retention of ^{3}H in the 1(3) position of glycerol.

Through the use of ^{32}P-phosphate, ^{14}C-mevalonate, and ^{14}C-glycerol administered as pulse-labels to *H. cutirubrum*, Moldoveanu and Kates[32] were able to construct the following tentative scheme for the diphytanyl ether analog of phosphatidylglycerophosphate (Figure 8). The alkylation step apparently involves dihydroxyacetone rather than glycerol itself. Based on the study of intermediates, the initial ether-linked side chains are believed to be unsaturated and are subsequently reduced to the phytanyl form. The other classes of diphytanyl ether glycerolipids are synthesized in an analogous fashion.

Because traces of long-chain fatty acids had been detected in *H. cutirubrum*, the ability of the organism to synthesize fatty acids from ^{14}C-acetate was tested.[33] Fatty acids could be formed *in vitro*, but the fatty acid synthetase system was inhibited by as much as 77% by the presence of even moderate levels of NACl or KCl. In sharp contrast, the enzymatic pathways governing the phytanyl group biosynthesis are not inhibited by 4 M salt.

Nishihara et al.[34] have examined the biosynthesis of tetraether lipids in *Methanobacterium thermoauto trophicum*, a methanogen which also contains diethers. On the basis of pulse-chase experiments conducted with [^{32}P] inorganic phosphate, it was concluded that the different tetraethers present were formed by a condensation of the isopranyl sidechains of two intact diether lipids which, if they contained polar groups, always contained different ones (Figure 9).

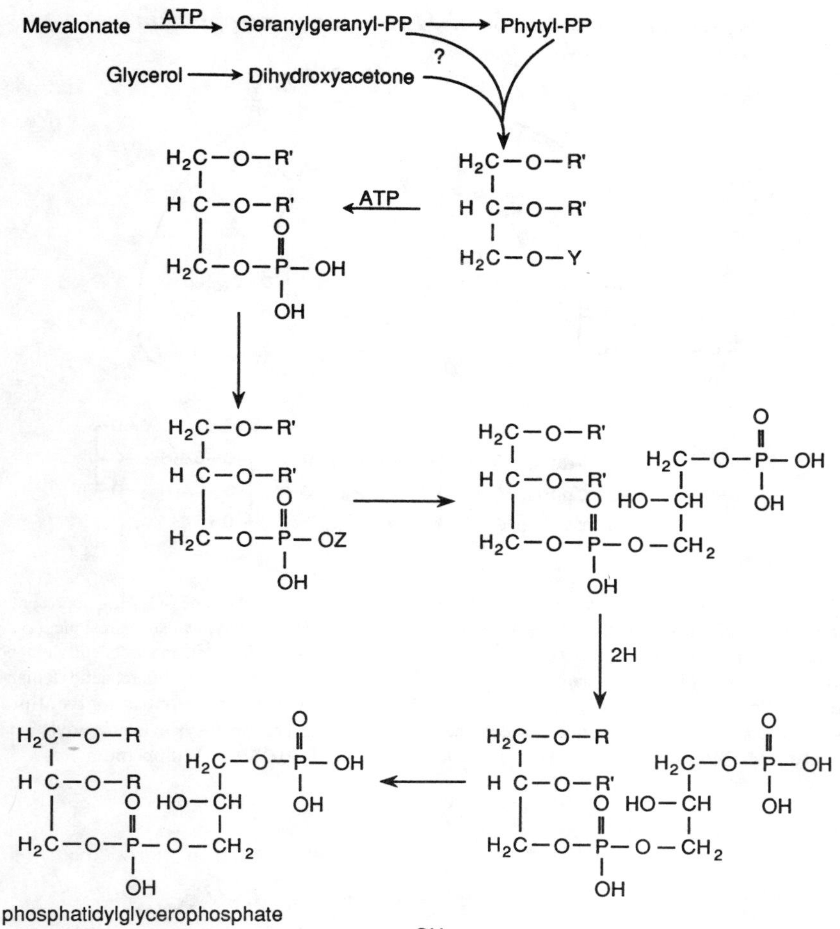

Mevalonate $\xrightarrow{ATP}$ Geranylgeranyl-PP → Phytyl-PP

Glycerol → Dihydroxyacetone ?

$H_2C{-}O{-}R'$ / $HC{-}O{-}R'$ / $H_2C{-}O{-}Y$ $\xrightarrow{ATP}$ $H_2C{-}O{-}R'$ / $HC{-}O{-}R'$ / $H_2C{-}O{-}P(O)(OH){-}OH$

↓

$H_2C{-}O{-}R'$ / $HC{-}O{-}R'$ / $H_2C{-}O{-}P(O)(OH){-}OZ$ → $H_2C{-}O{-}R'$ / $HC{-}O{-}R'$ / $H_2C{-}O{-}P(O)(OH){-}O{-}CH_2{-}CH(OH){-}CH_2{-}O{-}P(O)(OH){-}OH$

↓ 2H

$H_2C{-}O{-}R$ / $HC{-}O{-}R'$ / $H_2C{-}O{-}P(O)(OH){-}O{-}CH_2{-}CH(OH){-}CH_2{-}O{-}P(O)(OH){-}OH$ → $H_2C{-}O{-}R$ / $HC{-}O{-}R$ / $H_2C{-}O{-}P(O)(OH){-}O{-}CH_2{-}CH(OH){-}CH_2{-}O{-}P(O)(OH){-}OH$

phosphatidylglycerophosphate

R' = phytyl, $CH_3[CH(CH_3)(CH_2)_3]_3{-}C(CH_3){=}CH{-}CH_2{-}$, or

geranylgeranyl, $CH_3[C(CH_3){=}CH(CH_2)_2]_3{-}C(CH_3){=}CH{-}CH_2{-}$

R = phytanyl, $CH_3[CH(CH_3)(CH_2)_3]_3{-}CH(CH_3)CH_2{-}CH_2{-}$

FIGURE 8.

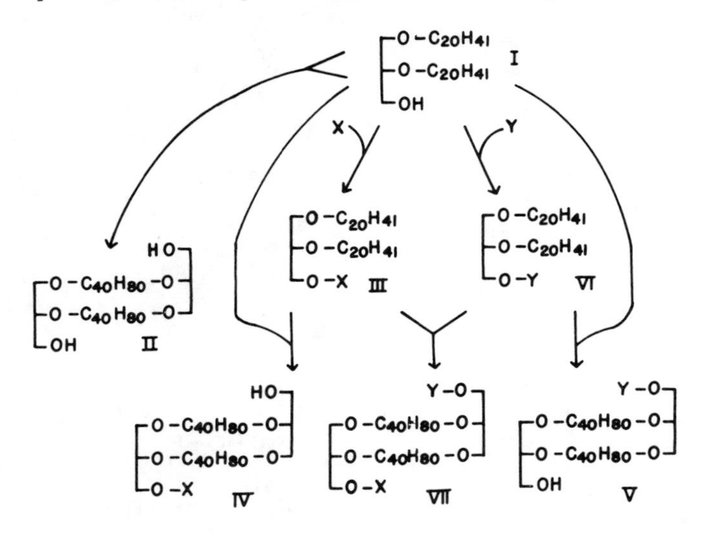

FIGURE 9. A proposed pathway of biosynthesis of tetraether polar lipids. $-C_{20}H_{41}$ and $-C_{40}H_{80}-$ represent phytanyl and biphytanediyl groups, respectively. (X) Water-soluble phosphoric ester moiety; (Y) sugar residue. (I) Archaeol; (II) caldarchaeol; (III) diether phospholipids, e.g., archaetidylserine, archaetidylinositol, and archaetidylethanolamine; (IV) tetraether phospholipids, e.g., caldarchaetidylserine, caldarchaetidylinositol, and caldarchaetidylethanolamine; (V) gentiobiosylcaldarchaeol; (VI) gentiobiosylarchaeol; (VII) Tetraether phosphoglycolipids, e.g., gentiobiosylcaldarchaetidylserine, gentiobiosylcaldarchaetidyl-inositol, and gentiobiosylcaldarchaetidylethanolamine. (From Nishihara, M., Morii, H., and Yosuke, K., *Biochemistry,* 28, 95–102, 1989. With permission.)

REFERENCES

1. **Gray, J. C.**, Control of isoprenoid biosynthesis in higher plants, *Adv. Bot. Res.*, 14, 25–91, 1987.
2. **Park, R. B.,** The chloroplast, in *Plant Biochemistry,* 3rd ed., Bonner, J. and Varner, J. E., Eds., Academic Press, New York, 1976, 115–145.
3. **Markwell, J. P., Thor, J. P., and Boggs, R. T.,** Higher plant chloroplasts: evidence that all the chlorophyll exists as chlorophyll-protein complexes, *Proc. Natl. Acad. Sci. U.S.A.*, 76, 1233–1235.
4. **Rüdiger, W. and Schock, S.,** Chlorophylls, in *Plant Pigments*, Goodwin, T. W. , Ed., Academic Press, New York, 1988, 1–59.
5. **Bogorad, L.,** Chlorophyll biosynthesis, in *Chemistry and Biochemistry of Plant Pigments,* Vol. 1, Goodwin, T. W., Ed., Academic Press, London, 1976, 64–148.
6. **Devlin, R. M. and Barker, A. V.,** *Photosynthesis,* Van Nostrand Reinhold, New York, 1971.
7. **Granick, S.,** Magnesium porphyrins formed by barley seedlings treated with δ-aminolevulinic acid, *Plant Physiol.*, 34, xviii, 1959.
8. **Fluhr, R., Havel, E., Klein, S., and Meller, E.,** Control of δ-aminolevulinic acid and chlorophyll accumulation in greening maize upon light-dark transitions, *Plant Physiol.,* 56, 497–501, 1975.
9. **Satter, R. L. and Galston, A. W.,** The physiological function of phytochrome, in *Chemistry and Biochemistry of Plant Pigments,* Vol 1., Goodwin, T. W., Ed., Academic Press, London, 1976, 680–735.
10. **Oelze-Karow, H. and Mohr, H.,** Control of chlorophyll b biosynthesis by phytochrome, *Photochem. Photobiol.,* 27, 189–193, 1978.
11. **De Greef, J. A. and Caubergs, R.,** Studies on greening of etiolated seedlings. II. Leaf greening by phytochrome action in the embryonic axis, *Physiol. Plant.,* 28, 71–76, 1974.
12. **Jones, O. T. G.,** Chlorophyll a biosynthesis, *Philos. Trans. R. Soc. London, Ser. B,* 273, 207–225, 1976.
13. **Castelfranco, P. A. and Jones, O. T. G.,** Protoheme turnover and chlorophyll synthesis in greening barley tissue, *Plant Physiol.,* 55, 485–490, 1975.

14. **Thomas, H., Bortlik, K., Rentsch, D., Schellenberg, M., and Matile, P.,** Catabolism of chlorophyll *in vivo:* significance of polar chlorophyll catabolites in a non-yellowing senescence mutant of *Festuca pratensis* Huds., *New Phytol.*, 111, 3–8, 1989.
15. **Peisker, C., Duggelin, T., Rentsch, D., and Matile, P.,** Phytol and the breakdown of chlorophyll in senescent leaves, *J. Plant Physiol.*, 135, 428–432, 1989.
16. **Rise, M., Cojocaru, M., Gottlieb, H. E., and Goldschmidt, E. E.,** Accumulation of α-tocopherol in senescing organs as related to chlorophyll degradation, *Plant Physiol.*, 89, 1028–1030, 1989.
17. **Trosper, T. and Allen, C. F.,** Carotenoid composition of spinach chloroplast grana and stroma lamellae, *Plant Physiol.*, 51, 584–585, 1973.
18. **Sitte, P.,** Functional organization of biomembranes, in *Lipids and Lipid Polymers in Higher Plants,* Trevini, M. and Lichtenthaler, H. K., Eds., Springer-Verlag, Berlin, 1977, 1–28.
19. **Britton, G.,** Biosynthesis of chloroplast carotenoids, in *Regulation of Chloroplast Differentiation,* Akoyunoglou, G. and Senger, H., Eds., Alan R. Liss, New York, 1986, 125–134.
20. **Nes, W. R. and McKean, M. L.,** *Biochemistry of Steroids and Other Isopentenoids,* University Park Press, Baltimore, 1977.
21. **Qureshi, A. A., Kim, M., Qureshi, N., and Porter, J. W.,** The enzymatic conversion of *cis*-[^{14}C] phytofluene, *trans* -[^{14}C] phytofluene, and *trans*- ζ–[^{14}C] carotene to poly-*cis*-acylic carotenes by a cell free preparation of tangerine tomato fruit plastids, *Arch. Biochem. Biophys.*, 162, 108–116, 1974.
22. **Kushwaha, S. C., Suzue, G., Subbarayan, C., and Porter, J. W.,** The conversion of phytoene-^{14}C to acyclic, monocyclic, and dicyclic carotenes and the conversion of lycopene-15,15′-^{3}H to mono-and dicyclic carotenes by soluble enzyme systems obtained from plastids of tomato fruits, *J. Biol. Chem.*, 245, 4708–4717, 1970.
23. **Goodwin, T. W.,** The prenyllipids of the membranes of higher plants, in *Lipids and Lipid Polymers in Higher Plants,* Trevini, M. and Lichtenthaler, H. K., Eds., Springer-Verlag, Berlin, 1977, 29–47.
24. **Goodwin, T. W.,** Terpenoids and chloroplast development, in *The Biochemistry of Chloroplasts,* Vol. 2, Goodwin, T. W., Ed., Academic Press, London, 1967, 721–733.
25. **Dietschy, J. M. and McGarry, J. D.,** Limitations of acetate as a substrate for measuring cholesterol synthesis in liver, *J. Biol. Chem.*, 249, 52–58, 1974.
26. **Wald, G.,** The visual function of vitamins A, *Vitam. Horm.*, 18, 417–430, 1960.
27. **Napoli, J. L.,** The biogenesis of retinoic acid: a physiologically significant promoter of differentiation, in *Chemistry and Biology of Synthetic Retinoids,* Dawson, M. I. and Okamura, W. H., Eds., CRC Press, Boca Raton, 1990, 230–249.
28. **Santi, B. P.,** Cellular retinoic acid-binding protein and the action of retinoic acid, in *Chemistry and Biology of Synthetic Retinoids,* Dawson, M. I. and Okamura, W. H., Eds., CRC Press, Boca Raton, 1990, 230–249.
29. **Kates, M.,** Ether-linked lipids in extremely halophilic bacteria, in *Ether Lipids: Chemistry and Biology,* Snyder, F., Ed., Academic Press, New York, 1972, 351–398.
30. **Kushwaha, S. C., Kates, M., and Stoeckenius, W.,** Comparison of purple membrane from *Halobacterium cutirubrum* and *Halobacterium halobium, Biochem. Biophys. Acta*, 426, 707–710, 1976.
31. **Goldfine, H. and Langworthy, T. A.,** A growing interest in bacterial ether lipids, *Trends Biochem Sci.*, 13, 217, 1988.
32. **Moldoveanu, N. and Kates, M.,** Biosynthetic studies of the polar lipids of *Halobacterium cutirubrum, Biochim. Biophys. Acta,* 960, 164–182, 1988.
33. **Pugh, E. L., Wassef, M. K., and Kates, M.,** Inhibition of fatty acid synthetase in *Halobacterium cutirubrum* and *Escherichia coli* by high salt concentrations, *Can. J. Biochem.*, 49, 953–958, 1971.
34. **Nishihara, M., Morii, H., and Yosuke, K.,** Heptads of polar ether lipids of an archaebacterium, *Methanobacterium thermoautotrophicum:* structure and biosynthetic relationship, *Biochemistry,* 28, 95–102, 1989.

Chapter 7

THE REGULATION OF GLYCOLIPID METABOLISM

I. INTRODUCTION

The glycolipids are members of an extremely diverse class of lipids distributed very widely in nature. Unlike the other major membrane lipid classes — phospholipids and sterols — the glycolipids are conspicuously localized in highly specialized membranes. In animals, glycolipids fashioned on a sphingoid base framework are tremendously enriched in the still mysterious membranes of brain and nerve tissues. The smaller amounts found elsewhere seem to function almost exclusively as components of the surface membrane of cells, or in the kidneys, as constituents of membranes engaged in salt transport. Plants have these same kinds of glycolipids in their plasma membranes and, in addition, have quite different types of glycolipids concentrated in the lamellar membranes of their chloroplasts, serving some as yet unrecognized function in photosynthesis. Bacterial glycolipids range from membrane components to elements of endotoxins.

In the interest of brevity, I shall devote the bulk of this chapter to the sphingosine-based glycolipids of animals and the glycerol-based glycolipids of green plants. The reader is referred to recent review by Ishizuka and Yamakawa[1] for information on the multitude of bacterial glycolipids.

II. GLYCOLIPID METABOLISM IN ANIMALS

Glycosphingolipids include a diverse group of lipids all composed of a long chain (sphingoid) base, a fatty acid, and a carbohydrate. The carbohydrate moiety, in particular, can be long and branched, giving rise to a variety of different structures.[2] Figure 1 illustrates in shorthand form the carbohydrate chains that distinguish the major complex glycosphingolipid series occurring in nature.

We shall consider here only representative samples of these lipids, namely the sialogangliosides and their simpler relatives the cerebrosides and cerebroside sulfates. These lipids are major components of the mammalian nervous system and for that reason have been singled out for extensive study.

A. CEREBROSIDE BIOSYNTHESIS

Animal glycolipids are almost exclusively derivatives of sphinogsine bases. The simplest quantitatively important representatives are the cerebrosides or monoglycosyl ceramides (Figure l).

Ceramide consists of a sphingosine base, usually sphingosine itself, joined to a long chain fatty acid through an amide linkage. The ceramides are considered to be key intermediates in the formation of all glycosphingolipids, providing from a heterogeneous, common pool the highly specific hydrophobic elements of gangliosides, cerebrosides, and sulfatides.

Cerebrosides result from the *O*-glycosidic bonding of the primary hydroxyl group of ceramide to the C_1 position of a monosaccharide. In nervous tissue, the sugar component is usually galactose. On the other hand, plasma cerebrosides have glucose as their major sugar. Cerebrosides of other tissues, such as lung and kidney, are mixtures of galactose- and glucose-containing molecules.

Our account of cerebroside biogenesis will begin with sphingosine, whose formation was discussed in Chapter 4.

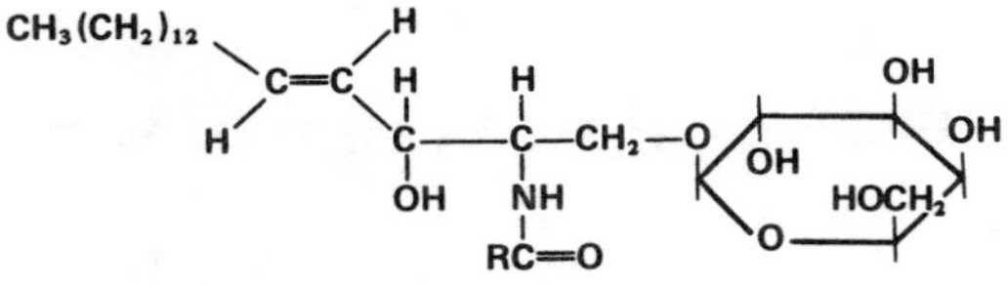

FIGURE 1. A typical cerebroside, *N*-acyl-sphingosine-1-β-D-galactopyranoside.

sphingosine + fatty acyl-CoA ⟶ ceramide + CoASH

FIGURE 2.

Morell and Radin[3] observed that brain microsomes can support the utilization of fatty acyl CoA for the acylation of sphingosine as follows (Figure 2): the acyltransferase involved exhibited a rather pronounced specificity towards certain fatty acids, utilizing stearoyl, lignoceroyl, palmitoyl, and oleoyl-CoAs in a molar ratio of 60:12:3:1. Since this ratio is close to those found in total brain sphingolipids, it seems likely that the fatty acid patterns are tailored at this enzymatic step.

High proportions of the cerebrosides and sulfatides of the vertebrate nervous system contain α-hydroxy fatty acids. These arise from hydroxyceramides (ceramides containing α-hydroxy fatty acids), shown by Ullman and Radin[4] to be synthesized in mouse brain microsomes from a sphingoid base and α-hydroxy fatty acid CoA esters.

Several laboratories have reported the formation of ceramide *in vitro* through the direct condensation of a sphingoid base with a free fatty acid (reviewed in reference 5). The physiological significance of this reaction, which requires neither ATP nor CoA and seems to represent a reversal of lysosomal ceramidase activity, is not known. Ceramide synthesis appears normal in the tissues of Farber's disease patients having defective ceramidase.[6]

Very long chain free fatty acids can be substrates for ceramide formation by a different brain enzyme system requiring certain undefined heat-stable and heat-labile factors as well as Mg^{2+} and reduced pyridine nucleotides.[5] The crude enzyme mixture also gives rise to hydroxyceramides, although free α-hydroxy fatty acids are themselves not effective substrates.

It has been postulated[5] that the substrate for α-hydroxylation is some activated form of a very long chain fatty acid. Conversely, Kaya et al.[7] surmised from studies of ceramide aminoethylphosphonate biosynthesis in *Tetrahymena* that fatty acid hydroxylation occurs after ceramide formation. This conclusion was reached after the transfer of cells containing [^{3}H]-palmitate prelabeled nonhydroxyceramide aminoethylphosphonates from 15 to 39°C triggered their rapid conversion to hydroxyceramide aminoethylphosphonates.

A final pathway for cerebroside biosynthesis is through the transfer of galactose from its UDP derivatives to sphingosine, yielding psycosine (galactosyl sphingosine). Cerebroside synthesis is then completed by amidation with a long chain fatty acyl CoA substrate. Both of the above reactions are reportedly catalyzed by brain microsomal enzymes, but there is some evidence that the latter reaction is nonenzymatic and therefore of little biological significance (see discussion by Kishimoto[5]).

The final step of cerebroside synthesis is the insertion of galactose from its nucleotide complex[8,9] (Figure 3), by a microsomal enzyme. Specificity exists at the level of ceramide utilization. The ceramides employed for cerebroside synthesis contain mainly longer chain (C_{22} to C_{26}) fatty acids, while those channeled into gangliosides and sphingomyelin have predominantly C_{18} and C_{20} species.

ceramide + UDP-galactose ⟶ cerebroside + UDP

FIGURE 3.

In developing brain, this enzyme preferentially transfers galactose to those ceramide species containing 2-hydroxy fatty acids and peaks in activity during myelination.[10] There appears to be a separate enzyme specific for ceramides containing nonhydroxy fatty acids.[11]

Whereas galactose-containing cerebrosides are localized primarily in the nervous system, a glucose-containing cerebroside is formed in a variety of tissues, where it may serve as an intermediate in the formation of other more complex sphingolipids. Glucocerebrosides are synthesized by a pathway generally resembling that involved in galactocerebroside formation.[5]

Cerebrosides are most plentiful in nervous tissue, where they increase from less than 0.02% of the dry weight in human fetal brain to adult levels of 2% in gray matter, 12% in white matter, and 8% in peripheral nerve.[12] On a molar basis, cerebrosides are second only to cholesterol as the most prevalent lipid component of myelin.

The dynamics of cerebroside formation in nervous tissue illustrates well the highly programmed but still poorly understood changes in lipid biosynthesis during myelination (Figure 4).

There is apparently a fairly rapid metabolic turnover of cerebroside constituents, at least in developing brain.[14] The sphingosine base of cerebrosides can be freed and used for the synthesis of other sphingolipids, and the fatty acid component of ceramide is also accessible to exchange into other lipids.[15] In adult animals, cerebroside turnover is slow, with a half-life of approximately 100 days.[12]

B. SULFATIDE BIOSYNTHESIS

After their completion, a fraction of the cerebroside molecules are further transformed into sulfatides (Figure 5). The donor of the sulfate moiety is the nucleotide, 3′-phosphoadenosine-5′-phosphosulfate (PAPS). The reaction is catalyzed by cerebroside sulfotransferase (3′phosphoadenosine-5′-phosphosulfate-galactosylceramidesulfo-transferase), a microsomal enzyme subsequently found to be localized in the Golgi apparatus.[16] As in the case of cerebrosides, sulfatide synthesis is very rapid during the myelination period in younger animals and decreases sharply during maturation. Siegrist et al.[17] have reported that the activity of mouse brain cerebroside sulfotransferase appears to vary during development as a function of the physical properties of its membrane environment. Adding back acetone-extracted lipids to delipidated (and inactivated) microsomes restores sulfotransferase activity to an extent determined by the age of the mice from which the lipids were taken. The drop in activity in older animals seems specifically to parallel the decreasing cholesterol to phospholipid molar ratio.

Hydrocortisone and its analog dexamethasone caused a sixfold increase in $^{35}SO_4^{-2}$ incorporation into sulfatides of cultured mouse glioblastoma cells.[18] Several lines of evidence suggested that the steroids acted to enhance sulfotransferase synthesis. This pattern of activity is compatible with the observed stimulation of myelination in intact rat brain by corticosteroids.

In addition to their presence in nervous tissue, sulfatides are also found in tissues that are very active in sodium transport, e.g., mammalian kidney and the salt gland of marine organisms.[19] Little information is available on the control of sulfatide metabolism with relation to its possible role in sodium transport.

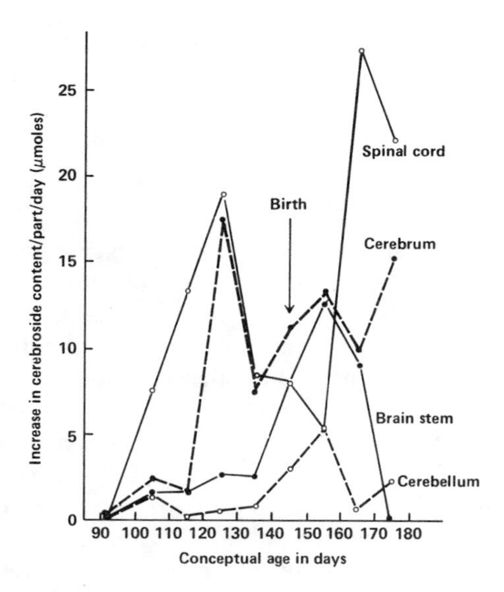

FIGURE 4. Changes in cerebroside concentration during fetal and postnatal development of the lamb central nervous system. (From Patterson, D. S. P., Sweasy, D., and Herbert, C. N., *J. Neurochem.*, 18, 2027, 1971. With permission.)

C. GANGLIOSIDE BIOSYNTHESIS

Another important structural lipid class derived from ceramide includes the gangliosides. These lipids are composed of ceramide linked by a glycosidic bond to the nonreducing end of an oligosaccharide containing hexose and *N*-acetylneuraminic acid units. Typical of the major gangliosides of brain is the following structure, (Figure 6) commonly referred to as G_{M1}, according to the shorthand nomenclature of Svennerholm.[20] A wide variety of gangliosides is found in nervous tissue. In gray matter, where gangliosides occur in highest concentrations, they account for approximately 6% of the lipid weight. Further information on the ganglioside content of various neural cell types may be found in a review by Bowen et al.[14] Closely related gangliosides, such as the structure shown in Figure 7, occur outside the brain.

Wherever gangliosides are found, they are localized primarily in the plasma membrane of the parent cell. In the case of nerve tissue, there is a pronounced enrichment of gangliosides in the surface membranes of the nerve ending, otherwise known as the synaptosomal plasma membranes. The hydrophobic ceramide moiety of the ganglioside anchors the molecule in the lipid bilayer, leaving the negatively charged carbohydrate chain exposed to the external aqueous milieu. The oligosaccharide tail is thus ideally positioned to fulfill its role as a receptor site for glycoprotein hormones of the thyroid-stimulating hormone type and possibly for other substances such as serotonin and interferon.[21]

Technological advances in ganglioside analysis have allowed investigators to detect a marked specificity dictating the types of oligosaccharide chain capable of interacting with a given stimulatory agent. For this reason, one suspects that control of the sequential carbohydrate addition reactions must be closely regulated.

Useful insights into ganglioside synthesis and intracellular movement have come from use of the monovalent cationic ionophore monensin.[22] This ionophore arrests the intracellular transport of newly synthesized materials through the Golgi apparatus and causes an accumulation of intermediates.

The proposed sequence of reactions leading to the formation of some of the common gangliosides is presented in Figure 8. Beginning with glucosylceramide, galactose is trans-

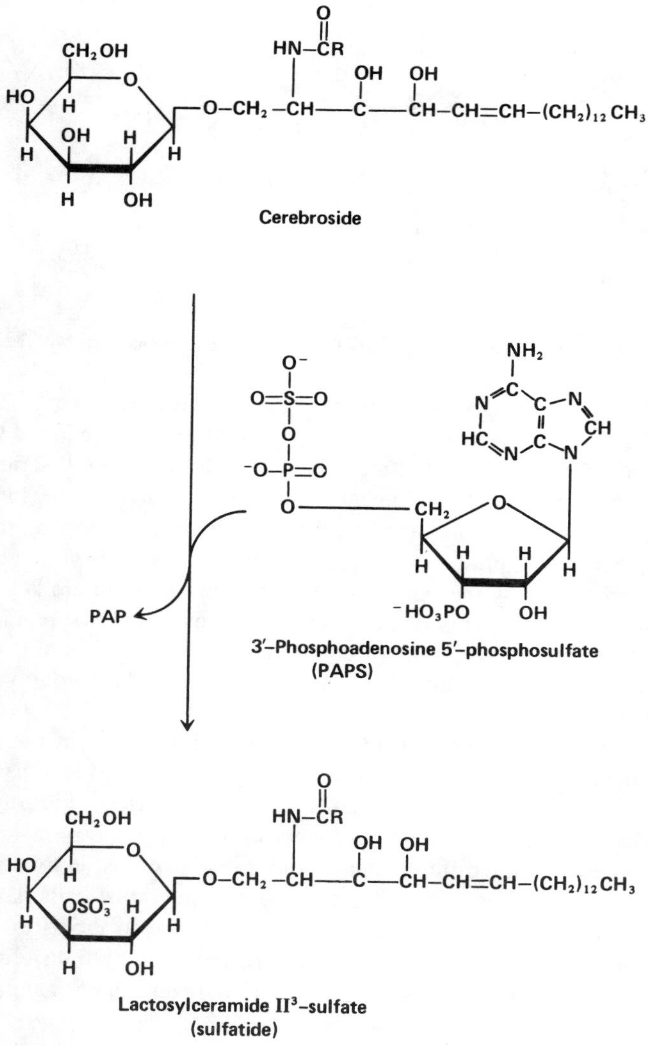

Cerebroside

PAP

3'-Phosphoadenosine 5'-phosphosulfate
(PAPS)

Lactosylceramide II^3-sulfate
(sulfatide)

FIGURE 5.

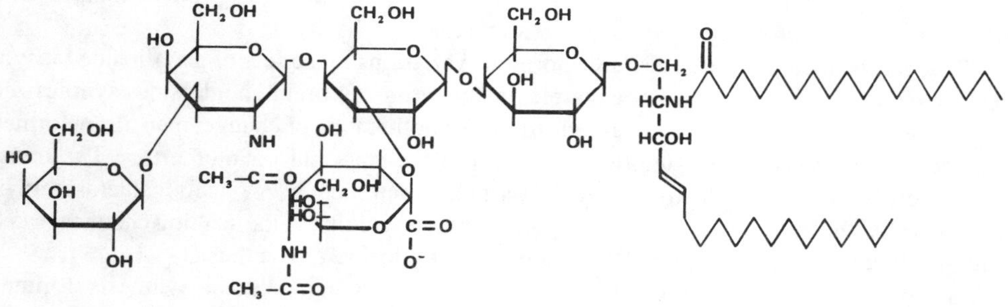

FIGURE 6. Monosialosyl-*N*-tetraglycosylceramide, often abbreviated G_{M1}, a common ganglioside of brain.

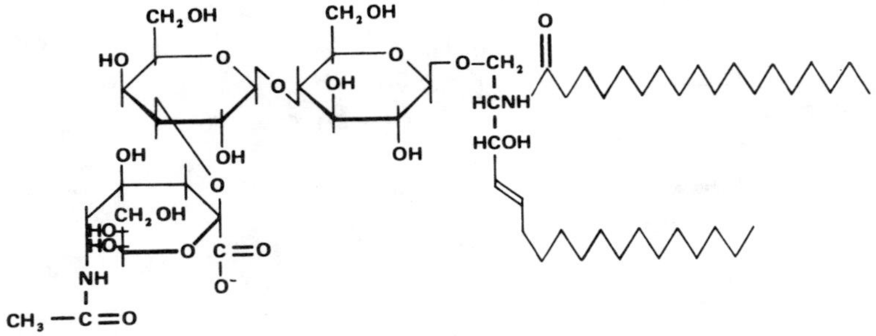

FIGURE 7. Monosialosyl-lactosylceramide of G_{M3}. This is the "hematoside" of visceral organs.

ferred via UDP-galactose to give lactosylceramide, which may have 1, 2, or 3 *N*-acetyl-neuraminic acid groups added by separate sialyltransferases. This series of Golgi apparatus-localized reactions produces the initial members of the asialo a and b series and possibly c series. As a result of competition experiments using LacCer, G_{M3}, and G_{D3} it was determined that the *N*-acetylgalactosamine addition in all the series is catalyzed by the same GalNac transferase.[23] Likewise, the initial *N*-acetyl-neuraminic acid residues placed on G_{A1}, G_{M1}, and G_{D1b} acceptors were added by the same sialyltransferase. Since certain key enzymes appear to operate equally well on all ganglioside series, it was postulated that regulation of ganglioside biosynthesis involves the distinctive sialyltransferases I and II (and perhaps III), which are involved in early steps. Modulation of these activities could selectively alter the relative proportions of the principal series.

Most of the later stages of ganglioside biosynthesis take place in the Golgi apparatus. Completed gangliosides are transported to the cell surface in vesicles that fuse with the plasma membrane, thereby localizing the gangliosides on its outer surface (Figure 9). Recycling of intact gangliosides or their degradation products can occur via endocytosis.

The relative amounts of the different gangliosides in a given tissue reflect the activities of the glycosyltransferases present. Apart from nervous tissue, most cells have only traces of gangliosides more complex than G_{M3} and G_{D3}. This is due to the virtual absence of the enzyme, UDP-*N*-acetylgalactosamine:G_{M3} *N*-acetylgalactosaminyltransferase. This enzyme is present in brain tissue and initiates the formation of several other ganglioside species in significant amounts.[21]

The proposal by Roseman[25] in 1970 that the ganglioside oligosaccharide chain is synthesized by a multienzyme complex has been generally supported by recent experimentation.[26] It is not yet clear whether the multiglycosyltransferase systems in some cases lack certain of the component enzymes or whether they contain one or more modifier proteins that, in turn, control the activity of specific glycosyl-transferases.[21]

Whatever the mechanisms, there are enormous variations in the rate of ganglioside biosynthesis in certain circumstances. For example, the quantities of brain gangliosides synthesized vary tremendously during development. Illustrating another type of change, many transformed cells lack the more complex gangliosides found in their normal counterparts.[27] The transformed cells exhibit a block in UDP-*N*-acetyl galactosamine: G_{M3} *N*-acetylgalactosaminyl-transferase. At the present time, few details are available to connect the striking changes in cell form and function with this altered regulation of ganglioside biosynthesis.

Much remains to be learned regarding the regulation of ganglioside biosynthesis. Limited evidence has been reported for control by feedback inhibition,[28] by altering the state of the Golgi cisternae where the transferase enzymes are localized,[29] or through regulation of gene expression.[30] As yet no one mechanism has been found to predominate.

Under circumstances where normal protein synthesis was blocked by inhibitors or by viral infection, HeLa cells experienced a decreased content of gangliosides and a concurrent large

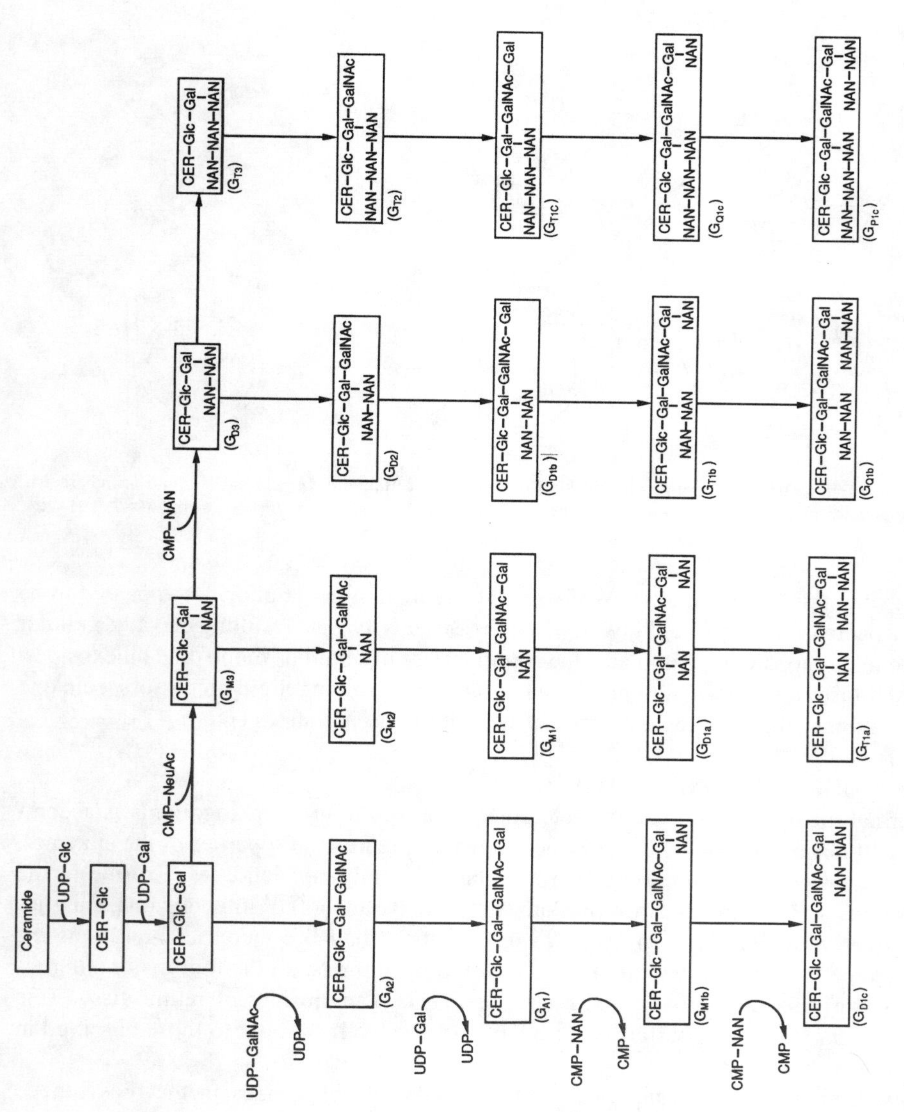

FIGURE 8. Precursor — product relationships among gangliosides.

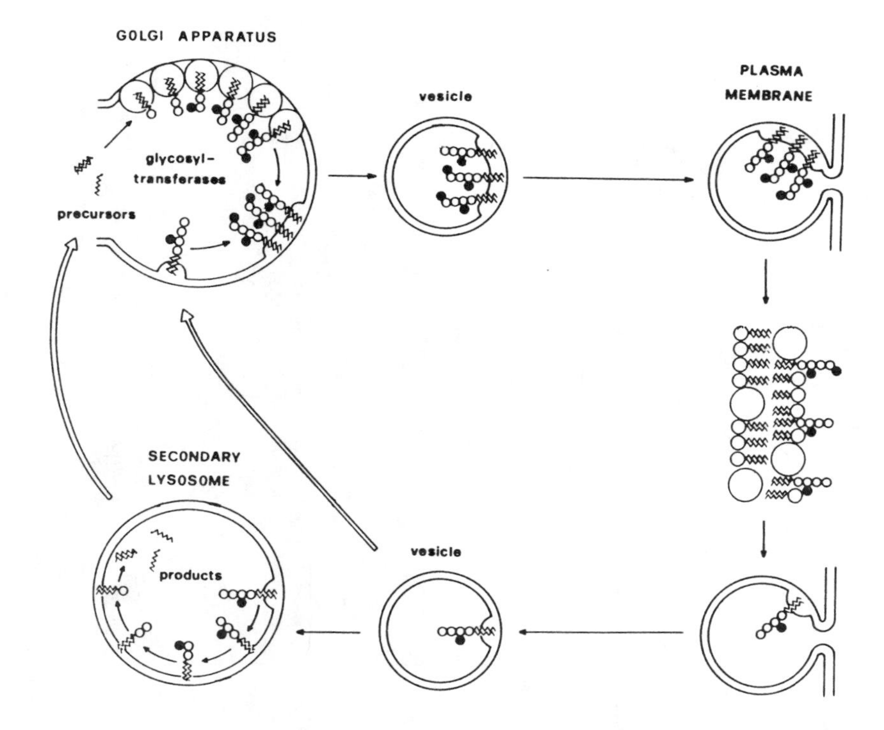

FIGURE 9. Intracellular trafficking of gangliosides. (Modified from Tettamanti, G., Chidoni, R., and Trinchera, M., *Gangliosides and Modulation of Neuronal Functions,* Rahmann, H., Ed., Springer-Verlag, Berlin, 1987, 191–204. With permission.)

increase in ceramide mono- and dihexosides.[31] The reduction in gangliosides appeared to be caused by a decrease in glycolipid glycosyltransferase activity. The resulting elevated cellular level of ceramides apparently stimulated the abnormal rise in ceramide mono- and dihexosides. This system illustrates on a small scale the shunting of glycosphingolipid precursors from one pathway to another that typifies a number of hereditary disease states (see p. 17).

D. GLYCOSPHINGOLIPID CATABOLISM

The normal metabolic turnover and interconversion of the glycosphingolipids is accomplished in part through action by the enzymes shown in Figure 10. Properties of the enzymes, which, with the possible exception of certain plasma membrane sialidases,[32] reside in the Golgi apparatus, have been described by Conzelmann and Sandhoff.[33] In young rats, gangliosides turn over with a half-life of 10 to 25 days, but the half-life increases rapidly as the animals mature.[34] During the period of brain development in rats, all the four major gangliosides undergo metabolic turnover at similar rates.[35] The gangliosides of retinal tissue (the major component being G_{D3}) turn over with half-lives generally similar to those observed in the brain proper.[36]

In addition to the altered ganglioside patterns induced by changes in the biosynthetic pathway, there exists a wide variety of abnormal conditions in which genetic deficiencies in catabolism lead to ganglioside accumulation[37] (see Chapter 1, Table 4, p. 17). In some cases, e.g., generalized gangliosidosis, G_{M1} accumulates to a level ten times the normal value in the central nervous system due to a deficiency at the degradative enzyme β-galactosidase. In other defects, such as Tay-Sachs and Sandhoff's diseases, G_{M2} builds up to 100 to 300 times its normal level in brain tissue because of a deficiency in hexosaminidase, blocking the cleavage of the terminal *N*-acetylgalactosamine moiety from G_{M2}.[38]

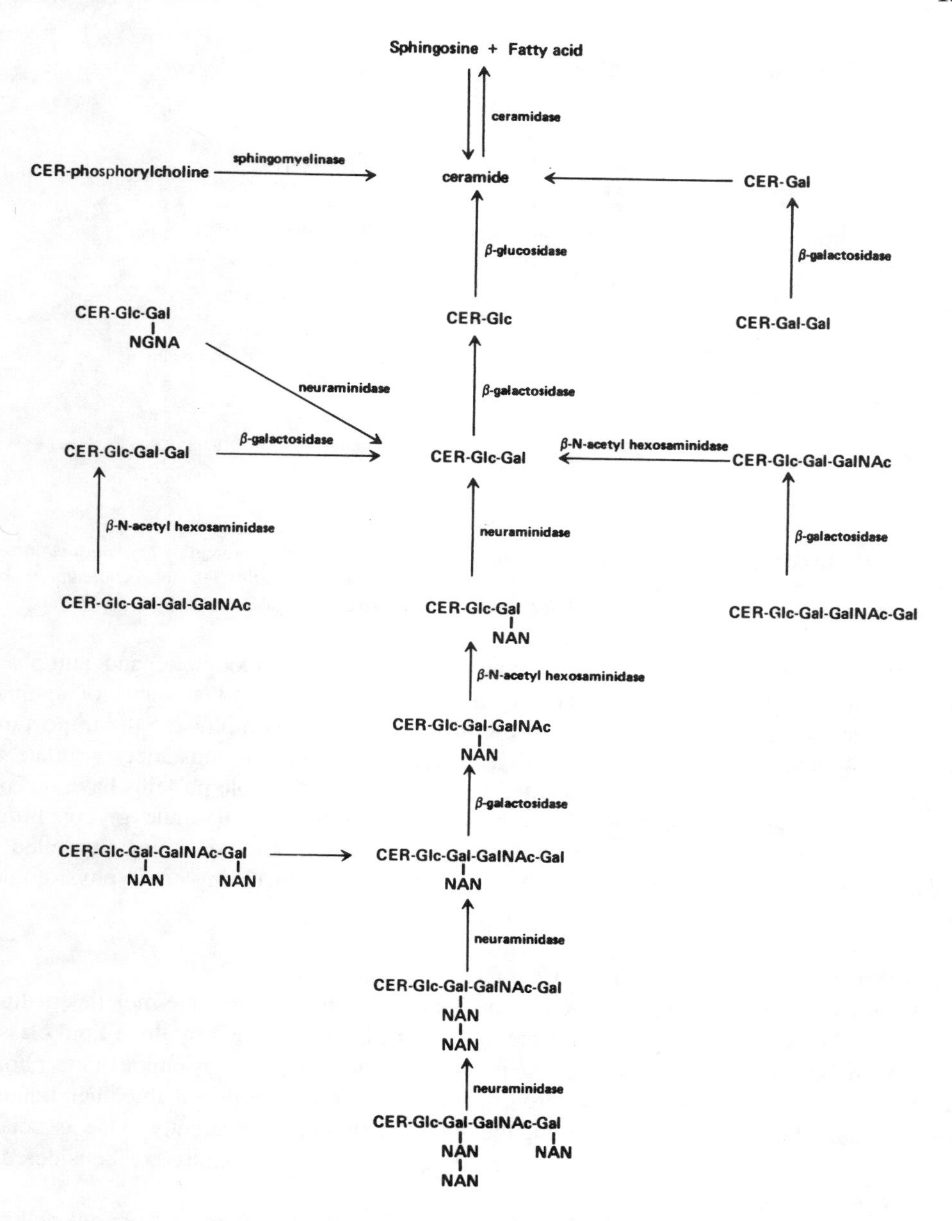

FIGURE 10. Pathways of sphingolipid degradation. Abbreviations: CER, ceramide; Glc glucose: Gal, galactose, GalNAc, *N*-acetylgalactosamine; NAN, *N*-acetylneuraminic acid; NGNA, *N*-glycolylneuraminic acid. (From Ramsey, R. B. and Nicholas, H. J., *Adv. Lipid Res.* 10, 144–232, 1972. With permission.)

Cerebroside storage in excessive amounts in the brain leads to a number of abnormal hereditary conditions, among them Gaucher's disease. This involves a large increase in glucocerebroside, mainly in nonneural tissues, due to a deficiency of glucocerebrosidase. Fabry's disease, on the other hand, causes an accumulation of dihexosyl ceramide and trihexosyl ceramide in the visceral organs.

In most instances, very little is known concerning the detailed nature of these enzymatic defects. It has recently become clear that sphingolipidoses are lysosomal storage diseases.[39] The build-up of specific sphingolipids occurs when a certain key degradative enzyme is

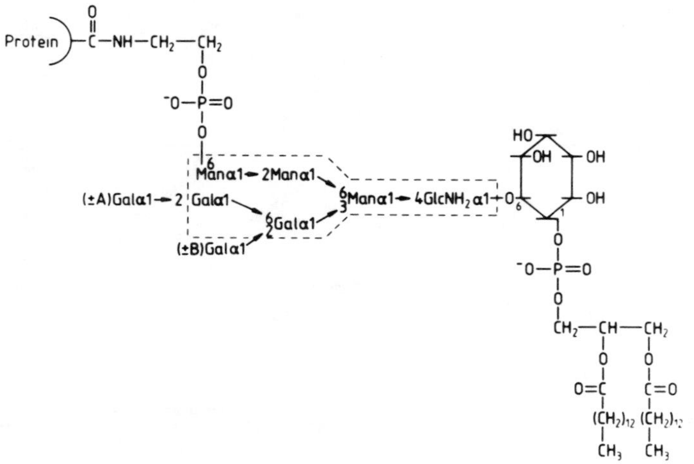

FIGURE 11. The phosphatidylinositol glycan anchor of the variant surface glycoprotein of *Trypanosoma brucei*, variant 117. The dashed line encloses the glycan backbone common to all phosphatidylinositol anchored proteins of this variant. The molecules are usually substituted with one or two additional galactose residues (A and/or B).

absent from the lysosomes that continuously participate in endocytotic and autophagic digestion of membranous organelles, which contain at least small amounts of sphingolipids. The disastrous consequences of the blocks (see p. 17) emphasize the importance of the little understood catabolic pathways that under normal circumstances regulate the levels of glycosphingolipids. Several small (8 to 13 kDa) heat stable proteins have recently been found to activate the specific lysosomal hydrolases that degrade glycosphingolipids. Although considerable structural and genetic information is being compiled on these "spingolipid activator proteins",[40] little is known concerning their physiological function.

E. PHOSPHATIDYLINOSITOL GLYCANS

A number of proteins are anchored in membranes by the hydrophobic moieties of lipids covalently bound to the proteins. This membrane linkage is provided by three lipid classes: long chain fatty acids, usually myristate or palmitate, which are joined by amide or ester bonds to the protein,[41] long chain prenyl groups attached to the protein via thioether bonds,[42] and phosphatidyl-inositol glycans having the glycan chain linked covalently to the associated protein.[43] The biosynthesis and catabolism of the protein-lipid bonds are considered in Chapter 8.

The structure of the phosphatidylinositol glycan can vary considerably from one cell type to another. Its biochemistry will be described only briefly since the glycolipid is present in small amounts as a nonprotein-linked membrane component. A relatively simple form of the membrane anchor is illustrated in Figure 11. In its fully assembled state the lipid is embedded in the outer leaflet of the plasma membrane, tethering a protein to the cell surface by the flexible glycan chain typically four hexoses in length.

So far, the direct link to the membrane has almost always involved phosphatidylinositol (ceramide-containing analogs have been reported from *Dictyostelium*[44]), but this component often has one or more unusual properties. For example, the anchor of a *Leishmania donovani* variant surface glycoprotein consists of a lysophosphatidylinositol containing an unbranched alkyl ether side chain 24 or 26 carbon in length.[45] In other cases, such as the surface anchored acetylcholinesterase of human erythrocytes,[46] the inositol moiety is linked covalently to the lipid anchor and to the glycan chain as usual, but in addition, one of the remaining hydroxyl

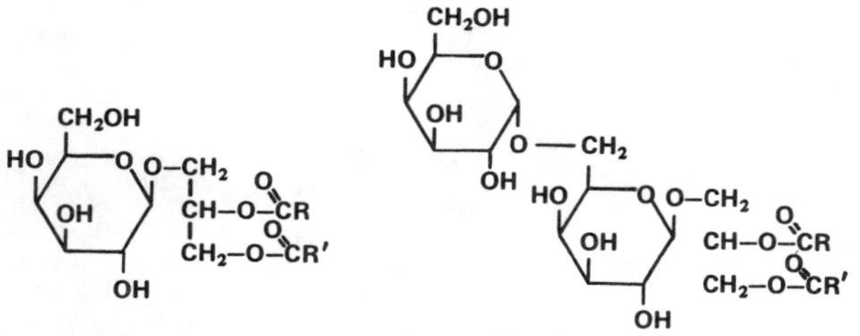

(A) **Monogalactosyl diglyceride (1,2-diacyl-3-*O*-β-D-galactopyranosyl-*sn*-glycerol)**

(B) **Digalactosyl diglyceride [1,2-diacyl-3-*O*-(α-D-galactopyranosyl-(1→6)-*O*-β-D-galactopyranosyl)-*sn*-glycerol]**

FIGURE 12.

groups of inositol is palmitoylated. This latter modification renders the phosphatidylinositol glycan resistant to cleavage by phosphatidylinositol-specific phospholipase C.

Small amounts of the completed or partially completed phosphatidylinositol glycans exist in cells in a nonprotein-associated form.[47] This pool of anchor precursors is used for the placement of a phosphatidylinositol glycan onto a carboxyl terminal amino acid newly created by the coordinated cleavage of a small peptide from the protein's original carboxyl end (see p. 172). Following the attachment of the phosphatidylinositol glycan anchor to protein, further modification, including additional branching, glycosylation and remodeling of the fatty acyl composition, appears to take place in certain instances. Virtually nothing is known regarding the regulation of these processes.

III. GLYCOLIPID METABOLISM IN PLANTS

A. OCCURRENCE OF GLYCOSYL GLYCERIDES

This is the other major class of membrane glycolipids. Biochemical properties of the more common representatives occurring in bacteria, higher plants, and animals have been reviewed by Sastry.[48] In our discussion, we shall also include the glycerol-based plant sulfolipid.

Plant tissues are by far the most significant sources of glycosyl glycerides. The principal lipids present are mono- and digalactosyldiacylglycerol (Figure 12).

These compounds are particularly concentrated in photosynthetic tissues, where along with a smaller amount of sulfolipid they sometimes account for as much as 40% of the total lipids. Other plant parts, including seeds, fruits, and roots, also contain galactosyl glycerides, usually in lesser amounts.[48] These lipids are always associated with chloroplasts or nonphotosynthetic types of plastids. In some instances, very small amounts of glucose analogs of the galactolipids have been identified.[50]

The fatty acid pattern of the plant galactolipids is unusually rich in polyunsaturates. In plant leaves, α-linolenic acid frequently accounts for more than 90% of the total fatty acids. This means, of course, that many of the mono- and digalactosyldiglycerides are the dilinolenoyl molecular species. However, the diglactosyl diglycerides are often less unsaturated than the monogalactosyl derivatives. When saturated fatty acids do occur, they are generally, but not invariably, esterified to the 1-position of the glycerol moiety.[48]

Monogalactosyldiglycerides have also been reported from mammalian nervous tissue, where they are present in amounts equivalent to approximately 2% of the glycerophosphatides. Unlike their plant analogs, they contain mainly palmitic, stearic, and oleic acids.[48] Based on the rapid increase of brain monogalactosyldiglycerides during myelination, it has been assumed that they are primarily constituents of myelin.

The plant sulfolipid, sulfoquinovosyldiglyceride (Figure 13), occurs with the mono- and

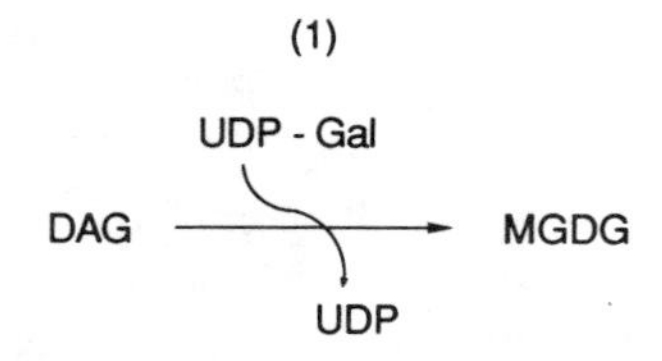

Sulfoquinovosyl diglyceride [1,2-diacyl-3-*O*-(6′—sulfo-α-D-quinovopyranosyl) *sn*-glycerol]

FIGURE 13.

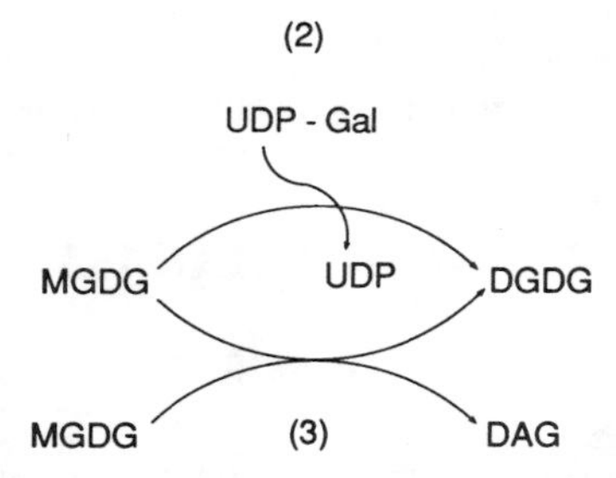

FIGURE 14. The formation of MGDG (1) and its conversion to DGDG by two alternative pathways (2 and 3).

digalactosyldiglycerides and often accounts for as much as 10% of the total glycolipid fraction. Palmitic and linolenic acids are characteristic fatty acids of sulfolipid, which may be thought of as being primarily a chloroplast structural component.

B. METABOLISM OF GLYCOSYL GLYCERIDES

The details of galactolipid biosynthesis have been reviewed by Joyard and Douce.[51] Galactolipid assembly always takes place in the chloroplast. The β-glycosidic bond between DAG and galactose is formed in the inner envelope membrane by the enzyme UDP-galactose:diacylglycerol galactosyltransferase (Figure 14, reaction 1). The resulting monogalactosyldiacylglycerol (MGDG) can be further converted to digalactosyldiacylglycerol (DGDG) by one of two potential pathways. Early evidence (reviewed by Joyard and Douce[50]) pointed to the addition of a second galactose moiety to MGDG via UDP-galactose (Figure 14, reaction 2). Later work, mainly with isolated chloroplasts, demonstrated a different pathway, involving the transfer of galactose directly from one MGDG molecule to another by a galactolipid:galactolipid galactosyltransferase[52] (Figure 14, reaction 3) This latter enzyme, now known to be located in the outer envelope membrane, is quite active in isolated chloroplasts but serious questions have been raised as to its importance in intact cells.[51]

The logistics of galactolipid synthesis is complicated because in many plants there are two distinct major sources of diacylglycerol (DAG) precursors.[53] One source is phosphatidylcho-

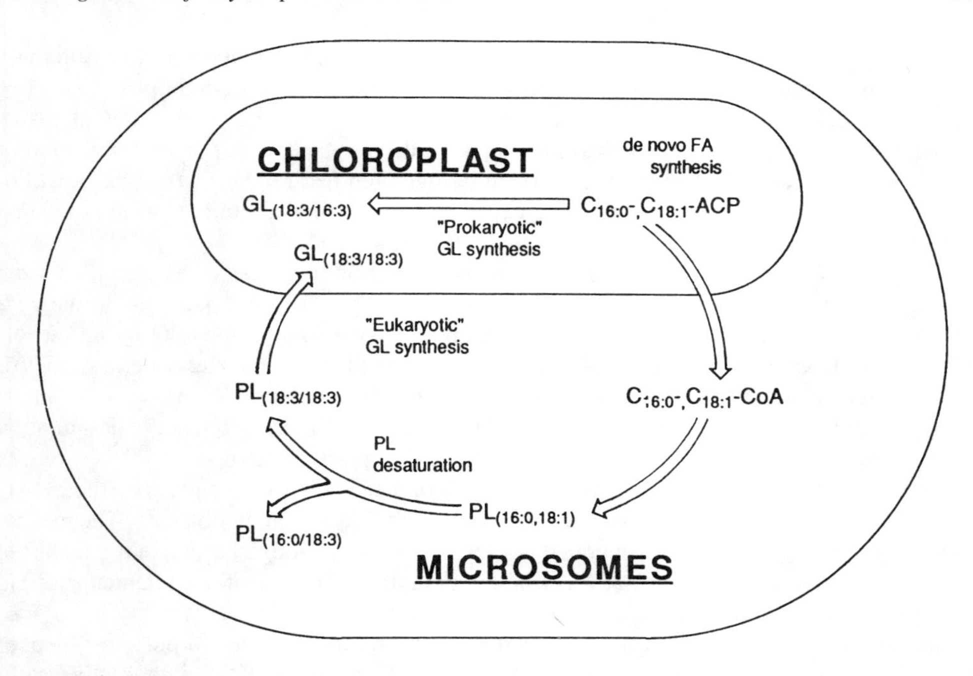

FIGURE 15. Galactolipid formation in chloroplasts via the "prokaryotic" pathway and the "eukaryotic" pathway.

line (and perhaps other phospholipids) made in the endoplasmic reticulum. The DAG moiety of phosphatidylcholine is generally constituted with oleate or linoleate at the *sn*-1 position and linoleate at the *sn*-2 position. This pattern, with a C_{18} fatty acid always present at the *sn*-2 position, is typically produced by endoplasmic reticulum acyltransferase. Such lipids are said to be formed by the eukaryotic pathway. In contrast, galactolipid synthesis can also draw upon a pool of DAG produced in the chloroplast from fatty acids made there and transferred to the glycerol-3-phosphate backbone directly from fatty acyl-ACP. The acyltransferases responsible for these acylations are situated in the inner envelope membrane and are similar in their positional specificities and other properties to bacterial enzymes. The resulting lipids are referred to as products of the prokaryotic pathway. The biosynthetic scheme is outlined in Figure 15.

The actual makeup of the monogalactosyldiacylglycerol mixture synthesized in a given plant is genetically predetermined — certain species, e.g., pea and all grasses are "18:3 plants", meaning that their galactolipids all have the eukaryotic pattern, with a C_{18} fatty acid in the *sn*-2 position. Other species, including spinach and *Arabidopsis thaliana,* contain approximately half eukaryotic and half prokaryotic type galactolipids and are termed "16:3 plants" because the C_{16} acyl chain at the *sn*-2 position usually becomes desaturated to hexadecatrienoic acid. Some algae, such as *Chlamydomonas reinhardtii* and *Dunaliella salina* contain almost exclusively prokaryotic type galactolipids.[54]

The galactosyltransferase responsible for monogalactosyldiacylglycerol synthesis appears to have a broad specificity for DAG molecular species[51] and can utilize either eukaryotic or prokaryotic DAG with equal facility. The molecular species composition of galactolipids in a particular plant is therefore dependent upon the supply of DAG. It has been reasoned that 18:3 plants are deficient in the chloroplast phosphatidic acid phosphatase capable of transforming the prokaryotic phosphatidic acid generated in the envelope membrane into a DAG substrate usable for galactolipid synthesis.[55]

Studies of an *Arabidopsis* mutant deficient in the glycerol-3-phosphate acyltransferase activity needed for prokaryotic type lipid synthesis in the chloroplast[56] have provided a striking example of the plant cell's capacity for regulating its galactolipid content. The greatly reduced acyltransferase activity still permitted an almost normal complement of one product, phosphatidylglycerol, but sharply diminished the level of prokaryotic galactolipids. However, a compensatory overproduction of eukaryotic galactolipids resulted in the mutant plants having a galactolipid content nearly as high as in wild-type *Arabidopsis*. This conversion of *Arabidopsis* from a 16:3 plant to an 18:3 plant by mutation had little effect on its growth or development under standard conditions, but chloroplast structural alterations were noted that could impair its function under environmental stress.[57]

The initial monogalactosyldiacylglycerol molecular species synthesized by the prokaryotic pathway is generally agreed to contain oleate at the *sn*-1 position and palmitate at the *sn*-2 position.[51] There follows a rapid desaturation of both fatty acids, culminating in high levels of linolenate, hexadecatrienoate and, in some species, hexadecatetraenoate. Kinetic studies of galactolipid metabolism have strongly suggested that the substrate for desaturation is the intact mono- or digalactosyldiacylglycerol molecule,[54] but neither the precise sequence of desaturation events nor the exact site of the activity within the chloroplast are known.

The eukaryotic type of galactolipids, which predominate in the 18:3 plants, draw their DAG precursors from a cytosolic pool constituted mainly with linoleate and linolenate. The mechanisms by which these DAG are generated (largely from phosphatidylcholine) and transferred into the chloroplast envelope are not yet clear, but desaturation may continue after galactolipid assembly, giving, in most species, a predominant dilinolenoyl product.

Although the plant sulfolipid, sulfoquinovosyldiacylglycerol, is a major lipid constituent of chloroplasts, surprisingly little is known regarding its biosynthesis.[58] Two alternative pathways for synthesis of the sulfoquinovose head group have been proposed. One envisions cysteic acid as a direct precursor of sulfoquinovose, but a series of competition and dilution experiments have made it seem more likely that cysteic acid simply provides sulfate that may be incorporated into sulfoquinovose.[58,59]

A second possible pathway involves the conversion of sulfate to adenosine 5′-phosphosulfate, which donates sulfur to the synthetic process either directly or after being processed to 3′-phosphoadenosine 5′-phosphosulfate.[58] Gupta and Sastry[59] detected an incorporation of $[^{35}S]SO^{2-}{}_4$, and, to a much smaller extent, $[^{35}S]$sulfoquinovose into sulfoquinovosyldiacylglycerol of groundnut leaves but did not find a detectable incorporation of $[^{35}S]$cysteic acid or $[^{35}S]$sulfoquinovosylglycerol into the lipid, although the latter compounds were taken up by the leaf discs. The authors favored an earlier proposal by Zill and Cheniae[60] that sulfonation of a glycosyldiacylglycerol may represent the major route of sulfoquinovosyldiacylglycerol formation.

REFERENCES

1. **Ishizuka, I. and Yamakawa, T.,** Glycoglycerolipids, in *Glycolipids*, Wiegandt, H., Ed., Elsevier, Amsterdam, 1985, 101–197.
2. **Makita, A. and Taniguchi, N.,** Glycosphingolipids, in *Glycolipids*, Wiegandt, H., Ed., Elsevier, Amsterdam, 1985, 1–99.
3. **Morrell, P. and Radin, N. S.,** Specificity in ceramide biosynthesis from long chain bases and various fatty acyl coenzymes A's by brain microsomes, *J. Biol. Chem.*, 245, 342–350, 1970.
4. **Ullman, M. D. and Radin, N. S.,** Enzymatic formation of hydroxy ceramides and comparison with enzymes forming nonhydroxy ceramides, *Arch. Biochem. Biophys.*, 157, 767–777, 1972.
5. **Kishimoto, Y.,** Sphingolipid formation, in *The Enzymes, VXI*, Boyer, P. D., Ed. Academic Press, New York, 1983, 357–407.
6. **Moser, H. W. and Chen, W. W.,** Ceramidase deficiency: Farber's lipogranulomatosis, in *The Metabolic Basis of Inherited Disease*, Stanbury, J. B., Wyngaarden, J. B., Frederickson, D. S., Goldstein, J. L., and Brown, M. S., McGraw-Hill, New York, 1983, 820–830.

7. **Kaya, K., Ramesha, C. S., and Thompson, G. A., Jr.,** On the formation of α-hydroxy fatty acids, *J. Biol. Chem.*, 259, 3548–3553, 1984.
8. **Morell, P. and Radin, N. S.,** Synthesis of cerebroside by brain from uridine diphosphate galactose and ceramide containing hydroxy fatty acid, *Biochemistry,* 8, 506–512, 1969.
9. **Basu, S., Schultz, A. M., Basu, M., and Roseman, S.,** Enzymatic synthesis of galactocerebroside by galactosyltransferase from embryonic chicken brain, *J. Biol. Chem.*, 246, 4272–4279, 1971.
10. **Mandel, P., Nussbaum, J. L., Neskovic, N. M., Sarlieve, L. L., and Kurihara, T.,** Regulation of myelinogenesis, *Adv. Enz. Reg.*, 10, 101–118, 1972.
11. **Morell, P. and Braun, P.,** Biosynthesis and metabolic degradation of sphingolipids not containing sialic acid, *J. Lipid Res.*, 13, 293–310, 1972.
12. **Davison, A. N.,** Lipid metabolism in nervous tissue, in *Comprehensive Biochemistry,* Vol. 18, Florkin, M. and Stotz, E. H., Eds., Elsevier, Amsterdam, 1970, 293–329.
13. **Patterson, D. S. P., Sweasey, D., and Hebert, C. N.,** Changes occurring in the chemical composition of the central nervous system during foetal and post-natal development of the sheep, *J. Neurochem.*, 18, 2027–2040, 1971.
14. **Bowen, D. M., Davison, A. N., and Ramsey, R. B.,** The dynamic role of lipids in the nervous system, in *MTP International Review of Science,* Vol. 4, Goodwin, T. W., Ed., Butterworths, London, 1974, 141–179.
15. **Okabe, H. and Kishimoto, Y.,** *In vivo* metabolism of ceramides in rat brain. Fatty acid replacement and esterification of ceramide, *J. Biol. Chem.*, 252, 7068–7073, 1977.
16. **Fleischer, B. and Zambrano, F.,** Localisation of cerebroside-sulfotransferase activity in the Golgi apparatus of rat kidney, *Biochem. Biophys. Res. Comm.*, 52, 951–958, 1973.
17. **Siegrist, H. P., Jutzi, H., Steck, A. J., Burkart, T., Wiesmann, U., and Herschkowitz, N.,** Age-dependent modulation of 3′-phosphoadenosine-5′-phosphosulfate-galactosylceramide sulfotransferase by lipids extracted from the microsomal membranes and artificial lipid mixtures, *Biochim. Biophys. Acta,* 489, 58–63, 1977.
18. **Dawson, G. and Kernes, S. M.,** Mechanism of action of hydrocortisone potentiation of sulfogalactosylceramide synthesis in mouse oligodendroglioma clonal cell lines, *J. Biol. Chem.*, 254, 163–167, 1979.
19. **Karlsson, K.-A., Samuelsson, B. E., and Steen, G. O.,** The lipid composition and Na^+-K^+-dependent adenosine-triphosphatase activity of the salt (nasal) gland of eider duck and herring gull, a role for sulphatides in sodium ion transport, *Eur. J. Biochem.*, 46, 243–258, 1974.
20. **Svennerholm, L.,** Ganglioside metabolism, in *Comprehensive Biochemistry,* Vol. 18, Florkin, M. and Stotz, E. H., Eds., Elsevier, Amsterdam, 1970, 201–227.
21. **Fishman, P. H and Brady, R. O.,** Biosynthesis and function of gangliosides, *Science,* 194, 906–915, 1976.
22. **Saito, M., Saito, M., and Rosenberg, A.,** Influence of monovalent cation transport on anabolism of glycosphingolipids in cultured human fibroblasts, *Biochemistry,* 24, 3054–3059, 1985.
23. **Pohlentz, G., Klein, D., Schwarzmann, G., Schmitz, D., and Sandhoff, K.,** Both G_{A2}, G_{M2} and G_{D2} synthases and G_{M1b}, G_{D1a}, and G_{T1b} synthases are single enzymes in Golgi vesicles from rat liver, *Proc. Natl., Acad. Sci. U.S.A.*, 85, 7044–7048, 1988.
24. **Tettamanti, G., Chidoni, R., and Trinchera, M.,** Fundamentals of brain ganglioside biosynthesis, in *Gangliosides and Modulation of Neuronal Functions,* Rahmann, H., Ed., Springer-Verlag, Berlin, 1987, 191–204.
25. **Roseman, S.,** The synthesis of complex carbohydrates by multiglycosyltransferase systems and their potential function in intercellular adhesions, *Chem. Phys. Lipids,* 5, 270–297, 1970.
26. **Fishman, P. H.,** Normal and abnormal biosynthesis of gangliosides, *Chem. Phys. Lipids,* 13, 305–326, 1974.
27. **Brady, R. O. and Fishman, P. H.,** Alterations of galactosaminyl- and galactosyltransferases in cultured mammalian cells and *in vivo,* in *The Enzymes of Biological Membranes,* Vol. 2, Martonosi, A., Ed., Plenum Press, New York, 1976, 421–442.
28. **Yusuf, H. K. M., Schwarzmann, G., Pohlentz, G., and Sandhoff, K.,** Oligosialogangliosides inhibit G_{M2}- and G_{D3} -synthesis in isolated Golgi vesicles from rat liver, *Hoppe-Seylers Z. Biol. Chem.*, 368, 455–462, 1987.
29. **Berger, E. G. and Hesford, F.,** Localization of galactosyl- and sialyltransferase by immunofluorescence: evidence for different sites, *Proc. Natl. Acad. Sci. U.S.A.* 82, 4736–4739, 1985.
30. **Hashimoto, Y., Otsuka, H., Sudo, K., Suzuki, K., Suzuki, A., and Yamakawa, T.,** Genetic regulation of G_{M2} expression in liver of mouse, *J. Biochem.*, 93, 895–901, 1983.
31. **Anderson, T.,** Regulation of glycolipid biosynthesis: effects of virus infection and drug-induced translational inhibition on glycolipid metabolism, *Biochemistry,* 18, 2395–2400, 1979.
32. **Sandhoff, K., Schwarzmann, G., Sarmientos, F., and Conzelmann, E.,** in *Fundamentals of Ganglioside Catabolism, Gangliosides and Modulation of Neuronal Functions,* Rahmann, H., Ed., Springer-Verlag, Berlin, 1987, 231–250.
33. **Conzelmann, E. and Sandhoff, K.,** Glycolipid and glycoprotein degradation, *Adv. Enzymol.* 60, 89–216, 1987.

34. **Suzuki, K.,** Formation and turnover of the major brain gangliosides during development, *J. Neurochem.,* 14, 917–925, 1967.
35. **Holm, M. and Svennerholm, L.,** Biosynthesis and biodegradation of rat brain gangliosides studied *in vivo, J. Neurochem.,* 19, 609–622, 1972.
36. **Holm, M.,** Biodegradation of the major rabbit retinal gangliosides studied *in vivo, FEBS Lett.,* 77, 225–227, 1977.
37. **Brady, R. O.,** The chemistry and control of hereditary lipid diseases, *Chem. Phys. Lipids,* 13, 271–282, 1974.
38. **Brady, R. O.,** Lipidoses, in *Handbook of Neurochemistry,* Vol. 10, Lajtha, A., Ed., Plenum Press, New York, 1985, 81–97.
39. **Desnick, R. J., Thorpe, S. R., and Fiddler, M. B.,** Toward enzyme therapy for lysosomal storage diseases, *Physiol Rev.,* 56, 57–99, 1976.
40. **O'Brien, J. S., Kretz K. A., Dewji, N., Wenger, D. A., Esch, F., and Fluharty, A. L.,** Coding of two sphingolipid activator proteins (SAP-1 and SAP-2) by same genetic locus, *Science,* 241, 1098–1101, 1988.
41. **Grand, R. J. A.,** Acylation of viral and eukaryotic proteins, *Biochem. J.,* 258, 625–638, 1989.
42. **Glomset, J. A., Gelb, M. H., and Farnsworth, C. C.,** Prenyl proteins in eukaryotic cells: a new type of membrane anchor, *Trends Biochem. Sci.,* 15, 139–142, 1990.
43. **Low, M. G.,** The glycosyl-phosphatidylinositol anchor of membrane proteins, *Biochim. Biophys. Acta,* 988, 427–454, 1989.
44. **Stadler, J., Keenan, T. W., Bauer, G., and Gerisch, G.,** The contact site A glycoprotein of *Dictyostelium discoideum* carries a phospholipid anchor of a novel type, *EMBO J.,* 8, 371–377, 1989.
45. **Orlandi, P. A. and Turco, S. J.,** Structure of the lipid moiety of the *Leishmania donovani* lipophosphoglycan, *J. Biol. Chem.,* 262, 10384–10391, 1987.
46. **Roberts, W. L., Myher, J. J., Kuksis, A., Low, M. G., and Rosenberry, T. L.,** Lipid analysis of the glycoinositol phospholipid membrane anchor of human erythrocyte acetylcholinesterase., *J. Biol. Chem.,* 263, 18766–18775, 1988.
47. **Cross, G. A. M.,** Glycolipid anchoring of plasma membrane proteins, *Annu. Rev. Cell Biol.,* 6, 1–39, 1990.
48. **Sastry, P. S.,** Glycosyl glycerides, *Adv. Lipid Res.,* 12, 251–310, 1974.
49. **Kates, M.,** Plant phospholipids and glycolipids, *Adv. Lipid Res.,* 8, 225–265, 1970.
50. **Jamieson, G. R. and Reid, E. H.,** The sugar components and the galactosyl diglycerides from green plants, *Phytochemistry,* 15, 135–136, 1976.
51. **Joyard, J. and Douce, R.,** Galactolipid synthesis, in *The Biochemistry of Plants,* Vol. 9, Stumpf, P. K., Ed., Academic Press, New York, 1987, 215–274.
52. **Van Besouw, A. and Wintermans, J. F. G. M.,** Galactolipid formation in chloroplast envelopes. I. Evidence for two mechanisms in galactosylation, *Biochim. Biophys. Acta,* 529, 44–53, 1978.
53. **Roughan, G. and Slack, R.,** Glycerolipid synthesis in leaves, *Trends Biochem. Sci.,* 9, 383–386, 1984.
54. **Cho, S. H. and Thompson, G. A., Jr.,** On the metabolic relationships between monogalactosyldiacylglycerol and digalactosyldiacylglycerol molecular species in *Dunaliella salina, J. Biol. Chem.,* 262, 7586–7593, 1987.
55. **Heinz, E. and Roughan, P. G.,** Similarities and differences in lipid metabolism of chloroplasts isolated from 18:3 and 16:3 plants, *Plant Physiol.,* 72, 273–279, 1983.
56. **Kunst, L., Browse, J., and Somerville, C.,** Altered regulation of lipid biosynthesis in a mutant of *Arabidopsis* deficient in chloroplast glycerol-3-phosphate acyltransferase activity, *Proc. Natl. Acad. Sci. U.S.A.,* 85, 4143–4147, 1988.
57. **Kunst, L., Browse, J., and Somerville, C.,** Altered chloroplast structure and function in a mutant of *Arabidopsis* deficient in plastid glycerol-3-phosphate acyltransferase activity, *Plant Physiol.,* 90, 846–853, 1989.
58. **Mudd, J. B. and Kleppinger-Sparace, K. F.,** Sulfolipids, in *The Biochemistry of Plants,* Vol. 9, Stumpf, P. K., Ed., Academic Press, New York, 1987, 275–289.
59. **Gupta, S. D. and Sastry, P. S.,** The biosynthesis of sulfoquinovosyl-diacylglycerol: studies with groundnut (*Arachis hypogaea*) leaves, *Arch. Biochem. Biophys.,* 260, 125–133, 1988.
60. **Zill, L. P. and Cheniae, G. M.,** Lipid metabolism, *Annu. Rev. Plant Physiol.,* 13, 225–264, 1962.

Chapter 8

LIPIDS COVALENTLY BOUND TO PROTEINS

I. FATTY ACYLATED PROTEINS

A. INTRODUCTION

The first protein containing covalently bound fatty acid was discovered in 1951 by Folch and Lees.[1] This myelin proteolipid was initially thought to be unique, but in recent years, beginning in the early 1980s, a wide variety of fatty acylated proteins has been described. These proteins appear to be ubiquitous, as judged from their reported presence in viruses, bacteria, plants, and animals.[2]

The fatty acyl chain can be linked by thioester, ester, or amide linkage to the protein (Figure 1). Ester-linked fatty acids can be experimentally distinguished from their amide-linked counterparts by the sensitivity of the former to hydrolysis in the presence of hydroxylamine. It is now clear that the two linkage types are found on entirely different subsets of proteins. Furthermore, the metabolism of the two acylated protein types differs in many respects, as described below, with ester-linked acyl chains often exhibiting a rapid metabolic turnover while amide-linked acyl chains become permanently attached to their designated protein during translation.

Fatty acylation promotes a tighter association of proteins with other subunits of a functional complex, such as a virus coat[3] or with membranes,[4] although in some instances, e.g., the low molecular weight GTP-binding proteins, attachment of a polyisoprenoid chain in addition to palmitoylation appears necessary for strong membrane anchoring.[5]

Much interest has recently been directed towards a number of fatty acylated proteins that participate in transmembrane signaling pathways.[6] In one such study, palmitoylation of a 64 kDa protein appeared to target its translocation from the cytosol to the plasma membrane. Mitogen stimulation of quiescent cells caused a deacylation of this protein, suggesting a possible role for protein fatty acyl chain turnover in signaling.

B. PROTEINS CONTAINING ESTER-LINKED FATTY ACIDS

Perhaps as many as 100 individual proteins are now known to have ester-linked fatty acids. It appears that in most cases the fatty acid is attached to cysteine and in a minority of cases, serine, although the type of linkage has not yet been determined in many instances. The principal fatty acid attached in this fashion is palmitate.

Insufficient experience has been gained to decide which factors predispose a particular cysteine for fatty acylation. The fusion glycoprotein of human respiratory syncytial virus contains palmitate linked to cysteine 550, which is situated in the cytoplasmic region of the hydrophobic anchor domain.[8] Rhodopsin contains two acyl chains, attached to adjacent cysteines of the protein. These cysteines are also located in a cytoplasmic segment near a membrane-spanning domain (Figure 2).[9]

The bulk of available evidence suggests that fatty acylation of proteins occurs in the endoplasmic reticulum as a posttranslational event.[2] Fatty acid attachment to protein continues at relatively high levels for a period of time following the inhibition of protein synthesis by cycloheximide.[10] In some systems palmitoylation appears to take place in the Golgi membranes (discussed in reference 2) and in the chloroplast.[11]

The palmitoylation of certain proteins is a transient event. In erythrocytes, where there is no active protein synthesis, several acylated proteins rapidly incorporated exogenous [^{3}H]palmitate, and the radioactivity in these various products turned over with half lives

A.

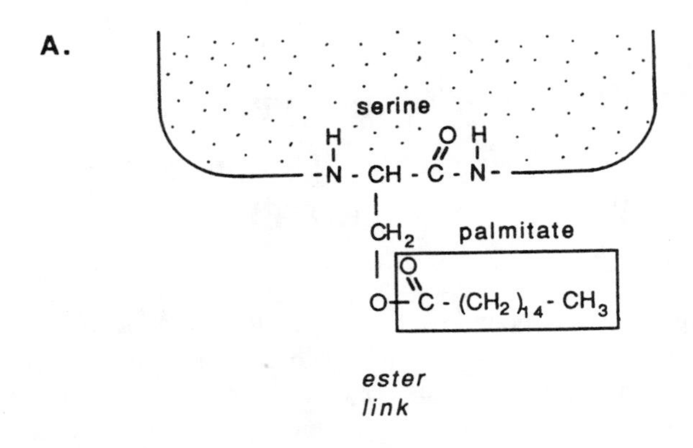

B.

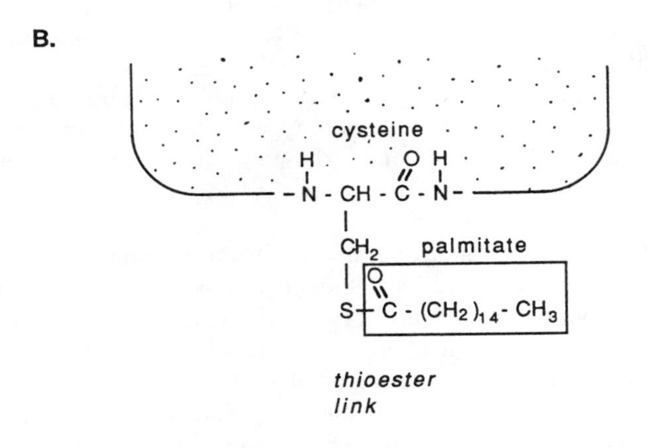

C.

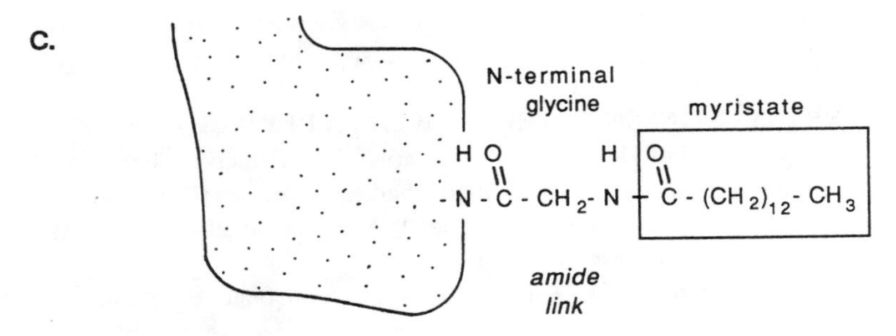

FIGURE 1.

ranging from less than 30 min to more than 3 h.[12] The hydrolase responsible for fatty acid deacylation has not been characterized.

Ester-type acylation of proteins requires fatty acyl-CoA as a substrate. In many cases tested palmitoyl-CoA is preferred over longer or shorter chain fatty acyl-CoAs or unsaturated derivatives. However, myristate and stearate are not uncommonly detected in fatty acids cleaved from acylated proteins, and stearate occasionally predominates.[2]

C. PROTEINS CONTAINING AMIDE-LINKED FATTY ACIDS

These derivatives are different in many respects from the fatty acid esters of proteins discussed above. Myristate is by far the most common fatty acid found in amide linkage, and

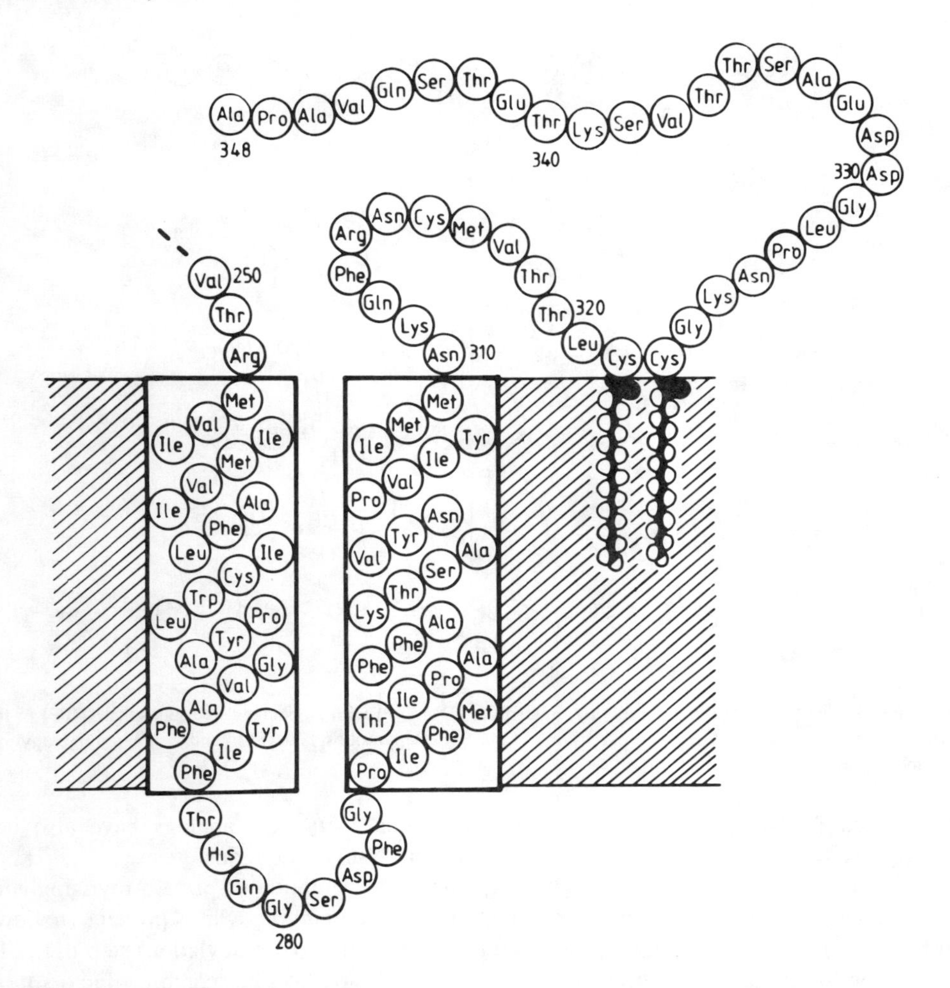

FIGURE 2. The location of the palmitoylated cysteines 322 and 323 near the seventh transmembrane domain of rhodopsin. (From Ovchinnikov, Y. A., Abdulaev, N. G., and Bogachuk, A. S., *FEBS Lett.*, 230, 1–5, 1988. With permission.)

the attachment of fatty acids in this way is generally referred to as myristoylation. Myristate is almost invariably attached to N-terminal glycine residue as a cotranslational modification, which apparently persists for the lifetime of the protein. Among the many proteins modified in this way are GTP-binding proteins, protein kinases, oncogene products, and viral coat proteins.[2]

Myristoylation is catalyzed by the enzyme myristoyl-CoA:protein N-myristoyltransferase (N-myristoyl-transferase). As the initiator methionine residue is removed from the growing peptide chain during translation to expose an amino-terminal glycine, the myristoyl group is transferred, forming a stable amide link.[14] The detection of myristate bound to nascent polypeptides isolated from ribosomes affirms the cotranslational nature of the myristoylation process.[13] While it is not known precisely what factors determine that an N-terminal glycine will be myristoylated, the 7-10 amino acids nearest the N-terminus are involved in substrate recognition, with a serine residue at position 5 being especially important for high affinity binding to the enzyme.[15]

In vitro studies found myristoyl-CoA to be strongly preferred as the acyl substrate over both shorter and longer chain fatty acyl derivatives, although decanoyl-CoA and lauroyl-CoA could

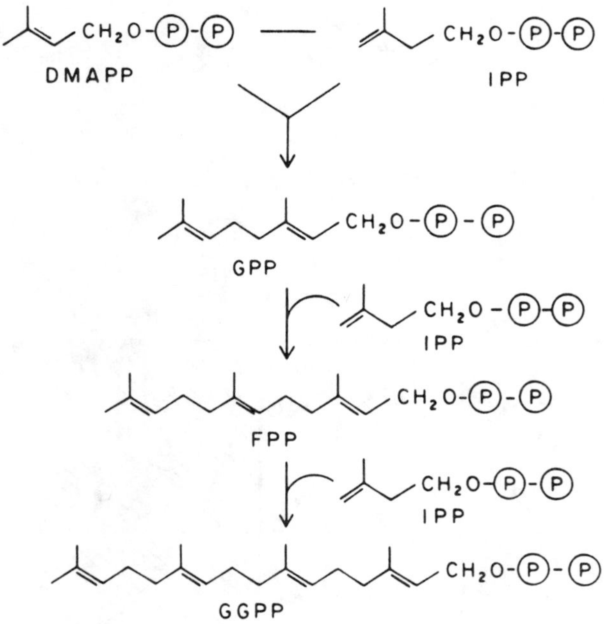

FIGURE 3. The fusion of dimethyl allylpyrophosphate (DMAPP) with isopentenyl pyrophosphate (IPP) to form geranylpyrophosphate, followed by its extension to form farnesylpyrophosphate (FPP) and then geranylgeranylpyrophosphate (GGPP).

be utilized.[16] Some unnatural heteroatom substituted fatty acid analogs have also been incorporated,[17] while other myristate analogs inhibit N-myristoyltransferase.[18]

Virtually no information is available regarding the regulation of protein myristoylation. Rapid changes in the cellular levels and functions of certain myristoylated proteins (reviewed in Schmidt[2]) may or may not be influenced by the acylation (or deacylation) step itself. The availability of the purified N-myristoyltransferase will facilitate the search for factors modifying its action.

II. ISOPRENYLATED PROTEINS

A. INTRODUCTION

Growing numbers of proteins are being shown to have polyisoprenyl groups attached to a modified cysteine at their carboxyl terminus. These groups, mainly containing either 15 or 20 carbon atoms, are widespread in yeast and mammalian cells.[20] Like fatty acyl groups, isoprenyl chains have been detected in GTP-binding proteins and other regulatory proteins, where they are thought to help stabilize the intracellular association of the proteins with membranes.[21]

B. BIOSYNTHESIS

As outlined in Chapter 6, isoprenoid chains grow by the sequential addition of 5 carbon isopentenyl groups. The formation of all-*trans* farnesyl pyrophosphate and all-*trans* geranylgeranyl pyrophosphate, the immediate precursors of protein-bound isoprenyl chains, is shown in Figure 3.

Based largely on studies of *ras* protein biosynthesis, Hancock et al.[5] proposed that isoprenyl chains are attached to target proteins as the initial step of posttranslational processing

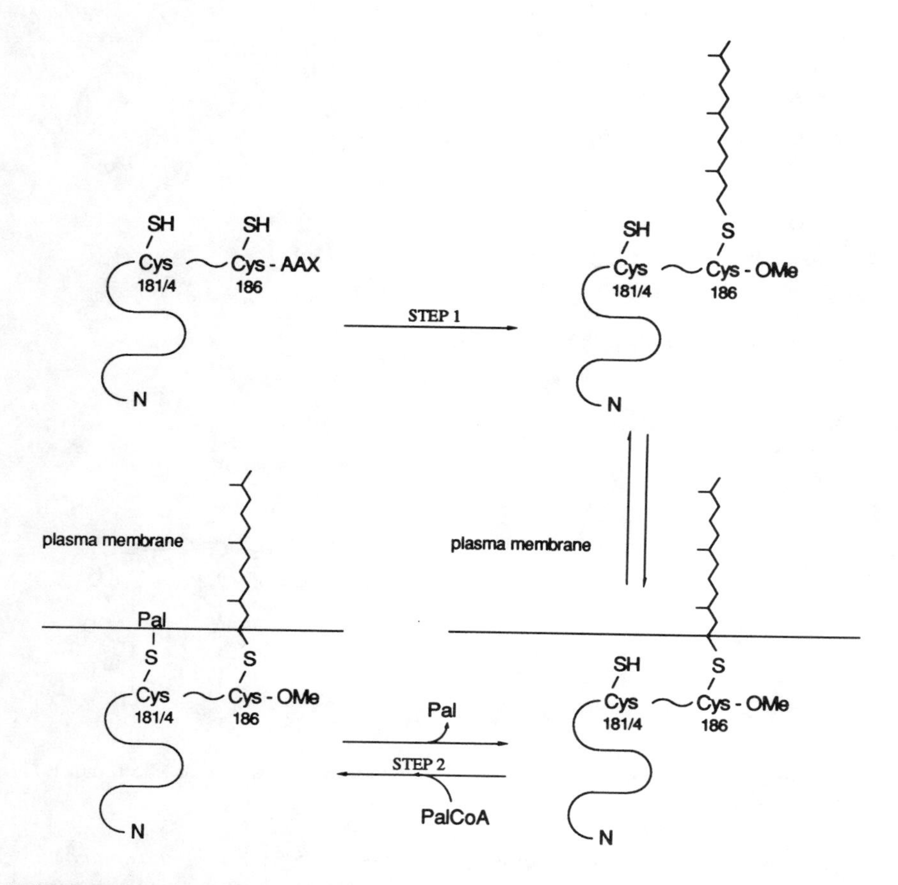

FIGURE 4. Farnesylation of a protein on a cysteine residue located at the fourth position from the carboxyl terminus (step 1), followed by hydrolysis of the terminal tripeptide and carboxylation of the farnesylated cysteine. After transfer to the membrane, the protein is then palmitoylated on a nearby cysteine.

(Figure 4). A soluble isoprenyl:protein transferase having two approximately 50 kDa subunits[22] catalyzes the transfer of a farnesyl group from the farnesyl pyrophosphate precursor to a cysteine residue 4 amino acids from the carboxyl terminus. This terminal motif is sometimes referred to as a *Caax* box because the second and third positions are typically occupied by aliphatic amino acids. The isoprenyl transferase, once thought to be very specific for this *Caax* box, now appears capable of isoprenylating *rab* proteins having the carboxyl terminal motif *GGCC*. [23]

Following isoprenylation, the terminal three amino acids are removed by a poorly characterized protease, and the isoprenylated cysteine, now C-terminal, is carboxymethylated. Many of the *ras* proteins modified in this way are also subsequently palmitoylated.

III. PHOSPHATIDYLINOSITOL-ANCHORED PROTEINS

A. INTRODUCTION

Since the discovery in the late 1970s that alkaline phosphatase and acetylcholinesterase could be released from the surface of intact cells by types of bacterial phospholipase C specific for phosphatidylinositol (PI), more than 50 proteins have been found linked to the extracellular face of the plasma membrane by a phosphatidylinositol glycan anchor.[24] The most intensively

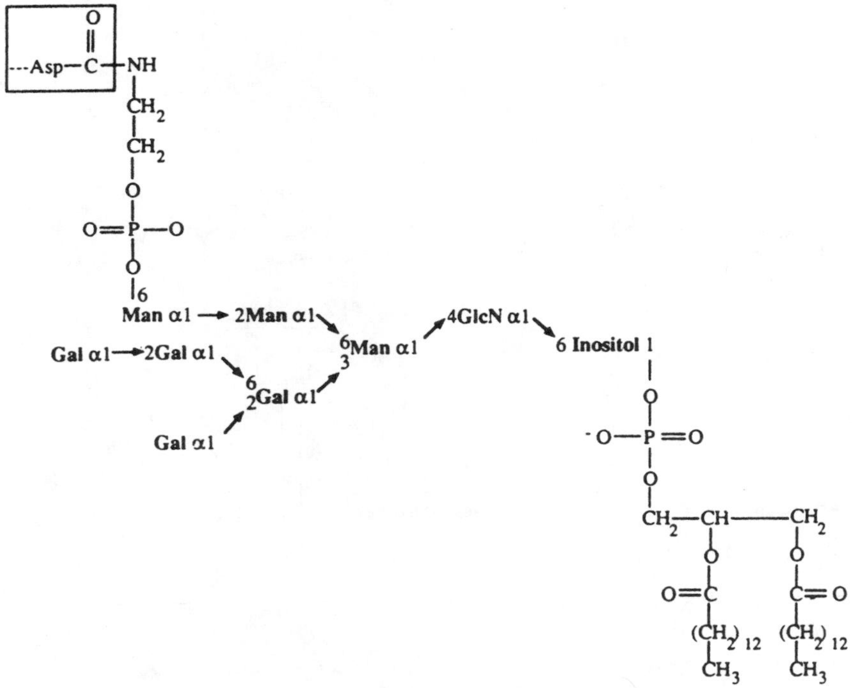

FIGURE 5. Structure of the PI glycan anchor of variant surface glycoprotein of *Trypanosoma brucei*, variant 117. The anchor is bound to the carboxyl terminus of the protein, upper box.

studied PI-anchored protein is the variant surface glycoprotein (VSG) of *Trypanosoma brucei*, the parasite responsible for sleeping sickness. This protein, which forms a 10 mm thick protective coating over the cell surface during certain stages of the organisms' life cycle, is attached to the membrane by the phosphatidylinositol glycan (PI glycan) chain shown in Figure 5. Several variations of this structure are known, featuring, for example, an additional phosphorylethanolamine or an absence of galactose units.[24] In *Dictyostelium* a ceramide derivative appears to replace phosphatidylinositol as the hydrophobic membrane-anchoring moiety.[25]

B. BIOSYNTHESIS

The formation of PI glycan anchor precursors has already been discussed in Chapter 7. Here we will review the process by which the anchor is attached to proteins, which, in most cases, are destined to reside on the outer face of the cell's plasma membrane.

Much of the available information on the assembly of the mature PI-anchored protein has been gained from work with trypanosomes, in which these components can account for as much as 10% of the total cellular protein.[24] But the general pattern of synthesis, as reviewed by Low[24] and by Cross,[26] appears to be similar in other cell types.

Preformed PI glycan anchor precursors exist in a pool available for attachment to the appropriate proteins within 1 to 5 min following synthesis of the proteins on the endoplasmic reticulum. Attachment of the PI glycan anchor occurs simultaneously with or immediately following cleavage of a peptide approximately 20 to 30 amino acids long from the carboxyl terminus of the preprotein. This signal domain must contain a short (15–20) sequence of hydrophobic residues, but no striking sequence homology has been found relating the PI-

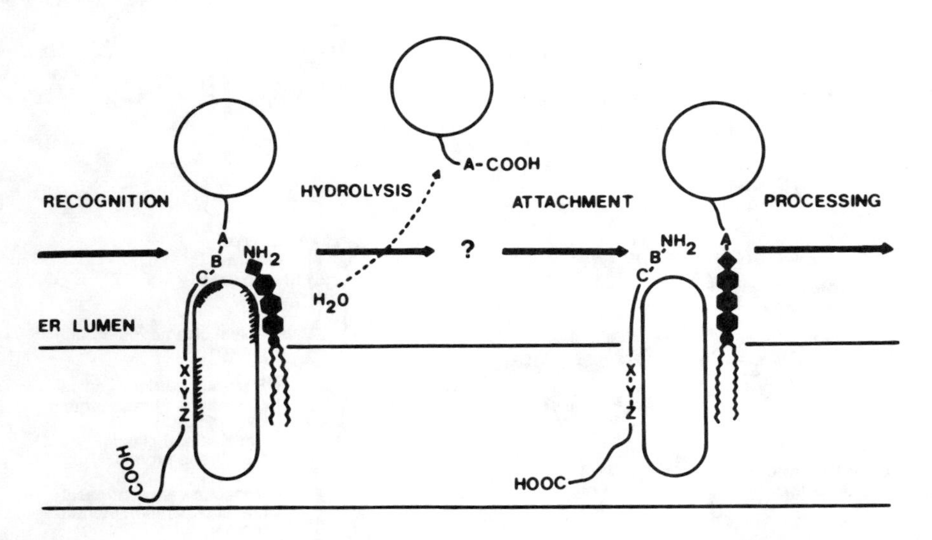

FIGURE 6. Attachment of protein to the GPI anchor. In this hypothetical representation of the attachment process at least three stages are envisaged. Immediately after translation is completed, the polypeptide interacts with an ER membrane protein that recognizes features of the C-terminal region. These features probably include at minimum the hydrophobic/hydrophilic balance (X-Y-Z) and the region immediately around the processing site (A-B-C) although in individual proteins upsteam regions of the polypeptide may also place steric or other constraints on the processing that cannot be generalized. The GPI anchor precursor is also recognized (presumably the polar head group) and attachment of the protein then occurs by a transamidase type of reaction involving the free amino group on the terminal ethanolamine residue of the GPI precursor and cleavage of the C-terminal region of the polypeptide. After attachment to the GPI-anchor the protein will be transported via the Golgi to the cell surface, possibly involving novel sorting and targeting mechanisms in addition to conventional posttranslational processing of N-linked glycans. Processing of the GPI anchor itself may take place after attachment as found for the attachment of galactose to VSG in the late ER or Golgi. (From Low, M. G., *Biochim. Biophys. Acta,* 988, 427–454, 1989. With permission.)

anchored proteins that have been analyzed. Following peptide removal, the PI glycan anchor is attached to the new terminal carboxyl group in a transamidase type of reaction involving the free amino group of the ethanolamine residue at the end of the PI glycan (Figure 6).

After its attachment to protein, the PI glycan anchor may be further embellished in Golgi vesicles by the addition of galactose or ethanolamine phosphate residues. Fatty acid remodeling of the PI glycan may also occur, probably before its linkage to protein, with the relatively long chain fatty acids of the original phosphatidylinositol precursor being replaced by myristate.[27] An additional and functionally very important modification detected on certain PI-anchored proteins is the attachment of palmitic acid to a free hydroxyl group of the inositol moiety. This fatty acyl group renders the PI glycan chain resistant to attack by phosphatidylinositol-specific phospholipase C.[28] Inositol acylation is not found on the VSG coating the *Trypanosoma brucei* bloodstream form, but its presence blocks phospholipase C cleavage of a related major protein from the surface of the insect midgut stage of *T. brucei*.[29] The factors determining which PI-anchored proteins will be thus modified by inositol acylation are not known, and there is no evidence that, once attached, this fatty acid can subsequently be removed from the anchor moiety.

Following its synthesis within the cell, the PI-anchored protein is disseminated to the cell surface via a highly specific process dictated by the presence of its PI glycan anchor. The role of the anchor in intracellular trafficking has been elegantly shown in epithelial type (MDCK) cells transfected to express the ectodomain of a herpes simplex protein (normally targeted to the basolateral surface) fused with the C-terminal 37 amino acids of decay accelerating factor

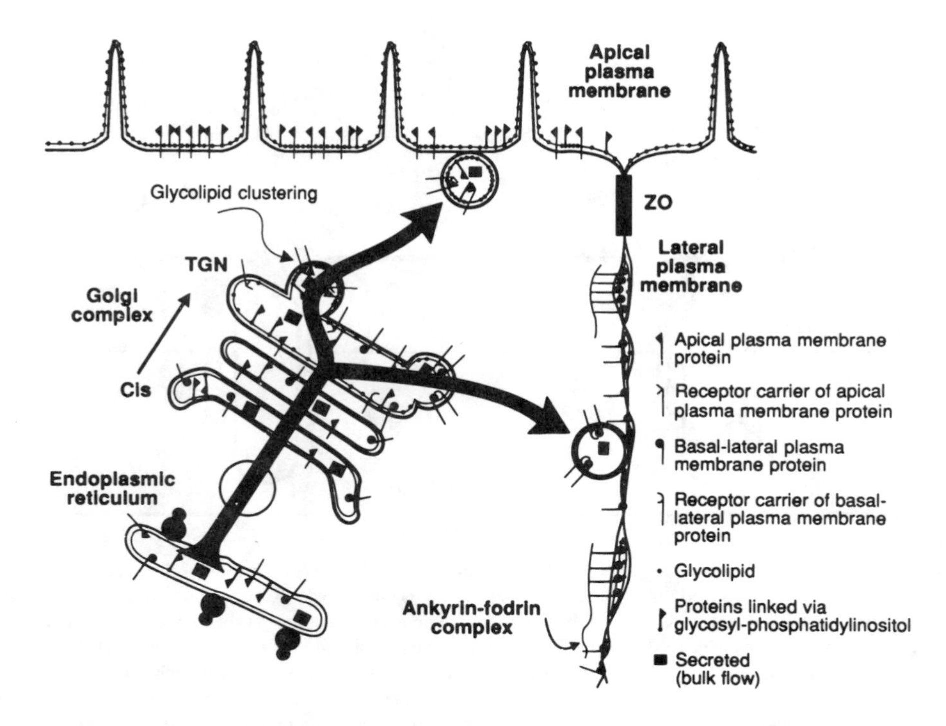

FIGURE 7. Proposed pathway for specific targeting of PI-anchored proteins to the apical plasma membrane of epithelial cells. (From Rodriguez-Boulan, E. and Nelson, W. J., *Science*, 245, 718–725, 1989. With permission.)

that constitute a PI glycan attachment site.[30] The resulting PI-anchored protein was efficiently transported to the apical membrane as are a variety of normal PI-anchored proteins. The mechanism governing this very specific targeting is not understood, but clustering of the PI-anchored proteins in post-Golgi vesicles destined for the plasma membrane may be facilitated by the apparent accumulation of glycosphingolipids in the same regions[31,32] (Figure 7).

C. CLEAVAGE OF THE PI GLYCAN ANCHORS

Lysis of *T. brucei* activates a hydrolase that converts the PI-anchored VSG from a membrane bound to a soluble form. This enzyme was shown to be a phospholipase C, specific for PI glycans and inactive towards phospholipids, including phosphatidylinositol (reviewed in Low[24]).

Immunolocalization studies have suggested that the *T. brucei* phospholipase C is associated with the cytosol-oriented face of internal membrane vesicles.[33] An enzyme having a similar activity has been purified from rat liver plasma membranes, and evidence for this type of phospholipase has also been reported for other cells (reviewed in Low[24]).

Most of the PI glycan-specific phospholipase Cs described to date have been intracellular, and their accessibility to the externally oriented PI-anchored proteins (see Figure 8) would seem to be limited. Therefore, great interest surrounded the discovery of a PI glycan-specific phospholipase D in mammalian plasma and serum.[34] The enzyme has been purified from bovine serum.[35] Although it exhibited little activity for the PI-anchored alkaline phosphatase on intact HeLa cells, it readily hydrolyzed that enzyme when the cells were lysed with detergent. Some as yet undetermined factors in plasma may be required for the hydrolysis of proteins anchored to living cells.

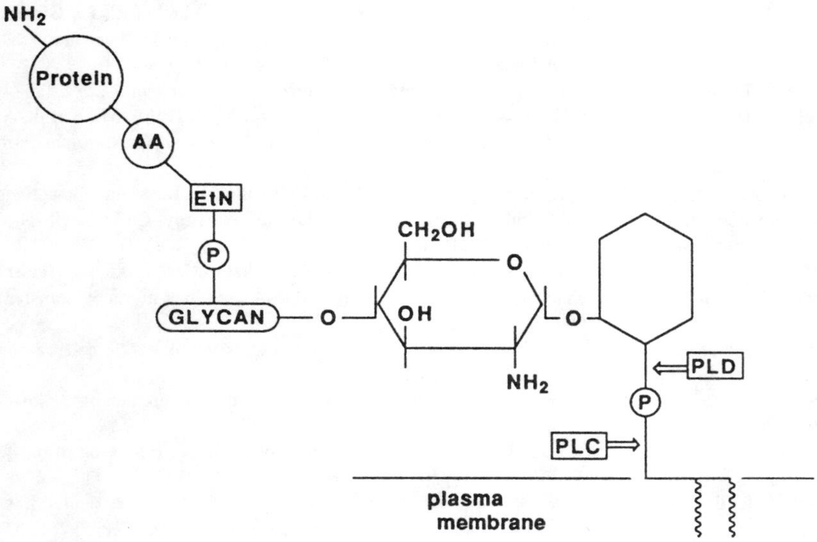

FIGURE 8. Sites of hydrolysis of externally oriented PI-anchored proteins by phospholipase C (PLC) and phospholipase D (PLD).

REFERENCES

1. **Folch, J. and Lees, M.,** Proteolipides, a new type of tissue lipoproteins, *J. Biol. Chem.,* 191, 807–817, 1951.
2. **Schmidt, M. F. G.,** Fatty acylation of proteins, *Biochim. Biophys. Acta,* 988, 411–426, 1989.
3. **Chow, M., Newman, J. F. E., Filman, D., Hogle, J. M., Rowlands, D. J., and Brown, F.,** Myristoylation of picornavirus capsid protein VP4 and its structural significance, *Nature,* 327, 482–486, 1987.
4. **Grand, R. J. A.,** Acylation of viral and eukaryotic proteins, *Biochem. J.,* 258, 625–638, 1989.
5. **Hancock, J. F., Magee, A. I., Childs, J. E., and Marshall, C. J.,** All *ras* proteins are polyisoprenylated but only some are palmitoylated, *Cell,* 57, 1167–1177, 1989.
6. **James, G. and Olson, E. N.,** Fatty acylated proteins as components of intracellular signaling pathways, *Biochemistry,* 29, 2623–2634, 1990.
7. **James, G. and Olson, E. N.,** Identification of a novel fatty acylated protein that partitions between the plasma membrane and cytosol and is deacylated in response to serum and growth factor stimulation, *J. Biol. Chem.,* 264, 20998–21006, 1989.
8. **Arumugham, R. G., Seid, R. C., Jr., Doyle, S., Hildreth, S. W., and Paradiso, P. R.,** Fatty acid acylation of the fusion glycoporotein of human respiratory syncytial virus, *J. Biol. Chem.,* 264, 10339–10342, 1989.
9. **Ovchinnikov, Y. A., Abdulaev, N. G., and Bogachuk, A. S.,** Two adjacent cysteine residues in the C-terminal cytoplasmic fragment of bovine rhodopsin are palmitylated, *FEBS Lett.,* 230, 1–5, 1988.
10. **Agrawal, H, C. and Agrawal, D.,** Effect of cycloheximide on palmitylation of PO protein of the peripheral nervous system myelin, *Biochem. J.,* 263, 173–177, 1989.
11. **Mattoo, A. K. and Edelman, M.,** Intramembrane translocation and posttranslational palmitoylation of the chloroplast 32-kDa herbicide-binding protein, *Proc. Natl. Acad. Sci. U.S.A.,* 84, 1497–1501, 1987.
12. **Staufienbiel, M.,** Fatty acids covalently bound to erythrocyte proteins undergo a differential turnover *in vivo*, *J. Biol. Chem.,* 263, 13615–13622, 1988.
13. **Wilcox, C., Hu, J.-S., and Olson, E. N.,** Acylation of proteins with myristic acid occurs cotranslationally, *Science,* 238, 1275–1278, 1987.
14. **Towler, D. and Glaser, L.,** Protein fatty acid acylation: enzymatic synthesis of an N-myristoylglycyl peptide, *Proc. Natl. Acad. Sci. U.S.A.,* 83, 2812–2816, 1986.

15. **Towler, D. A., Adams, S. P., Eubanks, S. R., Towery, D. S., Jackson-Machelski, E., Glaser, L., and Gordon, J. I.**, Myristoyl CoA: protein N-myristoyltransferase activities from rat liver and yeast possess overlapping yet distinct peptide substrate specificities, *J. Biol. Chem.*, 263, 1784–1790, 1988.
16. **Glover, C. J., Goddard, C., and Felsted, R. L.**, N-myristoylation of p60^{src}, *Biochem. J.*, 250, 485–491, 1988.
17. **Heuckenroth, R. O., Jackson-Machelski, E., Adams, S. P., Kishore, N. S., Huhn, M., Katoh, A., Lu, T., Gokel, G. W., and Gordon, J. I.**, Novel fatty acid substrates for myristoyl-CoA:protein N-myristoyltransferase, *J. Lipid Res.*, 31, 1121–1129, 1990.
18. **Paige, L. A., Zheng, G., DeFrees, S. A., Cassady, J. M., and Geahlen, R. L.**, Metabolic activation of 2-substituted derivatives of myristic acid to form potent inhibitors of myristoyl-CoA:protein N-myristoyltransferase, *Biochemistry*, 29, 10566–10573, 1990.
19. **Towler, D. A., Adams, S. P., Eubanks, S. R., Towery, D. S., Jackson-Machelski, E., Glaser, L., and Gordon, J. I.**, Purification and characterization of yeast myristoyl CoA:protein N-myristoyl transferase, *Proc. Natl. Acad. Sci. U.S.A.*, 84, 2708–2712, 1987.
20. **Glomset, J. A., Gelb, M. H., and Farnsworth, C. C.**, Prenyl proteins in eukaryotic cells: a new type of membrane anchor, *Trends Biochem. Sci.*, 15, 139–142, 1990.
21. **Maltese, W. A.**, Posttranslational modification of proteins by isoprenoids in mammalian cells, *FASEB J.*, 4, 3319–3328, 1990.
22. **Reiss, Y., Goldstein, J. L., Seabra, M. C., Casey, P. J., and Brown, M. S.**, Inhibition of purified p21^{ras} farnesyl:protein transferase by Cys-AAX tetrapeptides, *Cell*, 62, 81–88, 1990.
23. **Kinsella, B. T. and Maltese, W. A.**, *rab* GTP-binding proteins implicated in vesicular transport are isoprenylated *in vitro* at cysteines within a novel carboxyl-terminal motif, *J. Biol. Chem.*, 266, 8540–8544, 1991.
24. **Low, M. G.**, The glycosyl-phosphatidylinositol anchor of membrane proteins, *Biochim. Biophys. Acta*, 988, 427–454, 1989.
25. **Stadler, J., Keenan, T. W., Bauer, G., and Gerisch, G.**, The contact site A glycoprotein of *Dictyostelium discoideum* carries a phospholipid anchor of a novel type, *EMBO J.*, 8, 371–377, 1989.
26. **Cross, G. A. M.**, Glycolipid anchoring of plasma membrane proteins, *Annu. Rev. Cell Biol.*, 6, 1–37, 1990.
27. **Mayor, S., Menon, A. K., and Cross, G. A. M.**, Glycolipid precursors for the membrane anchor of *Trypanosoma brucei* variant surface glycoprotein, *J. Biol. Chem.*, 265, 6174–6181, 1990.
28. **Roberts, W. L., Santikarn, S., Reinhold, V. N., and Rosenberry, T. L.**, Structural characterization of the glycoinositol phospholipid membrane anchor of human erythrocyte acetylcholinesterase by fast atom bombardment mass spectrometry, *J. Biol. Chem.*, 263, 18776–18784, 1988.
29. **Field, M. C., Menon, A. K., and Cross, G. A. M.**, Developmental variation of glycosylphosphatidylinositol membrane anchors in *Trypanosoma brucei*, *J. Biol. Chem.*, 266, 8392–8400, 1991.
30. **Lisanti, M. P., Caras, I. W., Gilbert, J., Hanzel, D., and Rodriguez-Boulan, E.**, Vectorial apical delivery and slow endocytosis of a glycolipid-anchored fusion protein in transfected MDCK cells, *Proc. Natl. Acad. Sci. U.S.A.*, 87, 7419–7423, 1990.
31. **Montesano, R., Roth, J., Robert, A., and Orci, L.**, Non-coated membrane invaginations are involved in binding and internalization of cholera and tetanus toxins, *Nature*, 296, 651–653, 1982.
32. **Rodriguez-Boulan, E. and Nelson, W. J.**, Morphogenesis of the polarized epithelial cell phenotype, *Science*, 245, 718–725, 1989.
33. **Bulow, R., Griffiths, G., Webster, P., Sierhof, Y.-D., Opperdoes, F. R., and Overath, P.**, Intracellular localization of the glycosyl-phosphatidylinositol-specific phospholipase C of *Trypanosoma brucei*, *J. Cell Sci.*, 93, 233–240, 1989.
34. **Davitz, M. A., Hereld, D., Shak, S., Krakow, J., Englund, P. T., and Nussenzweig, V.**, A glycan-phosphatidylinositol-specific phospholipase D in human serum, *Science*, 238, 81–84, 1987.
35. **Huang, K.-S., Li, S., Fung, W.-J. C., Hulmes, J. D., Reik, L., Pan, Y.-C. E., and Low, M. G.**, Purification and characterization of glycosyl-phosphatidylinositol-specific phospholipase D, *J. Biol. Chem.*, 265, 17738–17745, 1990.

Chapter 9

THE SPECIFICITY AND RATES OF INTRACELLULAR LIPID MOVEMENT

I. INTRODUCTION

Although key steps in the biosynthesis of certain lipids take place in mitochondria, peroxisomes, plasma membranes, or chloroplasts, the vast majority of membrane lipid biosynthetic reactions of animal cells takes place in the endoplasmic reticulum. With the exception of the abundant glycolipids, this also holds true for plants. Studies on the topography of these enzymes shows them to be located exclusively on the cytoplasmic side of the E.R.[1] In recent years much research has been devoted to determining how the lipids produced on the E.R. (and also those produced elsewhere) are disseminated to other cellular locations and to the lumenal leaflet of the E.R. itself.

The process of intracellular lipid movement is complicated by the fact that each membrane, and even each leaflet of each membrane, maintains a characteristic mixture of lipid molecules drawn from a larger assortment of lipids available to it in other parts of the cell. Understanding how the cell preserves this dynamic balance, in which lipid molecules, although readily exchangeable, are equilibrated in unique, fixed proportions within each functionally distinct membrane, requires a consideration of several interrelated phenomena, as described below.

II. INTRACELLULAR MOVEMENT OF LIPIDS VIA "MEMBRANE FLOW"

The concept of "membrane flow", as developed in 1971 by Franke et al.[2] postulates that "the biogenesis of certain membrane is accomplished by the physical transfer of membrane material from one cell component to another in the course of their formation or normal functioning." The basis for this idea comes mainly from ultrastructural, cytochemical, and autoradiographic evidence gathered during study of the Golgi apparatus. Newly formed vesicles pinched off from the endoplasmic reticulum were found to be fused with the forming or *cis* face of the highly organized stack of flattened Golgi cisternae (Figure 1). The contents and membranes of each cisternum then underwent enzymatic modifications as it moved slowly across the stack to reach the maturing or *trans* face. From this latter membrane secretory vesicles were then released for movement through the cytoplasm to the plasma membrane or other membrane destinations.

Recent studies have shown the flow of membrane material to be more complex than illustrated in Figure 1. It is now clear that the Golgi cisternae are themselves relatively stationary and that material is fed progressively from one cisternum to the next by diffusable vesicular carriers.[3] These vesicles gradually and selectively ferry proteins and lipids to the *trans* face for further dissemination, mainly to the plasma membrane.

Multiple opportunities for the selective transport of both proteins and lipids exist in this system. Most study has been devoted to protein trafficking, but examples of highly directed lipid movement are also known. One particularly well characterized example in epithelial cells has been summarized by van Meer.[4] The plasma membrane of the epithelial cell is divided into an apical and a basolateral domain separated by an encircling band of tight junctions. The tight junctions prevent lateral diffusion of lipids in the externally oriented plasma membrane lipid monolayer but not those in the cytoplasmically oriented monolayer. The externally oriented

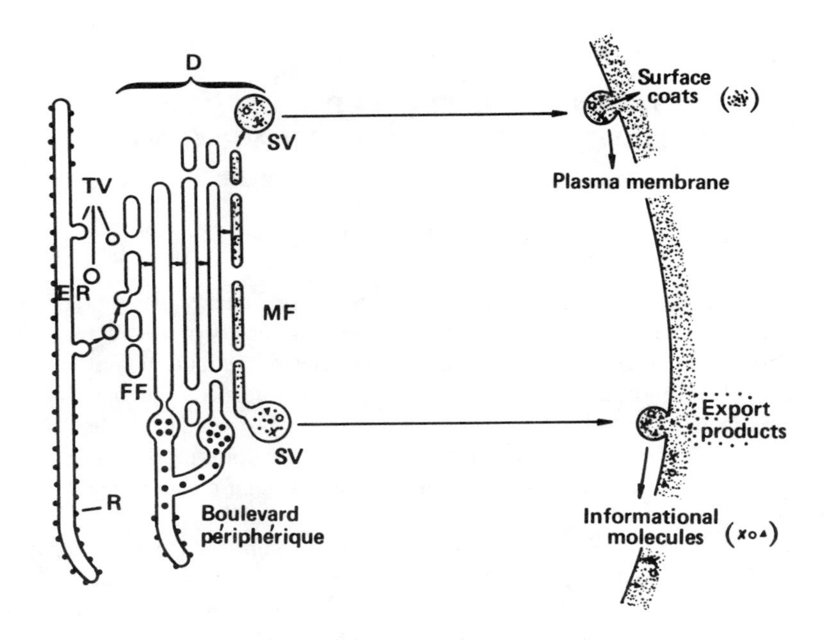

FIGURE 1. Diagrammatic representation of endomembrane functioning in membrane flow and differentiation. Transition vesicles (TV) derived from the endoplasmic reticulum (ER) migrate to the Golgi apparatus, where they fuse to add membrane to the forming face (FF) of the dictyosomes (D). The vesicle contents and membranes then mature and are ultimately discharged as secretory vesicles (SV) and/or fragments of cisternae from the maturing face (MF) of the dictyosomes. The vesicles migrate to the cell surface where their bounding membranes fuse with the plasma membrane to discharge the secretory product into the cell's exterior. (From Morré, D. J. and Ovtracht, L., *Int. Rev. Cytol.*, (Suppl. 5), 61, 1977. With permission.)

half of the apical membrane is greatly enriched in glycosphingolipids and, in some tissues, sphingomyelin, while the basolateral membrane contains a high proportion of phosphatidylcholine. It has been proposed that glycosphingolipids and phosphatidylcholine molecules undergo a lateral segregation in the luminal monolayer of the *trans* Golgi membrane and then bud separately into vesicles having different cellular destinations.[5] Glycosphingolipids do have a tendency to self-associate by hydrogen bonding.

Accompanying the glycosphingolipids as they cluster are certain proteins, particularly those linked to the membrane by a glycosylphosphatidylinositol anchor. The presence of these proteins in the glycosphingolipid enriched visicles may be influential in targeting them to apical membranes[6] (Figure 3).

The pronounced enrichment of sphingomyelin and glycosphingolipids in the plasma membrane's outwardly oriented lipid face can be traced back to the final steps of their biosynthesis on the luminal surface of the Golgi apparatus (reviewed by van Meer). Since there is apparently no transmembrane flip flop of these lipids, they remain highly oriented during transport of Golgi-derived vesicles to the plasma membrane and subsequent fusion with the plasma membrane, thereby exposing the sphingolipids on the cell surface.

Most (80 to 95%) of an animal cell's free cholesterol is localized in the plasma membrane.[7,8] Movement of cholesterol through the Golgi system has also been examined as a mechanism for achieving this pronounced enrichment at the cell periphery. Approximately 50% of the endoplasmic reticulum lipids have been estimated to cycle through the *cis* Golgi every 10 minutes,[9] and this has been assumed to be the main avenue for cholesterol trafficking.

More recently, Urbani and Simoni[10] compared the simultaneous movement of cholesterol

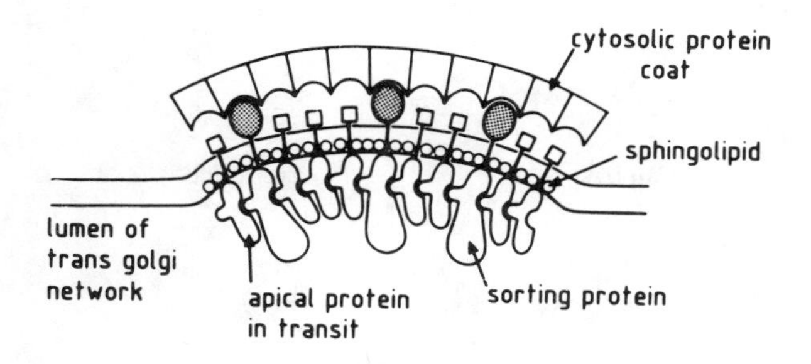

FIGURE 2. Schematic diagram of the proposed apical precursor microdomain in the membrane of the trans-Golgi network. The sphingolipids are clustered in the luminal leaflet of the bilayer. The putative sorting proteins form a surface on the cytosolic side, complementary to a cytosolic protein coat that fixes the domain spatially. The curving of the membrane into a vesicle is assumed to result from the structure of the coat protein. (From Simmons, K. and van Meer, G., *Biochemistry,* 27, 6197–6202, 1988. With permission.)

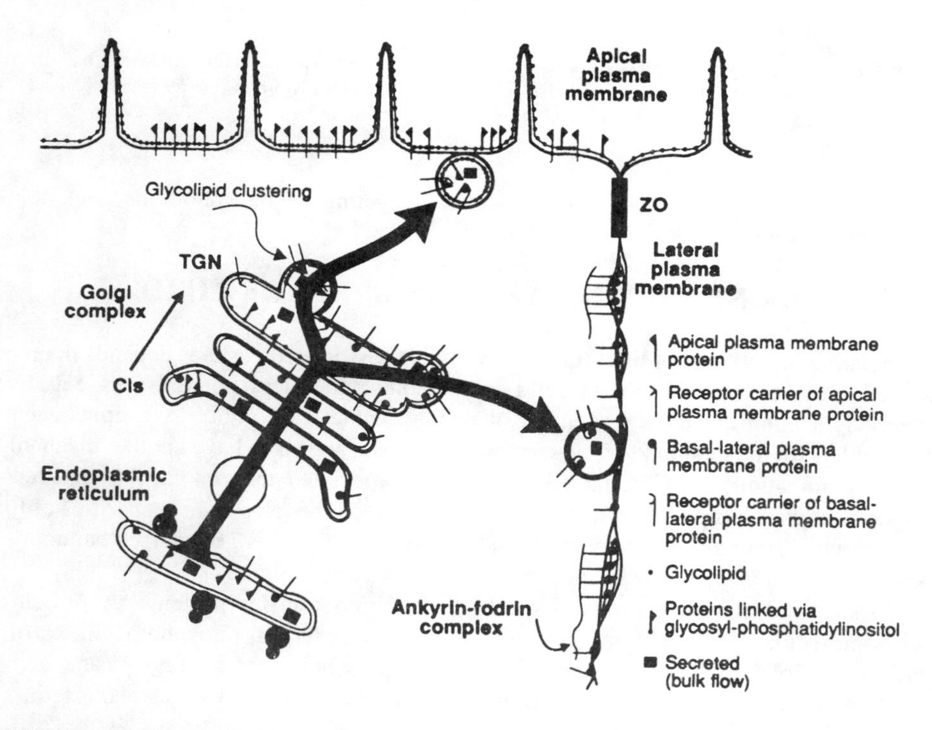

FIGURE 3. Postulated mechanism of glycolipid trafficking to the apical plasma membrane of epithelial cells. (From Rodriguez-Boulan, E. and Nelson, W. J., *Science,* 245, 718–725, 1989. With permission.)

and vesicular stomatitis virus G protein from their biosynthetic site in the endoplasmic reticulum to the plasma membrane of Chinese hamster ovary cells. By manipulating the transport rates using low temperature and the protein transport inhibitor Brefeldin A, the authors concluded that cholesterol and the G protein are both disseminated by vesicles but follow different paths to the plasma membrane, with cholesterol bypassing the Golgi apparatus as illustrated in Figure 4.

The mechanism leading to cholesterol enrichment in the plasma membrane is not understood. It may feature a selective binding of cholesterol to sphingomyelin, which is also highly concentrated there.[4] Alternatively, cholesterol making its way through the Golgi stacks may

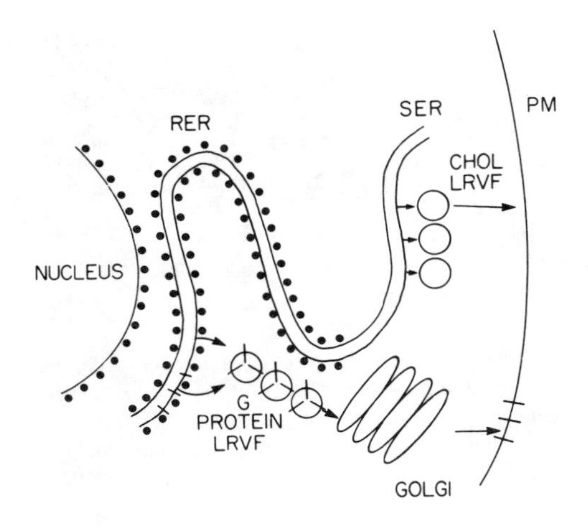

FIGURE 4. G protein and cholesterol (CHOL) are found in separate vesicles. G protein vesicle may bud off from the rough ER (RER), move through the Golgi, and be sorted to the PM. Cholesterol-rich vesicles may be derived from smooth ER (SER) and proceed directly to the PM. (From Urbani, L. and Simoni, R. D., *J. Biol. Chem.*, 265, 1919–1923, 1990. With permission.)

become concentrated there due to the selective recycling of phospholipids back to the endoplasmic reticulum.[8]

III. UNIMOLECULAR TRANSPORT OF LIPIDS

The relative importance of lipid transport via Golgi-associated vesicles depends upon cell structure and varies widely from one cell type to another. In at least some tissues, e.g., liver, the very rapid interorganellar movement of newly synthesized, radioactive lipids seemed incompatible with the membrane flow concept.[11] All cells appear to have another mechanism for the dissemination of lipids, namely, via protein-mediated monomolecular transfer. A multitude of studies has provided masses of information on the rates and specificities of lipid transfer proteins from animals,[12] plants[13] and microorganisms.[14] A generalized summary of the known unimolecular lipid transport pathways is shown in Figure 5.

Rat tissues contain three rather distinct phospholipid transfer proteins. A protein of 32 kDa molecular weight catalyzes the specific transfer of phosphatidylinositol, a second protein of 24.6 kDa molecular weight transfers phosphatidylcholine, and a third 14.5 kDa molecular weight protein transfers most phospholipids as well as cholesterol and glycolipids.[15]

The mechanisms of protein-mediated lipid transfer has been studied in considerable detail. The phosphatidylcholine-transfer protein appears to extract a single phospholipid molecule from the donor membrane and carry it through the aqueous milieu until an acceptor membrane is encountered.[15] The 14.5 kDa nonspecific lipid transfer protein, on the other hand, does not readily abstract a lipid molecule from the bilayer. Noting that inter-membrane transfer of phospholipids mediated by this latter protein was closely correlated with the rates of spontaneous transfer for the same lipids, Nichols and Pagano[16] proposed that the nonspecific lipid transfer protein may interact with the membrane interface, lowering the energy barrier to lipid monomer-interface dissociation and association. A similar proposal has been made to explain the enhanced transfer of cholesterol by this protein.[17] The primary structure of a protein capable of catalyzing the transfer of various glycosphingolipids and glyceroglycolipids but not phospholipids has recently been determined.[18] This protein has no homology with other known lipid transfer proteins.

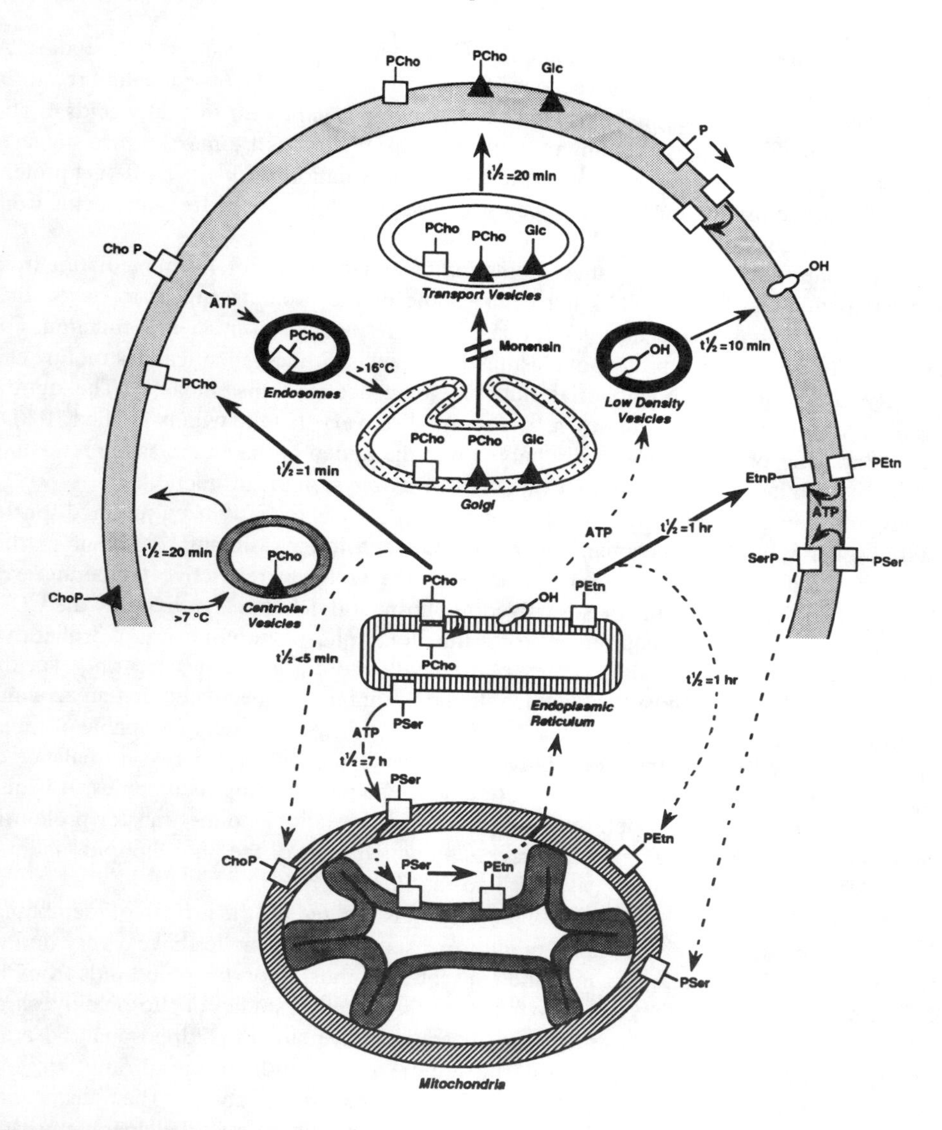

FIGURE 5. Schematic summary of unimolecular lipid translocations in animal cells. Major lipid translocation processes are shown: □, the diacylglycerol portion of phospholipids. ▲, the ceramide portion of sphingolipids. OH is the symbol for cholesterol. Abbreviations: Cho, choline; Etn, ethanolamine; Ser, serine; Glc, glucose. The dotted lines indicate that the translocation processes have been shown to occur but studies of mechanisms have not been conducted. The placement of the lipid in a given leaflet of the membrane bilayer is deliberate and indicates experimental evidence for the location. (From Voelker, D. R., *Experientia,* 46, 569–579, 1990. With permission.)

It is well established that lipid transfer from one bilayer to another can indeed occur without protein assistance in preparations of small unilamellar vesicles,[19] and in the large unilamellar vesicles more typical of cellular membranes.[20] Spontaneous phosphatidic acid exchange between rat liver microsomes and mitochondria has also been reported to occur in a nonprotein mediated fashion, apparently by diffusion of phosphatidic acid through the aqueous phase.[21] The enhanced lipid exchange normally observed between cell organelles may result from collisional exchange rendered more effective by the presence of lipid transfer proteins. In certain tissues, at least, the nonspecific lipid transfer protein has been shown to exist in a membrane-bound as well as a free form.[22]

Intracellular fatty acid binding proteins have been the subject of many recent studies. A group of 14 to 15 KDa proteins can account for as much as 2 to 5% of cytosolic protein in liver and adipose tissue, where they provide a high binding capacity for free fatty acids.[23] The physiological function of these binding proteins is not clear, although some evidence suggests possible roles in fatty acid transport, esterification, and oxidation. A distinct, 40 kDa protein associated with the plasma membrane appears to assist in the uptake of free fatty acids from extracellular sources.[23]

The quantitative significance of these lipid binding and transfer proteins in mediating lipid movement within the living cell is not clear. Some of the proteins appear to have other functions, and it is possible that in certain cases lipid transfer is an incidental function. For example, a fatty acid binding protein of plasma membrane is virtually identical in structure and apparently in function to mitochondrial glutamate-oxaloacetate transaminase.[24] The significance of this similarity is not yet known. The recent discovery that the widely studied 10 kDa plant lipid transfer protein is in fact discharged into the lumen of the endoplasmic reticulum following its synthesis has forced a reevaluation of its physiological function.[24a]

A new physiological role related to its transport ability has recently been reported for the yeast phosphatidylinositol/phosphatidylcholine transfer protein. Mutants deficient in this protein are incapable of protein transport through the Golgi at restrictive temperatures.[25] Mutations affecting several enzymes involved in phosphatidylcholine synthesis by the CDP-choline pathway bypass the requirement for a functional phosphatidylinositol/phosphatidylcholine transfer protein.[26] In these cases phosphatidylcholine is synthesized only via the sequential methylation of phosphatidylethanolamine. It might be speculated that an accumulation of ER-derived phosphatidylcholine molecules on transport proteins incapable of transferring them to elements of the Golgi prevents the establishment of a necessary balance of phosphatidylcholine and phosphatidylinositol among the participating membranes. It is also possible that the nonfunctional phosphatidylinositol/phosphatidylcholine transfer protein in some way disrupts the normal retrieval process by which most of the phospholipids shunted into the Golgi membranes are cycled back into the E.R. for reuse.

A substantial portion of the cell's lipid production is utilized for the growth of membranes within organelles such as mitochondria or chloroplasts. These organelles have a very limited selection of lipid biosynthetic enzymes and consequently must import certain lipids from the E.R. to supplement the ones they can form for themselves. Recent studies of phosphatidylserine translocation into yeast mitochondria have utilized the conversion of [^{3}H]phosphatidylserine to [^{3}H]phosphatidylethanolamine by the inner mitochondrial membrane enzyme phosphatidylserine decarboxylase as a means of assessing the rate of entry.[27] The translocation of phosphatidylserine does not require an electrochemical gradient across the inner membrane. It appears that the translocation of phosphatidylserine may involve the contact sites between outer and inner membranes previously shown to participate in protein import. Thus cell fractions enriched in contact sites were depleted in [^{3}H]phosphatidylserine and enriched in [^{3}H]phosphatidylethanolamine, consistent with localization of the decarboxylase there.[28]

IV. THE MAINTENANCE OF MEMBRANE LIPID ASYMMETRY

As described in Chapter 1, the evidence for asymmetry of certain cellular membranes with respect to their lipid molecules is now compelling.[29] Because establishing that asymmetry exists usually requires gaining selective access to one side of the membrane, most studies have involved cell surface membranes for the sake of convenience. Here the most frequently observed patterns have featured a predominance of choline-containing phospholipids and

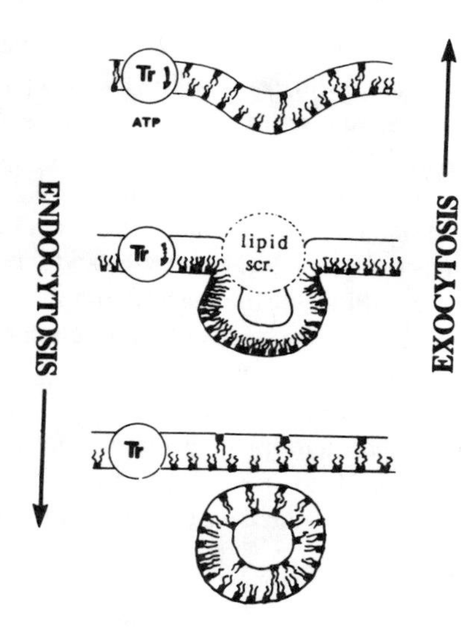

FIGURE 6. Schematic representation of a possible role of the aminophospholipid translocase (Tr) in the triggering of endocytosis and/or membrane budding (from top to bottom). Only the amino-phospholipids have been represented. The accumulation of amino-phospholipids on one side of the membrane due to the activity of the translocase is accompanied by membrane bending. At a certain threshold, endocytosis (or budding) takes place with possibly partial lipid scrambling (lipid scr) during the separation of the two membranes (i.e., between steps 2 and 3). The curvature does not necessarily take place near the translocase. Either the underlying cytoskeleton proteins or the clustering of receptors within the membrane determines the initiation sites for bending. Exocytosis may be visualized as the opposite scenario (from bottom to top). Parial lipid scrambling during endocytosis and exocytosis will supply the outer layer with the necessary aminophospholipids for a new endocytic process. Hence, a continuous ATP-driven lipid flow can take place. (Modified from Devaux, P. F., *Biochemistry*, 30, 1163–1173, 1991. With permission.)

glycosphingolipids in the externally oriented monolayer and an enrichment of phosphatidylethanolamine and phosphatidylserine in the internal monolayer. Membranes of the *trans*-cisternae of the Golgi apparatus and the vacuoles involved in the endocytotic pathway are plasma membrane-like in their lipid asymmetry.[4] However, there is much less, if any, lipid asymmetry in intracellular membranes such as the endoplasmic reticulum.[30]

The question to be addressed here is how lipid asymmetry is achieved and maintained amid all the dynamic lipid fluxes present in the cell. At least three quite different mechanisms for establishing membrane asymmetry have been considered experimentally.

The first possibility involves the establishment of a lipid asymmetry at the site of lipid biosynthesis, which means, in most cases, the endoplasmic reticulum. Phosphatidylcholine, phosphatidylethanolamine, and probably other phospholipids undergo a rapid transmembrane movement in the endoplasmic reticulum that does not require metabolic energy but seems to involve a membrane protein, popularly called a flippase (reviewed by Bishop and Bell[1]). Lipid asymmetry is therefore prevented from developing.

An asymmetric distribution of lipids has been shown to develop in artificial phospholipid membranes strictly through physical perturbations, e.g., by a transmembrane pH gradient[31] or through lipid phase separation,[20] but there is no strong evidence to support this mechanism *in vivo*.

It is likely that lipid asymmetry at the plasma membrane (and probably endocytotic system membrane) level is established primarily by an aminophospholipid translocase capable of establishing a predominantly inward orientation of phosphatidylserine and phosphatidylethanolamine (Figure 6). The lipid translocation mediated by this protein requires ATP and is inhibited by thiol-binding reagents and elevated intracellular levels of Ca[24] and vanadate.[32] While much of the early work was done with easily followed analogs of the aminophospholipids, endogenous phosphatidylserine was shown to appear spontaneously on the platelet surface following thiol group oxidation and then be returned to its normal, asymmetric position facing inward after reduction of the thiol groups with dithiothreitol.[33]

Once the phospholipid asymmetry is established, it may be stabilized to some extent by other structural elements, such as protein 4.1 of the erythrocyte membrane.[34] The appearance of phosphatidylserine on the external face of the plasma membrane leads to shape changes in platelets[35] and to sequestration of erythrocytes by the reticuloendothelial system.[36]

Other physiologically useful manifestations of membrane lipid asymmetry have been summarized by Devaux.[27] For example, the physical properties of the inner, aminophospholipid side would facilitate endocytosis and membrane fusion events while a sizable layer of water molecules associated with the choline head groups of the phosphatidylcholine/sphingomyelin-enriched side would discourage close contacts with adjacent cells.

REFERENCES

1. **Bishop, W. R. and Bell, R. M.,** Assembly of phospholipids into cellular membranes: biosynthesis, transmembrane movement and intracellular translocation, *Annu. Rev. Cell Biol.*, 4, 579–610, 1988.
2. **Franke, W. W., Morré, D. J., Deumling, B., Cheetham, R. D., Kartenbeck, J., Jarasch, E. D., and Zentgraf, H. W.,** Synthesis and turnover of membrane proteins in rat liver: an examination of the membrane flow hypothesis, *Z. Naturforsch, Teil B*, 26, 1031–1039, 1971.
3. **Rothman, J. E. and Orci, L.,** Movement of proteins through the Golgi stack: a molecular dissection of vesicular transport., *FASEB J.*, 4, 1460–1468, 1990.
4. **van Meer, G.,** Lipid traffic in animal cells, *Annu. Rev. Cell. Biol.*, 5, 247–275, 1989.
5. **Simmons, K. and van Meer, G.,** Lipid sorting in epithelial cells, *Biochemistry*, 27, 6197–6202, 1988.
6. **Rodriguez-Boulan, E. and Nelson, W. J.,** Morphogenesis of the polarized epithelial cell phenotype, *Science*, 245, 718–725, 1989.
7. **Lange, Y. and Ramos, B. V.,** Analysis of the distribution of cholesterol in the intact cell, *J. Biol. Chem.*, 258, 15130–15134, 1983.
8. **Reinhart, M. P.,** Intracellular sterol trafficking, *Experientia*, 46, 599–611, 1990.
9. **Wieland, F. T., Gleason, M. L., Serafini, T. A., and Rothman, J. E.,** The rate of bulk flow from the endoplasmic reticulum to the cell surface, *Cell*, 50, 289–300, 1987.
10. **Urbani, L. and Simoni, R. D.,** Cholesterol and vesicular stomatitis virus G protein take separate routes from the endoplasmic reticulum to the plasma membrane, *J. Biol. Chem.*, 265, 1919–1923, 1990.
11. **Vance, J. E.,** Compartmentalization of differential labeling of phospholipids of rat liver subcellular membranes, *Biochim. Biophys. Acta*, 963, 10–20, 1988.
12. **Voelker, D. R.,** Lipid transport pathways in mammalian cells, *Experientia*, 46, 569–579, 1990.
13. **Arondel, V. and Kader, J. C.,** Lipid transfer in plants, *Experientia*, 46, 579–585, 1990.
14. **Daum, G. and Paltauf, F.,** Lipid transport in microorganisms, *Experientia*, 46, 586–592, 1990.
15. **Wirtz, K. W. A. and Gadella, T. W. J., Jr.,** Properties and modes of action of specific and non-specific phospholipid transfer proteins, *Experientia*, 46, 592–599, 1990.
16. **Nichols, J. W. and Pagano, R. E.,** Resonance energy transfer of protein-mediated lipid transfer between visicles, *J. Biol. Chem.*, 258, 5368–5371, 1983.
17. **van Amerongen, A., Demel, R. A., Westerman, J., and Wirtz, K. W. A.,** Transfer of cholesterol and oxysterol derivatives by the nonspecific lipid transfer protein (sterol carrier protein): a study of its mode of action, *Biochim. Biophys. Acta.*, 1004, 36–43, 1989.
18. **Abe, A.**, Primary structure of glycolipid transfer protein from pig brain, *J. Biol. Chem.*, 265, 9634–9637, 1990.
19. **James, J. D. and Thompson, T. E.,** Spontaneous phosphatidylcholine transfer by collision between vesicles at high lipid concentration, *Biochemistry*, 28, 129–134, 1989.

20. **Wimley, W. C. and Thompson, T. E.,** Exchange and flip-flop of dimyristoylphosphatidylcholine in liquid-crystalline gel, and two-component, two-phase large unilamellar vesicles, *Biochemistry,* 29, 1296–1303, 1990.
21. **Baránska, J. and Wojtczak, L.,** Non-protein-mediated transfer of phosphatidic acid between microsomal and mitochondrial membranes, *Arch. Biochem. Biophys.,* 260, 301–308, 1988.
22. **van Amerongen, A., van Noort, M., van Beckhoren, J. R. C. M., Rommerts, J. O., and Wirtz, K. W. A.,** The subcellular distribution of the nonspecific lipid transfer protein (sterol carrier protein) in rat liver and adrenal gland, *Biochim. Biophys. Acta.,* 1001, 243–248, 1989.
23. **Clarke, S. D. and Armstrong, M. K.,** Cellular lipid binding proteins: expression, function, and nutritional regulation, *FASEB J.,* 3, 2480–2487, 1989.
24. **Berk, P. D., Wada, H., Horio, Y., Potter, B. J., Sorrentino, D., Zhou, S.-L., Isola, L. M., Stump, D., Kiang, C.-L, and Thung, S.,** Plasma membrane fatty acid binding protein and mitochondrial glutamic oxaloacetic transaminase of rat liver are related, *Proc. Natl. Acad. Sci. U.S.A.,* 87, 3484–3488, 1990.
24a. **Madrid, S.M. and von Wettstein, D.,** Reconciling contradictory notions on lipid transfer proteins in higher plants, *Plant Physiol. Biochem.* 29, 705–711, 1991.
25. **Bankaitis, V. A., Aitken, J. R., Cleves, A. E., and Dowhan, W.,** An essential role for a phospholipid transfer protein in yeast Golgi function, *Nature* 347, 561–562, 1990.
26. **Cleves, A. E., McGee, T. P., Whitters, E. A., Champion, K. M., Aitken, J. R., Dowhan, W., Goebl, M., and Bankaitis, V. A.,** Mutations in the CDP-choline pathway for phospholipid biosynthesis bypass the requirement for an essential phospholipid transfer protein, *Cell,* 64, 789–800, 1991.
27. **Simbeni, R., Paltauf, F., and Daum, G.,** Intramitochondrial transfer of phospholipids in the yeast, *Saccharomyces cerevisiae, J. Biol. Chem.,* 265, 281–285, 1990.
28. **Simbeni, R., Pon, L., Zinser, E., Paltauf, F., and Daum, G.,** Mitochondrial membrane contact sites of yeast. Characterization of lipid components and possible involvement in intramitochondrial translocation of phospholipids, *J. Biol. Chem.,* 266, 10047–10049, 1991.
29. **Devaux, P. F.,** Static and dynamic lipid asymmetry in cell membranes, *Biochemistry,* 30, 1163–1173, 1991.
30. **Hermann, A., Zachowski, A., and Devaux, P. F.,** Protein-mediated phospholipid translocation in the endoplasmic reticulum with a low lipid specificity, *Biochemistry,* 29, 2023–2027, 1990.
31. **Hope, M. J., Redelmeier, T. E., Wong, K. F., Rodrigueza, W., and Cullis P. R.,** Phospholipid asymmetry in large unilamellar vesicles induced by transmembrane pH gradients, *Biochemistry,* 28, 4181–4187, 1989.
32. **Morrot, G., Hervé, P., Zachowski, A., Fellmann, P., and Devaux, P. F.,** Aminophospholipid translocase of human erythrocytes: phospholipid substrate specificity and effect of cholesterol, *Biochemistry,* 28, 3456–3462, 1989.
33. **Bevers, E. M., Tilly, R. H. J., Senden, J. M. G., Comfurius, P., and Zwaal, R. F. A.,** Exposure of endogenous phosphatidylserine at the outer surface of stimulated platelets is reversed by restoration of aminophospholipid translocase activity, *Biochemistry,* 28, 2382–2387, 1989.
34. **Rybicki, A. C., Heath, R., Lubin, B., and Schwartz, R. S.,** Human erythrocyte protein 4.1 is a phosphatidylserine binding protein, *J. Clin. Inves.,* 81, 255–260, 1988.
35. **Sune, A. and Bienvenue, A.,** Relationship between the transverse distribution of phospholipids in plasma membrane and shape change of human platelets, *Biochemistry,* 27, 6794–6800, 1988.
36. **Schroit, A. J., Madsen, J. W., and Tanaka, Y.,** *In vivo* recognition and clearance of red blood cells containing phosphatidylserine in their plasma membranes, *J. Biol. Chem.,* 260, 5131–5138, 1985.

Chapter 10

THE EFFECTS OF EXOGENOUS LIPIDS UPON LIPID METABOLISM

I. INTRODUCTION

I reviewed in Chapter 1 evidence suggesting that each cellular membrane is maintained as closely as possible in a physical state that is optimal for performing its specific functions. This physical state is maintained to a large extent by regulating the membrane lipid composition. Whereas many organisms synthesize all of the necessary lipid components from their own metabolic pathways, others cannot produce the full spectrum of lipids needed. For example, a number of mammals, including humans, require that certain polyunsaturated fatty acids (called essential fatty acids) be provided from dietary sources. When the animals receive a diet deficient in these fats, they modify their own metabolic pathways in an effort to provide the increased quantity and the broad assortment of fatty acids needed.[1] As the term "essential" implies, compensation for the missing unsaturates is never complete, and physiological abnormalities result.

Another common situation is that in which the organism receives a variety of dietary lipids or lipid precursors in excess of its minimum needs. The surplus compounds are generally utilized fully, and endogenous lipid metabolism is usually altered accordingly so as to reduce the organism's own production of those dietary ingredients. This chapter provides several examples of how the balance between endogenous lipid synthesis and exogenous lipid supply is maintained.

II. EFFECTS OF EXOGENOUS FATTY ACIDS

Quite a number of reports have considered the responses of animals to prolonged diets free of lipids. This has two major effects upon lipid metabolism. First, the availability of excess carbohydrate and the lack of inhibitory fatty acyl-CoA combine to enhance the rate of fatty acid synthesis very greatly. Regulatory aspects of this phenomenon were discussed in Chapter 2. Second, the quantities and types of fatty acids formed *de novo* while the animal is on the fat-free diet may differ significantly from the varieties ingested in a more normal diet, especially if the organism is incapable of synthesizing certain required fatty acids. In such a case, metabolic compensation for the missing ingredients can sometimes be made; thus, optimal membrane physical properties are maintained.

Typical of early studies were the observations of Yamamoto et al.[2] Young rats were fed the following diets for 3 weeks: group A, fat-free; group B, 7% linolenate; group C, 7% linoleate. The latter diet, containing linoleate, which is an essential fatty acid for rats, was considered to be the control treatment. Differences in phospholipid fatty acid composition were widespread among the major components of isolated mitochondria and microsomes. An example of the observed changes is shown in Table 1. Animals ingesting no fatty acids elongated and desaturated 18:1 to 20:3 to offset the shortage of 18:2 and its product 20:4, which normally make up the bulk of polyunsaturated fatty acids. Animals fed linoleate converted that acid to 20:5 in sizable quantities.

The ability of rats to compensate for the deficiency of a normal fatty acid component depends upon the exact conditions of the experiment. In another study,[3] the ratio of liver phospholipid monounsaturates to polyunsaturates remained relatively high in rats fed a fat-

TABLE 1
The Effects that Feeding Young Rats Special Diets for 3 Weeks had on Fatty Acids of the Inositol-Containing Lipids from Liver Mitochondria

	Diet		
Fatty acid	Fat-free (%)	7% (18:3) Linolenate (%)	7% (18:2) Linoleate (%)
16:0	12.97 ± 1.33	13.10 ± 1.15	14.77 ± 0.77
16:1	7.36 ± 1.57	3.99 ± 0.35	4.01 ± 0.51
18:0	22.64 ± 1.27	24.03 ± 1.51	20.06 ± 1.22
18:1	16.69 ± 1.05	11.77 ± 0.88	11.55 ± 0.62
18:2	5.07 ± 1.57	4.62 ± 1.24	14.32 ± 2.47
18:3	—	3.65 ± 1.05	—
20:3	12.93 ± 1.22	1.74 ± 1.05	1.27 ± 0.83
20:4	15.83 ± 2.15	11.47 ± 0.65	26.39 ± 0.18
20.5	—	11.03 ± 1.19	—
22:4	0.94 ± 0.561	—	2.97 ± 0.79
22:5	—	4.89 ± 1.40	—
22:6	3.40 ± 0.46	8.47 ± 0.79	2.43 ± 0.86

From Yamamoto, A., Isozaki, M., Hirayama, K., and Sakai, Y., *J. Lipid Res.*, 6, 295, 1965. With permission.

free diet. This increase in the degree of fatty acid saturation has been shown to reduce the fluidity of the liver phospholipids, as estimated by calorimetry.[4]

Whereas liver is among the first tissues to reveal a lipid compositional change due to essential fatty acid deficiency, other tissues are also affected. Changes resembling those shown in Table 1 have been described in rat erythrocyte membranes after 15 weeks of fat-free feeding[5] and in mouse brain following a 6-month diet deficient in fatty acids.[6]

Rats on a fat-free diet dramatically increased their Δ9-fatty acid desaturase activity in order to desaturate the large quantities of 16:0 and 18:0 formed *de novo*.[7] The sensitivity of this increase to cycloheximide indicates that it is due largely to increased synthesis of one or more enzymes involved in desaturation. The only component showing a significant increase appears to be the terminal enzyme in the system, namely, the desaturase itself. Enhanced Δ9-desaturase activity is not necessarily a function of the *de novo* fatty acid synthesis rate, but rather is induced by accumulated saturated fatty acids — feeding saturated fatty acids can also raise the level of desaturase activity.[8] In this way excesses of dietary saturated fatty acids can rapidly be converted to a form more desirable for membrane synthesis.

Recent studies have focused on stearoyl-CoA desaturase mRNAs. Two closely related stearoyl-CoA desaturase genes have been described from mouse adipocytes.[9] By monitoring changes in mRNA levels, it was determined that one gene, SCD1, is expressed constitutively in adipose tissue, induced in liver and, to a lesser extent in kidney and lung, by a fat-free diet, and absent in brain, heart, and spleen under fat-free or fat-supplemented dietary regimes. In contrast, the SCD2 gene is constitutively expressed in brain, induced by refeeding a fat-free diet in adipose, kidney, and lung, and present at low levels in heart and spleen. SCD2 mRNA was not detected in liver. Although different control mechanisms are obviously participating, it is not yet clear whether regulation occurs mainly at the transcriptional level.

Other enzyme activities of lipid metabolism are also increased by a high fat diet. The pattern of change differs, depending upon the type of fat involved. Partially hydrogenated soybean oil was particularly effective in stimulating rat liver mitochondrial and microsomal palmitoyl-CoA synthetase, carnitine palmitoyltransferase, and microsomal glycerophosphate

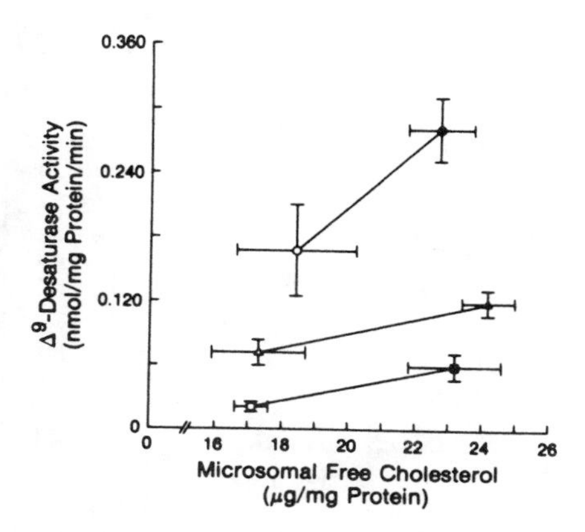

FIGURE 1. Effect of dietary cholesterol and/or *n*-3 fatty acid supplementation on palmitoylcoenzyme A desaturase activity of rat liver microsomes. Reaction rates are expressed as nmol per mg microsomal protein per min and are plotted against free cholesterol content of microsomal membranes. ○, △, □ represent low cholesterol (beef tallow, linseed oil and fish oil, respectively). ●, ▲, ■ represent cholesterol enriched (2%, w/w) diets (beef tallow and cholesterol, linseed oil and cholesterol, fish oil and cholesterol, respectively).

acyltransferase activities within 36 hours.[10] The enzyme activities associated with mitochondrial β-oxidation appeared to increase more rapidly than those for peroxisomal β-oxidation.

Perhaps as a result of the pronounced versatility many cells have in adjusting their pathways for lipid biosynthesis and utilization, the proportion of saturated to unsaturated fatty acids in membrane lipids remains remarkably constant despite large variations in the kind of exogenous fatty acids ingested. By thus avoiding major changes in the membrane fatty acid unsaturation index, membrane fluidity is maintained within a tolerable range. Some tissues are less resistant to lipid compositional change than others. Mammalian liver and heart respond relatively quickly to dietary lipids while brain is exceptionally slow to change.[11] Lipid alterations in the latter organ can be achieved only by long term deprivation or supplementation beginning before weaning.

Within the general pattern of stability, it is sometimes possible to achieve a substantial replacement of one polyunsaturated fatty acid by another. For example, diets rich in n-3 fatty acids result in increased proportions of 20:5 n-3, 22:5 n-3, and 22:6 n-3, often at the expense of 20:4 n-6.[12,13] The reverse relationship, e.g., a decrease in docosenoic n-3 polyunsaturated fatty acids after feeding 18:2 n-6, also has been observed.[14] Feeding rats a diet in which the 20% fat content was enriched in n-3 fatty acid-containing fish oil dramatically lowered the activity of fatty acid Δ9 desaturase, offsetting the activating effect that dietary cholesterol had on that enzyme (Figure 1).[15] The fish oil diet also lowered the cholesterol and arachidonic acid content of liver and serum, with the extent of change being dependent upon the other types of lipids also present in the diet (Figure 2).[16]

The high levels of n-3 fatty acids present in certain fish oils[17] compete with arachidonic acid for the *sn*-2-position of membrane phospholipids. Varying the n-3/n-6 polyunsaturated fatty acid balance in this way probably has a greater effect on the hormone-like activity of eicosanoids than on membrane fluidity per se. The prevalent n-6 C_{20} component of mammalian phospholipids, arachidonic acid (20:4), gives rise to classes of eicosanoids that differ both in structure and function from eicosanoids arising from n-3 acids such as eicosapentaenoic acid (20:5) (Figure 3).[18] Eicosapentaenoic acid also inhibits the formation of arachidonic acid from linoleic acid and competes with arachidonic acid as substrate for the cyclooxygenase that forms physiologically active eicosanoids (reviewed by Weaver and Holob[19]).

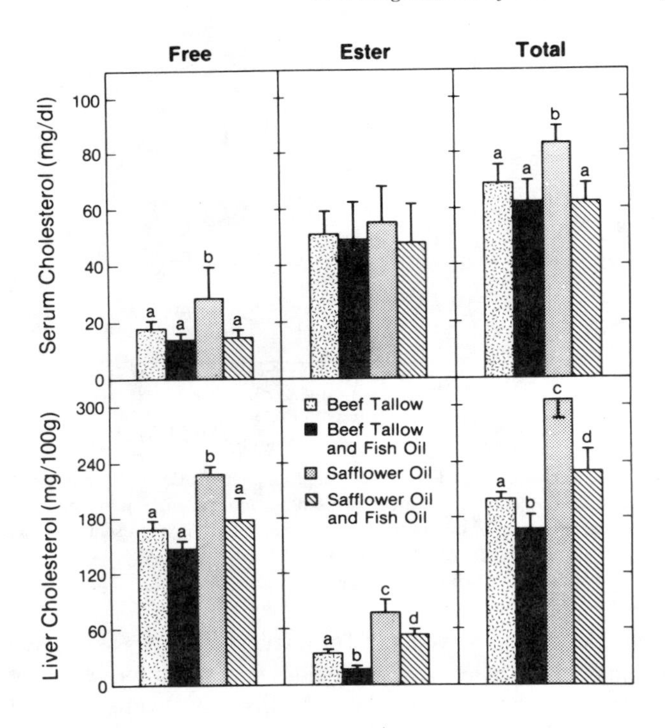

FIGURE 2. Effect of dietary fish oil on cholesterol content of rat serum and liver. Values without a common superscript are significantly different ($p < 0.05$). (From Garg, M. L., Wierzbicki, A. A., Thomson, A. B. R., and Clandinin, M. T., *Biochim. Biophys. Acta,* 962, 337–344, 1988. With permission.)

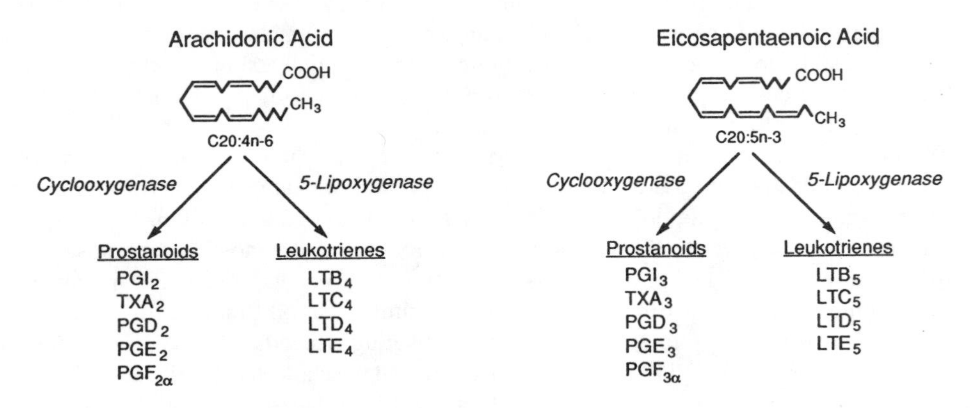

FIGURE 3. The major classes of eicosanoids derived from dietary *n*-6 and *n*-3 polyunsaturated C_{20} fatty acids. (From Leaf, A., and Weber, P. C., *New Engl. J. Med.,* 318, 549–557, 1988. With permission.)

Recent clinical research has explored the nutritional value of n-3 fatty acids. Evidence is accumulating for their beneficial effect in preventing cardiovascular disease[18] and a number of other disorders.[20,21]

Dietary cholesterol is capable of altering the fatty acid composition of membrane phospholipids (reviewed by McMurchie[22]). This may represent a compensatory response of the tissues to restore an optimal fluidity to lipids whose order was increased by the presence of additional sterols.

Although there are some exceptions,[23] generally speaking the quantitative distribution of phospholipid classes in a given membrane is not significantly altered by dietary fatty acid supplementation.

Based upon what we have learned from fundamental studies (see Chapter 1), it is logical to expect that membrane lipid compositional changes will engender alterations in the physical state of the membrane. These sometimes subtle changes may be inadequately measured by existing physical chemical techniques, and there is currently no way to predict their effects on membrane function. However, numerous empirical observations (reviewed by McMurchie[22]) have shown a correlation between lipid compositional changes, physical perturbations of the membrane, and function of membrane-associated enzymes. At least a few of these correlations are likely to be accounted for by direct fluidity-related enzyme activity changes.

A number of fatty acid supplementation experiments have been carried out with cultured animal cells.[24] The regulatory controls present in these cells appear capable of maintaining a relatively normal lipid pattern, even in the face of large supplements of a single fatty acid. In order to manipulate the fatty acid composition in LM and BHK_{21} cell phospholipids, Wisnieski et al.[25] found it necessary to inhibit fatty acid biosynthesis by growing the cells in the presence of a biotin analog and to administer exogenous fatty acids as the Tween® derivatives. Even under these circumstances, the cells were able to retain their characteristic fatty acid composition when fed derivatives of even-chain saturated or monounsaturated fatty acids.[26] Only when odd-chain saturated fatty acids or polyunsaturated fatty acids (which are not normally present) were supplied could appreciable changes be effected.

The modification of LM cell lipids by supplementation was further refined by Ferguson et al.,[27] who reduced the toxicity of the administered fatty acids by providing them as the bovine serum albumin complexes. It was clear that supplementation with polyunsaturated fatty acids under isothermal conditions produced a compensatory reduction in *de novo* synthesized monounsaturated acids. Thus, incorporation of exogenous 18:2 in amounts exceeding 30% of the total phospholipid fatty acids was almost exactly offset by a decrease in 16:1 and 18:1. A rather similar compensation has been reported by Cowen and Heydrick[28] to take place in cultured L cells.

The role of serum albumin in facilitating the entry of fatty acids into cells has received considerable study, and with good reason, since in higher animals, fatty acid transport from lipid depots to cellular sites of utilization occurs entirely by way of fatty acid-serum albumin complexes. Serum albumin possesses a variety of high, low, and medium affinity binding sites for fatty acids. Paris et al.[29] have shown that the uptake of palmitic acid by cultured chick embryo cardiac cells is greatly expedited if the palmitate is presented as the serum albumin complex, with the maximum effect being found at fatty acid:serum albumin molar ratios of 7 to 10. The effect is specific for saturated fatty acids and could be mimicked by the detergent, Tween® 40. Apparently, the stimulatory effect of serum albumin involves a conversion of metabolically inaccessible dimers and higher aggregates of palmitic acid to monomers of fatty acid bound to the serum albumin.

Unsaturated fatty acids are less prone to form aggregates and therefore fail to exhibit enhanced uptake in the presence of serum albumin. The presence of serum albumin in cell cultures does have a useful effect, however, in that it greatly diminishes the damaging detergent effect of free unsaturated fatty acid monomers. More recent work has indicated that fatty acids circulating in the plasma are extracted from their albumin binding sites by a hepatocyte plasma membrane — associated fatty acid binding protein, which also facilitates their movement into the cell.[30]

The response of the unicellular protozoan, *Tetrahymena pyriformis,* to fatty acid supplementation follows the same general pattern observed in mammalian cells. *Tetrahymena* can

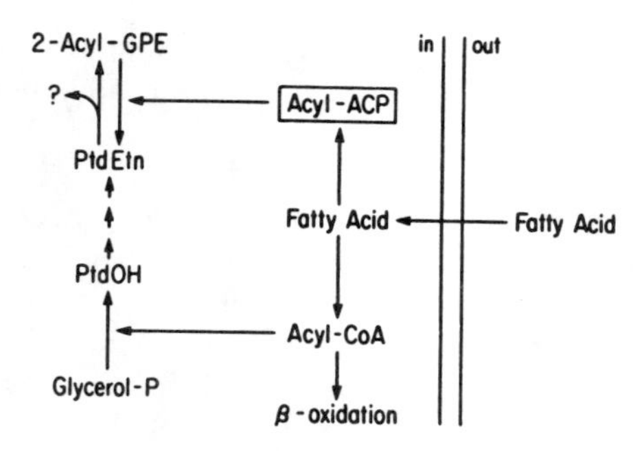

FIGURE 4. Various pathways for the utilization of exogenous fatty acids by *Escherichia coli*. The acyl-ACPs formed in this way (box) are segregated from those arising through *de novo* fatty acid biosynthesis and cannot be elongated or utilized by glycerol phosphate acyltransferase. (From Rock, C. O. and Jackowski, S., *J. Biol. Chem.*, 260, 12720–12724, 1985. With permission.)

assimilate massive amounts of exogenous lauric or palmitic acid with only minimal changes in its phospholipid fatty acid pattern.[31] This indicates a large increase in fatty acid elongation and desaturation in response to the fed acids. Temporary alterations in the fatty acid composition could be effected by the administration of large quantities of linoleic acid, but the increased unsaturation was quickly compensated for by a decreased desaturation of the saturated fatty acids synthesized *de novo*.[32]

Bacteria can readily utilize exogenous long chain fatty acids.[33] The pathways whereby these fatty acids are incorporated into phospholipids are not immediately obvious, because *Escherichia coli* and many other bacteria normally derive their phospholipid acyl chains by direct transfer of fatty acids from the ACP moieties to which they are attached throughout *de novo* synthesis. There would seem to be little need for an enzyme able to esterify preformed long chain acyl chains to ACP. Our present understanding of how bacteria, particularly *E. coli*, utilize exogenous fatty acids has been reviewed by de Mendoza and Farias.[34] Much recent progress has come through studies of *E. coli* mutants deficient in fatty acid metabolism.[35] Fatty acids are translocated across the bacterial inner membrane with the aid of an as yet uncharacterized gene product. Once inside the cell, these fatty acids can be activated by either acyl-CoA or acyl-ACP synthetase. The acyl-CoAs are substrates for the glycerol-3-phosphate acyltransferase-mediated formation of phosphatidic acid or for β-oxidation. The acyl-ACPs, on the other hand, are utilized specifically for the placement of an acyl chain into the *sn*-1 position of phosphatidylethanolamine (Figure 4). In this manner exogenous oleate can become incorporated into the position of phosphatidylethanolamine normally occupied by *de novo* synthesized palmitate. A number of questions remain unanswered regarding the mechanisms by which these two acylation substrates are selectively channeled into their individual products.

Certainly the most easily manipulated cells with respect to modifying membrane fatty acid composition are the mycoplasmas. Some species, such as *Mycoplasma mycoides*, completely lack the ability to synthesize or modify fatty acids, and all mycoplasmas have some nutritional requirement for fatty acids.[36] The much-studied species, *Acholeplasma laidlawii*, can synthesize saturated fatty acids *de novo*, but this synthesis is stopped if fatty acids are available in the medium.

McElhaney[37] reviewed a number of studies in which the effects of fatty acid supplementation on *de novo* fatty acid synthesis and membrane lipid composition were examined. Through supplementation coupled with the use of fatty acid synthesis inhibitors, it has been possible to produce mycoplasma membranes containing essentially only one type of acyl chain. This type of manipulation has been extremely useful in understanding the effects of particular fatty acids on membrane physical properties.[38]

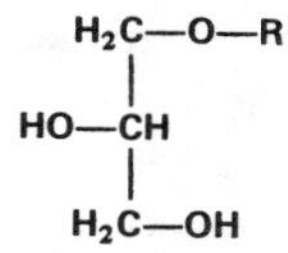

$$\begin{array}{l} H_2C{-}O{-}R \\ \quad | \\ HO{-}CH \\ \quad | \\ H_2C{-}OH \end{array}$$

FIGURE 5.

The capacity of *A. laidlawii* to compensate for these drastic lipid compositional and fluidity changes was tested by altering its growth temperature. It was concluded that *A. laidlawii* strain B completely lacks the ability to regulate the fluidity and physical state of its membrane lipids through temperature-induced alterations in fatty acid synthesis or differential utilization of mixed fatty acid supplements. This may not be entirely surprising since the organism normally resides in a homeothermic host. Unfortunately, the lack of regulatory powers diminishes its utility as a model for other cells that can undergo homeoviscous adaptation.

III. EFFECTS OF EXOGENOUS ETHER LIPIDS

In many tissues, the fatty acyl groups normally present at the l-position of some phospholipid glycerol moieties are replaced by ether-linked side chains, either having α, β-unsaturation, as in plasmalogens, or not, as in the alkyl ether phospholipids.[39] The alk-l-enyl side chains of plasmalogens are formed through the dehydrogenation of the corresponding alkyl ether derivatives, which are themselves synthesized *de novo* by the cells (see Chapter 4) or ingested in the diet.

Several studies have been carried out to determine the effect of massive supplementation of free alkyl glycerol ethers (Figure 5) to cultures of *T. pyriformis*, a cell that normally contains saturated ethers at the l-position of 20 to 30% of its phospholipids (there are no plasmalogens present). When small amounts (0.5 nmol/10^7 cells) of l-*O*-hexadecyl glycerol were fed, over 80% of the alkyl glyceryl ether was utilized for lipid synthesis without cleavage of the ether bond.[40] However, when higher levels (3000 nmol/10^7 cells) were administered, between 60 and 70% of the ether bonds were cleaved. The principal hydrophobic cleavage product, palmitic acid, was subsequently utilized for the synthesis of other lipids. The increased rate of ether-bond hydrolysis in the presence of excess substrate did not appear to stem from an induced synthesis of the cleavage enzyme.[41]

The growth of *Tetrahymena* in huge excesses of hexadecyl glycerol led to a surprisingly small rise in the percentage of ether-bound side chains in its membrane phospholipids.[42] The largest change was observed in the surface membrane fraction, which increased in ether lipid content from a value of 33 mol% of total phospholipids in control cells to 41 mol % in hexadecyl glycerol-fed cells. The increase was accompanied by a slight increase in certain fatty acids (16:0, 18:2, and 18:3) and a decrease in others (especially 18:1). These changes might have been caused in part by the large amounts of palmitic acid released by cleavage of the ether bond. The most unexpected effect of growth in the presence of hexadecyl glycerol was a significant elevation of one of the three major phospholipid classes, 2-aminoethyl phosphonolipid, in the cell surface membranes from the usual value of 22 to 31 mol% of total lipid phosphorus. This was offset by a decline in the structurally similar phosphatidylethanolamine.

IV. EFFECTS OF EXOGENOUS STEROLS

The effect of dietary sterols on lipid metabolism has been studied intensively over the past 20 years because of the possible role of cholesterol in human atherosclerosis. As described in Chapter 5, most organisms can synthesize an adequate supply of sterols. If sterols are absorbed from dietary sources, *de novo* synthesis is reduced in proportion, thus maintaining the combined supply of sterol from biosynthetic and dietary sources constant. However, what happens if the dietary intake is larger than the total metabolic need, as often happens in affluent human society? This problem has been studied in several animal species, with the rat being the most thoroughly examined.

High cholesterol diets appear to have little quantitative effect on membrane lipid composition. Rats fed a high (1.2%) cholesterol diet differ from controls in exhibiting (1) a marked inhibition of hepatic cholesterol synthesis, (2) an enhanced conversion of cholesterol to bile acids, and (3) an accumulation of cholesterol in the liver.[43] Cholesterol feeding does not raise the cholesterol level of any other tissue besides liver.[44,45] Specific tissue analyses have been reported, particularly for erythrocytes, which are directly exposed to high levels of cholesterol circulating in blood. The evidence suggests that rats fed a high cholesterol diet show no increase in the cholesterol level of their erythrocytes.[46,47] This is perhaps not too surprising in view of the observation that plasma cholesterol levels are a poor indicator of cholesterol balance. In humans, absorption of as much as 1 g/day of dietary cholesterol caused little change in the plasma cholesterol content.[48] Although the eight patients studied were remarkably variable in some of the parameters measured, most responded to increased cholesterol absorption by two compensatory mechanisms. They increased the excretion of cholesterol (but, unlike the rat and the dog, not bile acids) in bile, and they decreased *de novo* cholesterol synthesis. In a few patients, this compensation was not adequate to prevent an accumulation of cholesterol in body pools, but even these individuals did not show elevated plasma cholesterol.

The transfer of circulating cholesterol from plasma to cells is, as implied above, quite efficient. The mechanisms for interchanging cholesterol from plasma lipoproteins to cells has been studied extensively, employing the erythrocyte as a model. Using plasma that had been depleted of cholesterol through action by the enzyme, lecithin:cholesterol acyltransferase, Lange and D'Alessandro[49] determined that ^{3}H-cholesterol of exchange-labeled human erythrocytes was available for reequilibration with the plasma pool. Previous reports of a nonexchangeable cholesterol pool could not be ruled out by this technique, but the sizable amount (as much as 40%) of the cholesterol that was drawn out of the cells by depleted plasma was kinetically similar to the cellular pool available for equimolar exchange. Studies describing the kinetics of cholesterol exchange between plasma lipoproteins and cells have been reviewed by Phillips et al.[50]

As yet, little is known regarding the specificity of this exchange process. Studies with artificial lipid membranes[51] raise the possibility that the fluidity of a potential acceptor cell's plasma membrane could determine the amount of cholesterol transferred from circulating plasma.

Coronary heart disease in man is associated with the accumulation of cholesterol and cholesterol esters in the intima and media of arteries. Exposing weanling rabbits to a high cholesterol diet for 6 weeks induced an increase of acyl-CoA-cholesterol-acyltransferase activity in aorta.[52] This enhanced capacity for forming cholesterol esters persisted throughout a 9-week recovery period on a normal diet, as evidenced by the elevated formation of cholesterol esters when the animals were rechallenged with high cholesterol. It is not clear just how long this increase in enzyme activity might persist following normalization of plasma cholesterol.

Not all sterols are metabolically or nutritionally equivalent. It has long been recognized that certain plant sterols are very poorly absorbed by the animal intestine.[53] More recent studies have shown that a diet containing 0.8% β-sitosterol inhibited cholesterol absorption by the rat.[43] The reduction in absorbed cholesterol was sufficiently pronounced to stimulate an increased *de novo* cholesterol synthesis.

Whereas cholesterol and β-sitosterol were perceived in quite different ways by the system described above, the specificity is not so great in some organisms. As described on p. 131, the protozoan *T. pyriformis* can utilize dietary ergosterol as a substitute for its own sterol-like pentacyclic triterpenoid tetrahymanol.[54]

The membrane role of sterols has been studied in some detail using cultured mammalian cells.[24] One much-used technique has been the suppression of cholesterol synthesis by the use of compounds such as 25-hydroxycholesterol, which inhibits the key regulatory enzyme, 3-hydroxy-3-methylglutaryl coenzyme-A reductase. Chen et al.[55] followed this approach to block the formation of desmosterol, the principal sterol of L cells. Under these conditions, cell growth was resumed only when desmosterol, its precursor, mevalonic acid, or certain other sterols or sterol derivatives were added to the growth medium. This system is well suited for the analysis of the detailed structural requirements of sterol molecules enabling them to function in membranes.

In addition to studies of sterol-depleted cells, the metabolic effects of excessive sterols on cultured mammalian cells have been examined. There was a limit to the amount of cholesterol that could be utilized by rat hepatoma cells for incorporation into membranes.[56] As the level of exogenous cholesterol exceeded that needed for membrane growth, increasing amounts of cholesterol were converted to cholesterol esters, a nonmembrane storage form of the sterol. The regulatory mechanism for shunting excess cholesterol into an esterified form remains obscure.

In the course of studies with *Saccharomyces cerevisiae* sterol auxotrophs, Parks and associates discovered that the cells were unable to grow on cholestanol as a bulk sterol unless minute quantities (1 to 10 ng/ml) of ergosterol were also provided. Many different sterols and stanols could satisfy the bulk requirement for growth, but only those having unsaturation in the C-5,6 position or capable of being desaturated at the C-5 position could serve as the "sparking" component required for the resumption of growth.[57] Although sparking by effective combinations of ergosterol and cholesterol has been correlated with activation of a $pp60^{v\text{-}src}$-related protein kinase, the precise nature of the sparking function remains to be determined.

V. EFFECTS OF EXOGENOUS PRECURSORS OF PHOSPHOLIPID HEAD GROUPS

By and large, higher plants and animals appear fairly resistant to an alteration of their phospholipid polar head group distribution through dietary excesses of certain normal constituents. However, rats fed a diet supplemented with 1% *N*-monomethylethanolamine for 24 h showed quite appreciable concentrations of this amino alcohol and its dimethyl derivatives as polar head groups in liver and lung phospholipids.[58] The overall fatty acid composition of the tissues was altered somewhat by the changes, perhaps by way of compensation. Lee et al.[59] observed an incorporation of the unnatural base-analog, *N*-isopropylethanolamine, into the phospholipids of various rat liver organelles following its intraperitoneal injection in daily doses of 20 mg/100 g of body weight for 4 d. As much as 9% of the phospholipids contained this unusual analog.

Growth of tomato seedlings for 5 d in the presence of 5 m*M* ethanolamine led to an absolute increase of tissue phosphatidylethanolamine and phosphatidylserine.[60] Phospholipids of the ethanolamine-treated seedlings were enriched in linoleic acid.

Compared to the changes observed in intact organisms, much greater perturbation of phospholipid polar head groups can be achieved using cells grown in culture. Horwitz[24] has reviewed the types of modifications that are possible through supplementation or the use of specific inhibitors. Many mammalian cells grown in culture have a nutritional requirement for one or more lipid precursors. Mouse fibroblast L cells (LM cells) normally require choline supplementation. When the choline was replaced by monomethylethanolamine, dimethylethanolamine, 1-2-aminobutanol, or 3-aminopropanol, a large-scale alteration of lipid polar head groups was noted.[61] From 35 to 50 mol% of the phospholipids contained the administered base analog, usually at the expense of phosphatidylcholine and, to a lesser extent, phosphatidylethanolamine. Only supplementation with dimethylethanolamine allowed the cells to continue growing indefinitely; growth supported by the other additives ceased after several days.

Manipulation of LM cell phospholipid head groups by exogenous ethanolamine has also been studied. Substitution of 40 μg ethanolamine/ml medium for the normal choline supplement resulted, after 36 hr, in cells whose plasma membranes contained 27% phosphatidylcholine and 43% phosphatidylethanolamine vs. 58 and 20% of those lipids, respectively, in normal choline-supplemented cells.[62] Small changes in fatty acid composition accompanied the polar head group alterations. Both cell types grew well during the experimental period, but fluorescence polarization measurements indicated that the membranes of ethanolamine-fed cells were more viscous than those of the controls.

Changes in phospholipid distribution can also be shown in free-living eukaryotic microorganisms. *Tetrahymena* responded to growth in the presence of nitrogenous bases, such as ethanolamine, choline, or monomethylethanolamine (all administered in a final concentration of 8.1 m*M*), by modifying its phospholipid polar head group composition.[63] Monomethylethanolamine supplementation caused the greatest alteration, leading to the synthesis of the unnatural lipid, phosphatidylmonomethylethanolamine, in amounts as high as 34% of the total cellular phospholipids. This new lipid mainly replaced phosphatidylethanolamine.

By growing *Tetrahymena* in the presence of 2-aminoethylphosphonate (AEP) and 3-aminopropylphosphonate (APP), phosphonate analogs of the more familiar nitrogen bases, Smith[64] was able to replace much of the cell's phosphatidylethanolamine with phosphonolipids. This replacement sharply reduced the contribution of the phosphatidylethanolamine N-methyltransferase pathway for phosphatidylcholine synthesis and raised the contribution of the diacylglycerol:CDP-choline choline phosphotransferase-mediated pathway for phosphatidylcholine synthesis from 40% of the total in controls cells to 75 and 90%, respectively, for the AEP- and APP-supplemented cultures. A similarly increased reliance on the phosphotransferase pathway for phosphatidylcholine formation was observed following replacement of about 50% of the ethanolamine in *Tetrahymena* phosphatidylethanolamine by added 3-aminopropane-1-ol.[65]

Neurospora crassa was one of the first organisms to reveal an altered distribution of phospholipids due to manipulation. In 1964, Crocken and Nyc[66] observed that certain cholineless mutants accumulated phosphatidylmonomethylethanolamine even in the presence of exogenous choline. More recently, Hubbard and Brady[67] discovered that supplementing the medium of a choline-deficient *Neurospora* mutant with 720 μ*M* of monomethylethanolamine or dimethylethanolamine instead of choline produced dramatic alterations in lipid composition. For example, the level of phosphatidylcholine, normally 45% of the total phospholipids, was reduced to <1% in dimethylethanolamine-supplemented cells. This decrease was offset by the rise of phosphatidyldimethylethanolamine from 0 to 56%. The morphology and growth rate of the altered cells were similar to those of wild-type cells.

Bacteria seem less sensitive to factors that alter the lipid distribution in eukaryotic cells. The wild-type strains have proved rather unresponsive to supplementation with exogenous lipids and polar lipid precursors, and therefore investigators have largely resorted to the creation of mutants deficient in some aspect of lipid synthesis.[68] Beebe[69] isolated a mutant of *Bacillus subtilis* deficient in phosphatidylethanolamine. The growth of the mutant in the absence of exogenous phosphatidylethanolamine or glycerylphosphorylethanolamine was very slow, and its protoplasts were more fragile than those of the wild-type cells. The uptake of certain metabolites was significantly reduced.[70] Although the quantitative distribution of radioactivity from ^{14}C-glucose was measured in the various lipids, it was not possible to decipher with certainty which membrane lipids increased in the virtual absence of phosphatidylethanolamine.

Ohta et al.[71] isolated a temperature-sensitive mutant of *E. coli* that was unable to synthesize phosphatidylethanolamine at 42°C. Cell growth at the nonpermissive temperature was greatly reduced. The block in phosphatidylethanolamine formation triggered an increased synthesis of cardiolipin, one of the major membrane components, but the resultant change in lipid composition did not restore cell growth to the normal rate.

In some cases, the synthesis of a specific bacterial membrane lipid has been selectively inhibited by administering analogs of lipid precursors. An analog of glycerol-3-phosphate, namely, 3,4-dihydroxybutyl-1-phosphonate, fairly specifically blocks the formation of phosphatidylglycerol in *E. coli* without markedly affecting cell growth.[72] The analog itself appears to be incorporated into phospholipids of the cells.

REFERENCES

1. **McMurchie, E. T.,** Dietary lipids and the regulation of membrane fluidity and function, in *Physiological Regulation of Membrane Fluidity,* Aloia, R. C., Curtain, C. C., and Gordon, L. M., Eds., Alan R. Liss, New York, 1988, 189–237.
2. **Yamamoto, A., Isozaki, M., Hirayama, K., and Sakai, Y.,** Influence of dietary fatty acids on phospholipid fatty acid composition in subcellular particles of rat liver, *J. Lipid Res.,* 6, 295–300, 1965.
3. **Darsie, J., Gosha, S. K., and Holman, R. T.,** Induction of abnormal fatty acid metabolism and essential fatty acid deficiency in rats by dietary DDT, *Arch. Biochem. Biophys.,* 175, 262–269, 1976.
4. **Mabrey, S., Powis, G., Schenkman, J. B., and Tritton, T. R.,** Calorimetric studies of microsomal membranes, *J. Biol. Chem.,* 252, 2929–2933, 1977.
5. **Bloj, B., Morero, R. D., Farias, R. N., and Trucco, R. E.,** Membrane lipid fatty acids and regulation of membrane-bound enzymes. Allosteric behavior of erythrocyte Mg^{2+}-ATPase, $(Na^+ + K^+)$-ATPase and acetylcholinesterase from rats fed different fat supplemented diets, *Biochim. Biophys. Acta,* 311, 67–79, 1973.
6. **Sun, G. Y. and Sun, A. Y.,** Synaptosomal plasma membranes: acyl group composition of phosphoglycerides and $(Na^+ + K^+)$-ATPase activity during fatty acid deficiency, *J. Neurochem.,* 22, 15–18, 1974.
7. **Oshino, N. and Sato, R.,** The dietary control of the microsomal stearyl CoA desaturation enzyme system in rat liver, *Arch. Biochem. Biophys.,* 149, 369–377, 1972.
8. **Mercuri, O., Peluffo, R. O., and De Tomás, M. E.,** Effect of different diets on the Δ9-desaturase activity of normal and diabetic rats, *Biochim. Biophys. Acta,* 369, 264–268, 1974.
9. **Kaestner, K. H., Ntambi, J. M., Kelly, T. J., Jr., and Lane, M. D.,** Differentiation-induced gene expression in 3T3-LI Preadipocytes, *J. Biol. Chem.,* 264, 14755–1476l, 1989.
10. **Berge, R. K., Nilsson, A., and Husoy, A.-M.,** Rapid stimulation of liver palmitoyl-CoA synthetase, carnitine palmitoyltransferase, and glycerophosphate acyltransferase compared to β-oxidation and palmitoyl-CoA hydrolase in rats fed high-fat diets, *Biochim. Biophys. Acta,* 960, 417–426, 1988.
11. **Stubbs, C. D. and Smith, A. D.,** The modification of mammalian membrane polyunsaturated fatty acid composition in relation to membrane fluidity and function, *Biochim. Biophys. Acta,* 779, 89–137, 1984.
12. **Kurata, N. and Privett, O. S.,** Effect of dietary fatty acid composition on the biosynthesis of unsaturated fatty acids in rat liver microsomes, *Lipids,* 15, 512–518, 1980.
13. **Hølmer, G. and Beare-Rogers, J. L.,** Linseed oil and marine oil as sources of (n-3) fatty acids in rat heart, *Nutr. Res.,* 5, 1011–1014, 1985.
14. **Kramer, J. K. G., Farnworth, E. R., Thompson, B. K., Corner, A. H., and Tranholm, H. L.,** Reduction of mycardial necrosis in male albino rats by manipulation of dietary fatty acid levels, *Lipids,* 17, 372–382, 1982.

15. **Garg, M. L., Wierzbicki, A. A., Thomson, A. B. R., and Clandinin, M. T.,** Dietary cholesterol and/or n-3 fatty acid modulate Δ^9-desaturase activity in rat liver microsomes, *Biochim. Biophys. Acta*, 962, 330–336, 1988.
16. **Garg, M. L., Wierzbicki, A. A., Thomson, A. B. R., and Clandinin, M. T.**, Fish oil reduces cholesterol and arachidonic acid content more efficiently in rats fed diets containing low linoleic acid to saturated fatty acid ratios, *Biochim. Biophys. Acta*, 962, 337–344, 1988.
17. **Stansby, M. E.,** Fatty acids in fish, in *Health Effects of Polyunsaturated Fatty Acids in Seafoods,* Simopoulos, A. P., Kifer, R. R., and Martin, R. E., Eds. Academic Press, New York, 1986, 389–401.
18. **Leaf, A. and Weber, P. C.,** Cardiovascular effects of n-3 fatty acids, *New Engl. J. Med.,* 318, 549–557, 1988.
19. **Weaver, B. J. and Holub., B. J.,** Health effects and metabolism of dietary eicosapentaenoic acid, *Prog. Food Nutr. Sci.,* 12, 111–150, 1988.
20. **Simopoulos, A. P.,** Summary of the NATO Advanced Research Workshop on Dietary ω3 and ω6 Fatty Acids: Biological Effects and Nutritional Essentiality, *J. Nutr.,* 119, 521–528, 1989.
21. **Lees, R. S. and Karel, M., Eds.** *Omega-3 Fatty Acids in Health and Disease,* Marcel Dekker, New York, 1990, 240.
22. **McMurchie, E. J.,** Dietary lipids and the regulation of membrane fluidity and function, in *Physiological Regulation of Membrane Fluidity,* Aloia, R. C., Curtain, C. C., and Gordon, L. M., Eds., Alan R. Liss, New York, 1988, 189–237.
23. **Innis, S. M. and Glandinin, M. T.,** Mitochondrial-membrane polar-head-group composition is influenced by diet fat, *Biochem. J.,* 198, 231–234, 1981.
24. **Horwitz, A. F.,** Manipulation of the lipid composition of cultured animal cells, in *Dynamic Aspects of Cell Surface Organization,* Vol. 3, Poste, G. and Nicholson, G. L., Eds., North-Holland, Amsterdam, 1977, 295–305.
25. **Wisnieski, B. J., Williams, R. E., and Fox, C. F.,** Manipulation of fatty acid composition in animal cells grown in culture, *Proc. Natl. Acad. Sci. U.S.A.,* 70, 3669–3673, 1973.
26. **Williams, R. E., Wisnieski, B. J., Rittenhouse, H. G., and Fox, C. F.,** Utilization of fatty acid supplements by cultured animal cells, *Biochemistry,* 13, 1969–1977, 1974.
27. **Ferguson, K. A., Glaser, M., Bayer, W. H., and Vagelos, P.R.,** Alteration of fatty acid composition of LM cells by lipid supplementation and temperature, *Biochemistry,* 14, 146–151, 1975.
28. **Cowen, W. F. and Heydrick, F. P.,** Incorporation of C_{18} polyunsaturated fatty acids into L cell phospholipids under normal conditions and during infection with Venezuelan equine encephalitis virus, *Exp. Cell Res.,* 72, 354–360, 1972.
29. **Paris, S., Samuel, D., Jacques, J., Gache, C., Franchi, A., and Ailhaud, G.,** The role of serum albumin in the uptake of fatty acids by cultured cardiac cells from chick embryo, *Eur. J. Biochem.,* 83, 235–243, 1978.
30. **Clarke, S. D and Armstrong, M. K.,** Cellular lipid binding proteins: expression, function, and nutritional regulation, *FASEB J.,* 3, 2480–2487, 1989.
31. **Kitajima, Y. and Thompson, G. A., Jr.,** Self-regulation of membrane fluidity. The effect of saturated normal and methoxy fatty acid supplementation on *Tetrahymena* membrane physical properties and lipid composition, *Biochim. Biophys. Acta,* 468, 73–80, 1977.
32. **Kasai, R., Kitajima, Y., Martin, C. E., Nozawa, Y., Skriver, L., and Thompson, G. A., Jr.,** Molecular control of membrane properties during temperature acclimation. Membrane fluidity regulation of fatty acid desaturase action, *Biochemistry,* 15, 5228–5233, 1976.
33. **Silbert, D. F., Ruch, F., and Vagelos, P. R.,** Fatty acid replacements in a fatty acid auxotroph of *Escherichia coli, J. Bact.,* 95, 1658–1665, 1968.
34. **de Mendoza, D. and Farias, R. N.,** Effect of fatty acid supplementation on membrane fluidity in microorganisms, in *Physiological Regulation of Membrane Fluidity,* 119–148, Aloia, R. C., Curtain, C. C., and Gordon, L. M., Eds., Alan R. Liss, New York, 1988, 119–148.
35. **Rock, C. O. and Jackowski, S.,** Pathways for the incorporation of exogenous fatty acids into phosphatidylethanolamine in *Escherichia coli, J. Biol. Chem.,* 260, 12720–12724, 1985.
36. **Razin, S.,** Physiology of mycoplasmas, *Adv. Microb. Physiol.,* 10, 1–80, 1973.
37. **McElhaney, R. N.,** The structure and function of the *Acholeplasma laidlawii* plasma membrane, *Biochim. Biophys. Acta,* 779, 1–42, 1984.
38. **McElhaney, R. N.,** The influence of membrane lipid composition and physical properties of membrane structure and function in *Acholeplasma laidlawii, CRC Crit. Rev. Micro.,* 17, 1–32, 1989.
39. **Saito, Y., Silvius, J. R., and McElhaney, R. N.,** Membrane lipid biosynthesis in *Acholeplasma laidlawii* B: elongation of medium- and long-chain exogenous fatty acids in growing cells, *J. Bacteriol.,* 133, 66–74, 1978.
40. **Snyder, F., Ed.,** *Ether Lipids, Chemistry and Biology,* Academic Press, New York, 1972, 433.
41. **Kapoulas, V. M., Thompson G. A., Jr., and Hanahan, D. J.,** Metabolism of α-glyceryl ethers by *Tetrahymena pyriformis.* I. Characteristics of the *in vivo* degradation system, *Biochim. Biophys. Acta,* 176, 237–249, 1969.

42. **Kapoulas, V. M., Thompson, G. A., Jr., and Hanahan, D. J.,** Metabolism of a α-glyceryl ethers by *Tetrahymena pyriformis*. II. Properties of a cleavage system *in vitro*, *Biochim. Biophys. Acta,* 176, 250–264, 1969.
43. **Fukushima, H., Watanabe, T., and Nozawa, Y.,** Studies on *Tetrahymena* membranes. In vivo manipulation of membrane lipids by 1-O-hexadecyl glycerol feeding in *Tetrahymena pyriformis, Biochim. Biophys. Acta,* 436, 249–259, 1976.
44. **Raicht, R. F., Cohen, B. I., Shefer, S., and Mosbach, E. H.,** Sterol balance studies in the rat. Effects of dietary cholesterol and sitosterol on sterol balance and rate-limiting enzymes of sterol metabolism, *Biochim. Biophys. Acta,* 388, 374–384, 1978.
45. **Beher, W. T., Baker, G. D., and Penney, D. G.,** A comparative study of the effects of bile acids and cholesterol on cholesterol metabolism in the mouse, rat, hamster, and guinea pig, *J. Nutr.,* 79, 523–530, 1963.
46. **Quintão, E., Grundy, S. M., and Ahrens, E. H., Jr.,** Effects of dietary cholesterol on the regulation of total body cholesterol in man, *J. Lipid Res.,* 12, 233–247, 1971.
47. **Monsen, E. R., Okey, R., and Lyman R.,** Effect of diet and sex on the relative lipid composition of plasma and red blood cells in the rat, *Metabolism,* 11, 1113–1124, 1962.
48. **Bloj, B., Moreno, R. D., and Farias, R. N.,** Membrane fluidity, cholesterol and allosteric transitions of membrane-bound Mg^{2+}-ATPase, $(Na^+ + K^+)$-ATPase and acetylcholinesterase from rat erythrocytes, *FEBS Lett.,* 38, 101–105, 1973.
49. **Lange, Y. and D'Alessandro, J. S.,** Characterization of mechanisms for transfer of cholesterol between human erythrocytes and plasma, *Biochemistry,* 16, 4339–4343, 1977.
50. **Phillips, M. C., Johnson, W. J., and Rothblat, G. H.,** Mechanisms and consequences of cellular cholesterol exchange and transfer, *Biochim. Biophys. Acta,* 906, 223–276, 1987.
51. **Bloj, B. and Zilversmit, D. B.,** Complete exchangeability of cholesterol in phosphatidylcholine/cholesterol vesicles of different degrees of unsaturation, *Biochemistry,* 16, 3943–3948, 1977.
52. **Subbiah, M. T. R., Sprinkle, J. D., Rymaszewski, Z., and Yunker, R. L.,** Short-term exposure to high dietary cholesterol in early life: arterial changes and response after normalization of plasma cholesterol, *Am. J. Clin. Nutr.,* 50, 68–72, 1989.
53. **Borgström, B.,** Quantitative aspects of the intestinal absorption and metabolism of cholesterol and ß-sitosterol in the rat, *J. Lipid Res.,* 9, 473–481, 1968.
54. **Conner, R. L., Mallory, F. B., Landrey, J. R., Ferguson, K. A., Kaneshiro, E. S., and Ray, E.,** Ergosterol replacement of tetrahymanol in Tetrahymena membranes, *Biochem. Biophys. Res. Commun.,* 44, 995–1000, 1971.
55. **Chen, H. W., Kandutsch, A. A., and Waymouth, C.,** Inhibition of cell growth by oxygenated derivatives of cholesterol, *Nature* (London), 251, 419–421, 1974.
56. **Rothblat, G. H., Arbogast, L., Kritchevsky, D., and Naftulin, M.,** Cholesterol ester metabolism in tissue culture cells. II. Source of accumulated esterfied cholesterol in FU5AH rat hepatoma cells, *Lipids,* 11, 97–108, 1976.
57. **Rodriguez, R. J. and Parks, L. W.,** Structural and physiological features of sterols necessary to satisfy bulk membrane and sparking requirements in yeast sterol auxotrophs, *Arch. Biochem. Biophys.,* 225, 861–871, 1983.
58. **Katyal, S. L. and Lombardi, B.,** Quantitation of phosphatidyl N-methyl and N,N-dimethylaminoethanol in liver and lung of N-methylaminoethanol-fed rats, *Lipids,* 9, 81–85, 1974.
59. **Lee, T. C., Blank, M. L., Piantadosi, C., Ishaq, K. S., and Snyder, F.,** Incorporation and subcellular distribution of an unnatural phospholipid base-analog, N-isopropylethanolamine, in rat liver, *Biochim. Biophys. Acta,* 409, 218–224, 1975.
60. **Waring, A. J., Breidenbach, R. W., and Lyons, J. M.,** *In vivo* modification of plant membrane phospholipid composition, *Biochim. Biophys. Acta,* 443, 157–168, 1976.
61. **Glaser, M., Ferguson, K. A., and Vagelos, P. R.,** Manipulation of phospholipid composition of tissue culture cells, *Proc. Natl. Acad. Sci. U.S.A.,* 71, 4072–4076, 1974.
62. **Esko, J. D., Gilmore, J. R., and Glaser, M.,** Use of fluorescent probe to determine the viscosity of LM cell membranes with altered phospholipid compositions, *Biochemistry,* 16, 1881–1890, 1977.
63. **Kasai, R. and Nozawa, Y.,** unpublished.
64. **Smith, J. D.,** Phosphatidylcholine homeostasis in phosphatidyl-ethanolamine-depleted Tetrahymena, *Arch. Biochem. Biophys.,* 246, 347–354, 1986.
65. **Smith, J. D. and Barrows, L. J.,** Use of 3-aminopropanol as an ethanolamine analogue in the study of phospholipid metabolism in Tetrahymena, *Biochem. J.,* 254, 301–302, 1988.
66. **Crocken, B. J. and Nyc, J. F.,** Phospholipid variations in mutant strains of Neurospora crassa, *J. Biol. Chem.,* 239, 1727–1730, 1964.
67. **Hubbard, S. C. and Brody, S.,** Glycerophospholipid variation in choline and inositol auxotrophs of Neurospora crassa, *J. Biol. Chem.,* 250, 7173–7181, 1975.

68. **Cronan, J. E., Jr.,** A new method for selection of *Escherichia coli* mutants defective in membrane lipid synthesis, *Nature (London) New Biol.,* 240, 21–22, 1972.
69. **Beebe, J. L.,** Isolation and characterization of a phosphatidylethanolamine-deficient mutant of *Bacillus subtilis, J Bacteriol.,* 107, 704–711, 1971.
70. **Beebe, J. L.,** Transport alterations in a phosphatidylethanolamine-deficient mutant of *Bacillus subtilis, J. Bacteriol.,* 109, 939–942, 1972.
71. **Ohta, A., Okonogi, K., Shibuya, I., and Maruo, B.,** Isolation of *Escherichia coli* mutants with temperature-sensitive formation of phosphatidylethanolamine, *J. Gen. Appl. Microbiol.,* 20, 21–32, 1974.
72. **Shopsis, C. S., Engel, R., and Tropp, B. E.,** The inhibition of phosphatidylglycerol synthesis in *Escherichia coli* by 3,4-dihydroxybutyl-1-phosphonate, *J. Biol. Chem.,* 249, 2473–2477, 1974.

Chapter 11

THE EFFECTS OF ENVIRONMENTAL FACTORS ON LIPID METABOLISM

I. INTRODUCTION

The metabolism of membrane lipids, as described in the previous 10 chapters, appears to be regulated in such a way as to create for each membrane a very special assortment of lipids designed to provide an equally special physical environment within that membrane. Increasingly numerous indications point to the desirability of maintaining the fluidity of each functional type of membrane within rather narrow limits.

However, it is now quite obvious that the fluidity of a given membrane depends not only upon the makeup of its lipids, but also upon a host of environmental parameters, ranging from temperature, salinity, and other natural variables to the presence of ethanol or drugs administered by accident or design. Many environmental perturbations that upset the optimal physical properties of one or more cell membranes also evoke a compensatory or adaptive change seemingly designed to restore the membrane state to its functionally most desirable condition. This type of response is typified by the acclimation of a cell to temperature change. The actions of inorganic salts and certain drugs may also be of this type, although direct evidence is often lacking. On the following pages I provide a number of examples to illustrate the capacity of cells to compensate for changes in their environment. A more detailed review of this subject has been published by Hazel and Williams.[1]

II. ADAPTIVE RESPONSES TO ENVIRONMENTAL STRESS

Understanding the response of organisms to environmentally induced membrane fluidity changes has become possible only since the recent advances in physical chemical techniques described in Chapter 1. The picture tentatively emerging from these and related studies is one in which cells of all types possess a remarkable capacity for restoring environmentally altered membrane fluidity from a functionally unsuitable value to an optimal range through compensatory metabolic activity.

While pronounced environmental changes typically occur relatively slowly, over a period of hours, they can take place with dramatic suddenness. Thus organisms residing in saline ponds may experience a rapid influx of fresh water during a rainstorm. Likewise the membranes of dry seed undergo abrupt physical changes during the imbibition of water preparatory to germination or if desiccation occurs during the germination process.[1a] Such extreme stresses can not only engender simple alterations in the fluidity of existing membrane lipid bilayers but also short lived reorganization of membrane lipids into nonbilayer structures.[2] The extent of nonbilayer lipid orientation is dependent upon the proportion of molecules prone to assume that conformation, and this is often determined by the prior environmental history of the tissue.

The variable most extensively studied to date is temperature. Not only is its effect of immense practical significance, for example, in agriculture, but it is in some respects one of the most straightforward variables to analyze in basic studies. The temperature of most experimental subjects — certainly in poikilothermic organisms — is easily measured and is usually constant in all parts of the system. The more common metabolic responses of cells to temperature changes are summarized below.

A. CELLULAR RESPONSES TO ALTERATIONS IN ENVIRONMENTAL TEMPERATURE

1. Temperature Responses in Prokaryotes

Free living prokaryotes often live in habitats subject to significant variations in temperature. Because of their small size, these cells have few options but to compensate for the direct and indirect effects of temperature on the physical state of their membranes by making lipid alterations.[3] If compensation is prevented, for example, by withholding the exogenous fatty acids that *Acholeplasma laidlawii* requires or mutating *Escherichia coli* to block unsaturated fatty acid synthesis, growth continues only so long as at least half of the membranes exists in a liquid-crystalline state.[4]

Many scenarios have evolved for acclimation, most of them resulting in some degree of homeoviscous adaptation,[3] i.e., an alteration of lipid composition such that membrane physical properties are restored to or nearly to the state existing before the change in temperature. Examples of microbial responses illustrate their capacity for acclimation.

Anaerobic bacteria, exemplified by *E. coli,* regulate the unsaturation of their fatty acids at a step during *de novo* fatty acid biosynthesis. As described in Chapter 3, two forms of a key enzyme, ß-ketoacyl-ACP synthase, exist in the cells. Of these, synthase II shows a preference for elongating the unsaturated product palmitoleoyl-ACP. Since the other enzyme form, synthase I, which prefers elongating saturated substrates, is more easily inactivated at low temperature, the main product of synthase II, *cis*-vaccenic acid, accumulates. Because of these changes in the activity of existing enzyme molecules, the ratio of unsaturated to saturated fatty acids in *E. coli* can vary from 2.3 at 40 to 25 at 10°C.[5] Enhanced production of the C_{18} *cis*-vaccenate also means an increase in the average fatty acid chain length.

Another factor contributing to the temperature modification of *E. coli* fatty acid composition is the changing pattern of fatty acid supply and demand. During a short period after the cells have been chilled, phospholipid synthesis is depressed sufficiently to relieve the normally high demand for fatty acids. Consequently, the nascent fatty acids are retained for a longer time on the fatty acid synthase, where many of them are elongated by two or four additional carbon atoms.[6]

A second example is *Bacillus megaterium*, an aerobic organism whose lipid metabolism was exhaustively studied by Fulco and co-workers. *B. megaterium* membrane lipids contain no unsaturated fatty acids at all when the cells are grown at 35°C. However, transferring the organisms to 20°C leads to a rapid formation of monounsaturated fatty acids.[7] The sudden capacity for desaturation results from the induced biosynthesis of a Δ^5-fatty acid desaturase enzyme that was not present in the 35°C cells (Figure 1) (see p. 51 for details of induction). If the cells are returned to the higher temperature, desaturase synthesis gradually ceases and the existing Δ^5-desaturase is inactivated. Consequently, desaturation stops, and the unsaturated fatty acids present are rapidly diluted by *de novo* synthesis of saturated components. Induction of the desaturase rather than activation of a preexisting inactive enzyme was demonstrated by using inhibitors of protein synthesis.

Cyanobacteria (blue green algae) such as *Anabaena variabilis* also increase the degree of unsaturation in their membrane-bound fatty acids when exposed to low temperature.[8] The major chilling induced change in *A. variabilis* involves first a rapid desaturation of palmitate linked to the *sn*-2 position of glycolipids and then a slower increase in the desaturation of fatty acids linked to the *sn*-1 position of the same glycolipids. These enhanced rates of desaturation are prevented by inhibitors of protein synthesis, suggesting that the synthesis of additional desaturase is induced by low temperature.

The response of certain other cyanobacteria, in particular the chilling resistant species *Synechocystis* PCC 6803, to low temperature differs in some details but is basically similar to that described above for *A. variabilis*. It depends upon the induced synthesis of desaturases

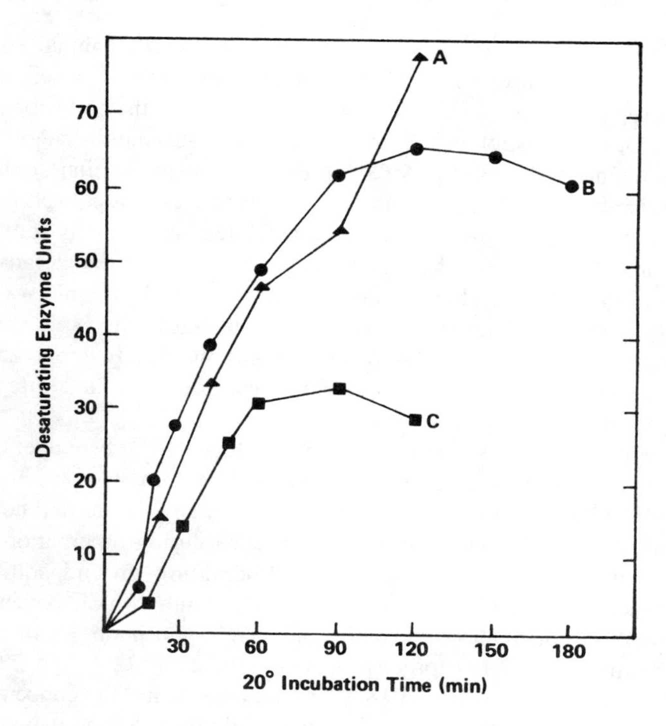

FIGURE 1. Temperature-mediated induction of Δ^5-desaturating enzyme in *B. megaterium* 14581. In three separate experiments, cultures of *B. megaterium* were grown at 35°C in the usual way to various densities and then cooled rapidly in ice to 20°C and transferred to a 20°C shaker. At appropriate time intervals after transfer 5-ml volumes of culture were withdrawn and assayed by the usual procedure to determine the levels of induced Δ^5-desaturating enzyme. In the first experiment (Curve A), transfer from 35°C took place at a cell concentration of 6.7 g/l. After 120 min at 20°C, the cell density was 10.5 g/l. In the second experiment (Curve B), transfer took place at 11.2 g/l and reached 13.8 g/l after 180 min at 20°C. In the last experiment (Curve C), the cells were transferred at 13.0 g/l and there was no further growth after 120 min at 20°C. (From Fulco, A. J., *J. Biol. Chem.*, 247, 3511–3519, 1972. With permission.)

capable of converting glycerolipid-bound oleate to linoleate and linolenate. On the other hand *Anacystis nidulans* is able to synthesize only monounsaturated fatty acids and accordingly is very sensitive to low temperature. Interestingly, when the gene for the Δ^{12} desaturase of *Synechocystis* was cloned and introduced into *A. nidulans,* the transformant not only contained lipids with polyunsaturated fatty acids, but also was more resistant than the wild type to low temperature damage.[9]

Unlike the bacteria described above, which regulate their membrane fluidity by adjusting the level of fatty acid unsaturation, the psychrophilic bacterium *Micrococcus cryophilus* responds to chilling from 20 to 0°C by decreasing the C_{18}/C_{16} ratio of its fatty acids from 3 to 1.[10] The organism achieves this by removing C_2 units from stearate and using these selectively to elongate a C_{14} intermediate to palmitate. The membrane-bound elongase is apparently regulated by the physical properties of its host membrane.

2. Temperature Responses in Animals

As the name implies, homeotherms are generally considered to maintain a constant body temperature. It therefore seems rather surprising that warm-blooded animals should alter their tissue lipids in acclimation to temperature extremes. However, a large number of reports have confirmed that exposure to low temperature does lead to an increased proportion of tissue unsaturated fatty acids. This is due to the fact that portions of an animal body cannot be maintained at the same high temperature of the body core. Arctic animals such as reindeer have been shown to store lower melting fats in their extremities, where the superficial temperature can drop to as much as 30°C below that measured in the body core.[11] Lipids from the subcutaneous fat of pigs kept at 0°C are much more highly unsaturated than equivalent lipids of animals maintained at 30 to 35°C.[12] In these and many similar studies, the effects measured were primarily associated with the neutral glycerides, but phospholipids have also been observed to respond to lowered environmental temperatures in some species. For example, Eybel and Simon[13] observed large increases in unsaturated fatty acids in the phospholipids as well as the neutral glycerides of cold-sensitive Arctic mice raised at 5°C and compared to 18°C controls. Other species of mice examined in the same study were cold-resistant, meaning that they were capable of maintaining their body temperature at a nearly normal value even when exposed to –40°C for 2 hr. These mice exhibited little change of fatty acid composition with temperature.

Hibernating mammals are an exception to the rule of sensitive body temperature homeostasis in that their cells can function at the normal body temperature of approximately 37°C and also at a body temperatures as low as 1°C.[14] Analysis of many tissues and cell fractions has made it clear that the torpor, which animals experience during hibernation, is sometimes accompanied by lipid compositional changes. Hibernating ground squirrels (*Citellus tridecemlineatus*) actually had a lower unsaturation index (sum of mole percent of each fatty acid times number of double bonds) than active animals in phospholipids of cerebral cortex, heart, and liver.[15] In the latter two organs, decreases in 20:4, 22:4, and 22:6 were offset primarily by elevated levels of 18:1 and 18:2. On the other hand, the concentration of most unsaturated fatty acids (and aldehydes) rose in phospholipids of Syrian hamster brain during the transition from full activity to hibernation. Although fatty acid unsaturation changes of this type have been frequently observed, they are not necessarily correlated with alterations in membrane fluidity.[14] While the possibility of fluidity-modifying lipid changes in an as yet unexamined organelle or membrane domain during entry into hibernation cannot be strictly ruled out, it seems unlikely at this point. Since metabolic activity is greatly reduced at low temperature, it is reasonable to expect that acclimation completely restoring the membrane physical state existing at the original high temperature is not necessary. The chilled cells seem to retain sufficient capacity even at low temperature to carry out their limited functions.

The ability of mammalian cells to adapt to temperature extremes is illustrated more directly in tissue cultures. Mouse LM cells grown in a lipid-free medium contained a significantly higher level of phospholipid fatty acid unsaturation at 28°C than at 37°C (Table 1).[16] If the cultures were supplemented with 20 μg/ml of linoleate, this was extensively incorporated into phospholipids, but increased unsaturation was still evident in the 28°C cells.

The response of LM cell endogenous lipid metabolism, although clearly detectable, appears to be of limited physiological capacity in overcoming the effects of low temperature. At 28°C, the cells grew more slowly than 36°C control cells and reached a much lower maximal density.[17] They were capable of increasing their maximal density only by supplementation with linolenic or oleic acid.

It is difficult to conduct definitive studies of the regulatory processes governing fatty acid composition in temperature-acclimating mammals because of the temperature gradient in the tissues. Cultured cells also have limitations. Accordingly, most studies of membrane alteration during temperature acclimation have utilized poikilotherms. A popular and typical poikilothermic subject is the goldfish (*Carassium auratus* L.). Johnston and Roots [18] compared the brains of goldfish acclimated to several temperatures, ranging from 30 to 5°C. The composition of the nonhydroxylated fatty acids from total brain lipid (Table 2) revealed an increase in unsaturated fatty acids with decreasing acclimation temperature. This tendency towards more fatty acid unsaturation at lower temperatures held in both the major phospholipids,

TABLE 1
Fatty Acid Composition of Phospholipids of LM Cells Grown at Different Temperature and with the Addition of Linoleate[a,b]

Fatty acid supplement	Temp. (°C)	Fatty acid composition (% by wt.) 14:0	16:0	16:1	18:0	18:1	18:2	% Unsaturated fatty acids
None	37	1.9	20.9	11.2	8.2	57.8		69.0
None	28	1.6	14.2	12.5	9.2	62.3		74.8
Differences[b]		–0.3 ± 0.3	–6.7 ± 2.6	+1.0 ± 2.4	+1.0 ± 0.9	+4.5 ± 3.5		
18:2 $\Delta^{9,12}$	37	1.8	19.5	4.5	11.6	29.7	32.8	67.0
18:2 $\Delta^{9,12}$	28	2.0	13.3	6.0	11.1	33.4	34.2	73.6
Differences		+0.2 ± 0.6	–6.8 ± 1.8	+1.5 ± 1.6	–0.5 ± 0.6	+3.7 ± 3.2	+1.4 ± 5.1	

[a] Cells were incubated in medium ± 20 μg/ml of linoleate (complexed to bovine serum albumin) for 16 hr at the indicated temperature.
[b] Differences in each of five experiments were averaged and the standard error was calculated by the formula $\sigma = (\Sigma \bar{X}^2 - N(X)^2)/(N - 1)$.

TABLE 2
Gas-Liquid Chromatography of the Methyl Esters of Nonhydroxylated Fatty Acids of Brain Lipid from Fish Acclimated to Different Temperatures

Fatty acid	% Fatty acid (peak area) 5°C	15°C	25°C	30°C
12–16	6.16	6.41	5.73	5.93
16:0[a]	24.44 ± 0.49[b]	24.36 ± 0.46	25.49 ± 0.26	24.78 ± 0.84
16:1	9.76 ± 0.64	8.84 ± 0.46	8.48 ± 0.57	7.83 ± 1.84
18:0	13.01 ± 2.42	14.41 ± 0.56	16.61 ± 0.47	18.33 ± 1.0
18:1	25.55 ± 3.41	21.48 ± 0.41	22.94 ± 0.62	24.51 ± 1.29
18:2	1.61 ± 0.74	2.68 ± 1.86	2.05 ± 0.22	0.87 ± 0.15
18:3, 30:0	0.93	0.91	0.62	0.77
20:4	4.13 ± 0.25	4.25 ± 1.12	2.51 ± 1.06	1.29 ± 0.42
22:5 or 6[c]	2.45 ± 0.19	1.99 ± 0.42	1.76 ± 1.45	0.47 ± 0.43
Others [d]	10.48	14.51	13.91	13.26

[a] Figure preceding colon indicates chain length, figure after colon the number of double bonds.
[b] Standard deviation of the mean (3 to 5 samples).
[c] No standard fatty acid available, whether 5 or 6 double bonds uncertain.
[d] Minor fatty acids including branched chain and unidentified acids of chain length beyond 22.

phosphatidylcholine and phosphatidylethanolamine.[19] The fatty aldehydes recovered from the ethanolamine plasmalogens were also more unsaturated in the low-temperature adapted fish, and the percentage of the total phosphoglycerides found as plasmalogens was significantly higher (45.73 ± 3.51) at 30°C than at 5°C (35.19 ± 1.37).[20]

A sizable number of temperature acclimation studies have now been carried out using fish.[21] The response to low temperature routinely includes an increase in fatty acid unsaturation and, as illustrated in Figure 2, a highly ordered revamping of the phospholipid molecular species composition. Another response, namely a reciprocal change in the relative proportions of phosphatidylcholine and phosphatidylethanolamine, is among the first chilling induced lipid alterations observed (Figure 3).

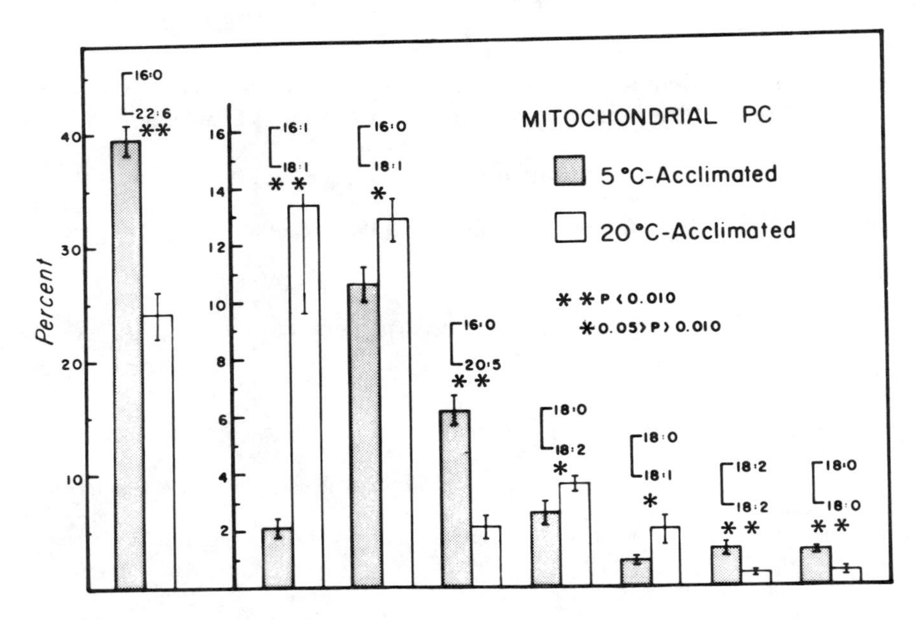

FIGURE 2. The effect of acclimation history on the molecular species composition of phosphatidylcholine isolated from liver mitochondria of 5 and 20°C-acclimated rainbow trout. Only molecular species experiencing significant changes in proportions are reported. Presented as mean ± S.E.M. (N = 8). (From Hazel, J. R., *Physiological Regulation of Membrane Fluidity,* Alan R. Liss, New York, 1988, 149–188. With permission.)

Eukaryotic microorganisms appear to be more complex in their response to altered temperature, but this apparent complexity may merely reflect the relative ease with which such lifeforms may be studied in detail. The unicellular protozoan, *Tetrahymena pyriformis,* has recently been subjected to a rather exhaustive analysis during adaptation to temperature change.[22] Protozoa are regularly subjected to temperature extremes in nature, and they are capable of surviving at 0°C.[23]

I shall illustrate the temperature response of these organisms by describing the changes in *T. pyriformis* strain NT-1, a thermostable variety isolated from a hot spring.[24] Cells grown at 39.5°C (near the upper limit of survival) had a pattern of phospholipid fatty acids significantly different from that found in cells grown at 15°C (Table 3).[24] Differences were also observed in the phospholipid polar head group distribution (Table 4).[24] Cell fractionation studies revealed that these differences occurred, to a greater or lesser degree, in membranes of intracellular organelles as well as in membranes of the cell surface.

Following a shift of 39.5°C-grown cells to 15°C, there was an extremely rapid rise in fatty acid unsaturation in microsomal phospholipids followed, after a lag of approximately 0.5 h, by a similar rise in unsaturation of phospholipids in the surface membranes.[25] These changes were accompanied by increased fluidity of, first, the microsomal lipids and, later, the surface membrane lipids, as inferred from fluorescence polarization measurements (Figure 4). This kinetic pattern was dictated by the fact that the cellular fatty acid desaturases, localized in the microsomal membranes, released their products initially to that membrane. Subsequent equilibration of the more unsaturated lipids to other membranes required the participation of intermembranous exchange reactions (see p. 180).

The close correlation of microsomal fatty acid unsaturation with the membrane physical state, as estimated by electron microscopic observations of lipid phase separation as well as fluorescence polarization measurements[25] is illustrated in Figure 5. As indicated on p. 59, the activation of fatty acid desaturation relative to fatty acid synthesis is thought to occur because

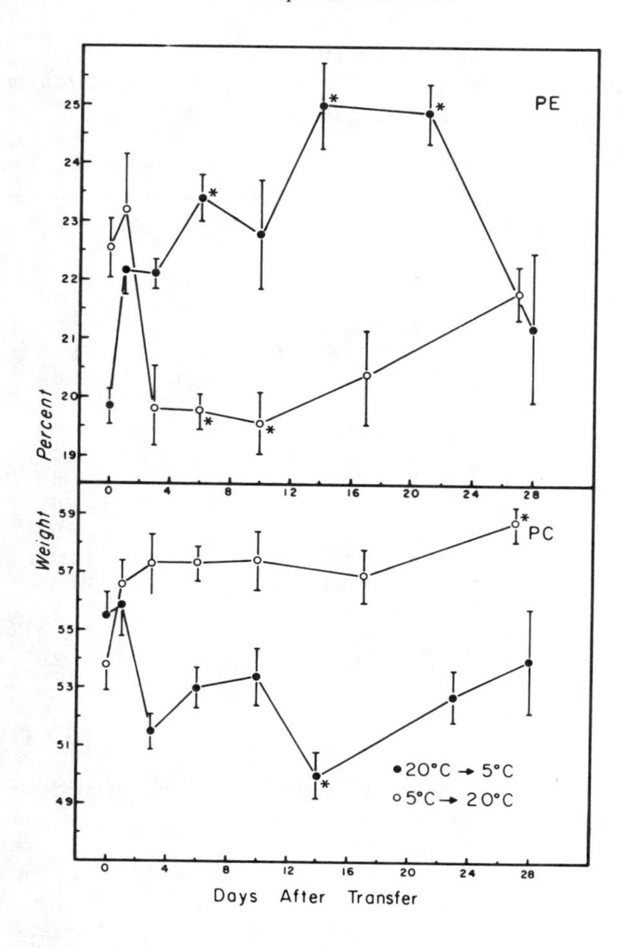

FIGURE 3. Changes in the weight percentages of phosphatidylethanolamine (top) and phosphatidylcholine (bottom) in trout gill during acclimation from 5 to 20°C (open symbols) and from 20 to 5°C (closed symbols). Presented as mean ± S.E.M. (N = 5). Asterisks indicate a statistically significant difference ($p < 0.05$) compared with the day 0 mean. (From Hazel, J. R., *Physiological Regulation of Membrane Fluidity,* Alan R. Liss, New York, 1988, 149–188. With permission.)

of membrane fluidity changes in the microsomal membrane environment occupied by the fatty acid desaturases. An increasing rigidity of this membrane promotes enhanced activity by the desaturase enzymes, with the products of the reactions being utilized to restore greater fluidity to the microsomal membranes and, ultimately, other membranes of the cell. In order to confirm the role of membrane fluidity (rather than temperature per se), as the prime regulatory factor, a number of studies have been conducted to measure the impact of fluidity varied independently of temperature and vice versa.[26] In each case, fatty acid desaturase activity was much more responsive to changes in fluidity than to temperature change.

In addition to the above-mentioned tendency of decreased membrane fluidity at low temperature to preserve relatively high activity of existing desaturase molecules, there is a chilling-induced synthesis of certain desaturases. By 1 h following the chilling of *Tetrahymena* from 39.5 to 15°C, palmitoyl-CoA desaturase activity had increased by nearly 50% and stearoyl-CoA desaturase activity by 27%.[27] This rise in activity, which could be inhibited by cycloheximide, persisted for only a few hours. No evidence was found for an induced synthesis of monoenoic or dienoic fatty acid desaturases.

TABLE 3
The Fatty Acid Composition of Total Phospholipids from *Tetrahymena* Grown at Different Temperatures

Component	Retention time (min)	Whole cells 15°C [6]	Whole cells 24°C [6]	Whole cells 39.5°C [5]
12:0	1.4	0.6 ± 0.3	1 .0 ± 0.5	0.9 ± 0.3
14:0	2.7	6.9 ± 0.7	6.6 ± 0.9	6.5 ± 1.0
ante-iso 15:0	3.2	1.5 ± 0.3	1 .5 ± 0.3	3.4 ± 0.4
15:0	3.7	tr[a]	0.8 ± 0.4	1.2 ± 0.1
16:0	5.1	8.9 ± 0.5	10.4 ± 1.2	12.6 ± 1.6
16:1	6.0	8.7 ± 0.9	8.9 ± 0.8	8.7 ± 1.1
17:0 + 16:2	7.1	2.2 ± 0.2	3.3 ± 0.4	4.4 ± 0.3
17:1	8.1	0.8 ± 0.4	1.2 ± 0.2	1.4 ± 0.2
?		0.9 ± 0.3	0.9 ± 0.2	1.2 ± 0.3
18:0	9.9	0.6 ± 0.3	1.4 ± 0.7	2.1 ± 0.4
?		1.3 ± 0.4	1.5 ± 0.1	1.1 ± 0.2
18:1	11.2	9.6 ± 1.6	7.4 ± 0.6	10.6 ± 3.9
?	13.3	7.0 ± 0.3	5.1 ± 0.4	3.4 ± 1.0
18:2	13.9	20.2 ± 1.6	18.2 ± 1.6	14.5 ± 0.9
18:3	16.3	31.1 ± 0.9	32.6 ± 2.0	24.5 ± 2.4
19:0	19.2	tr	tr	2.6 ± 2.2
20:1	21.0	tr	tr	tr
?	22.7	tr	tr	tr

Note: Results are expressed as the weight percent of total fatty acids as determined by gas-liquid chromatography. The numbers of separate experiments are shown in brackets.

[a] Less than 0.5%.

TABLE 4
Lipid Composition of Tetrahymena Whole Cells Adapted to Different Temperatures[a]

Lipid	15°C	24°C	39.5°C
μmol lipid phosphorus/10^6 cells	0.14 ± 0.01	0.13 ± 0.02	0.14 ± 0.01
Tetrahymanol/phosphorus molar ratio	0.081	0.078	0.083
Alkyl ether (mol% of phospholipids)	27.1 ± 6.6	—	20.9 ± 2.8
Individual phospholipids (mol%)			
Cardiolipin	5.2 ± 0.6	5.6 ± 0.4	7.8 ± 1.9
2-Aminoethyl phosphonolipid	29.0 ± 0.6	25.4 ± 2.0	15.6 ± 1.7
Ethanolamine glycerophosphatides	25.6 ± 4.4	34.4 ± 1.6	42.5 ± 1.3
Lysophosphatidylethanolamine, lyso-2-aminoethylphosphonolipid, ceramide aminoethylphosphonate	8.3 ± 0.6	5.4 ± 0.9	2.7 ± 1.0
Choline glycerophosphatides	27.2 ± 1.3	26.2 ± 0.5	26.8 ± 3.4
Lysophosphatidylcholine	2.8 ± 1.9	2.6 ± 1.8	2.8 ± 1.6

[a] Values represent averages ± S.D. of three or more experiments.

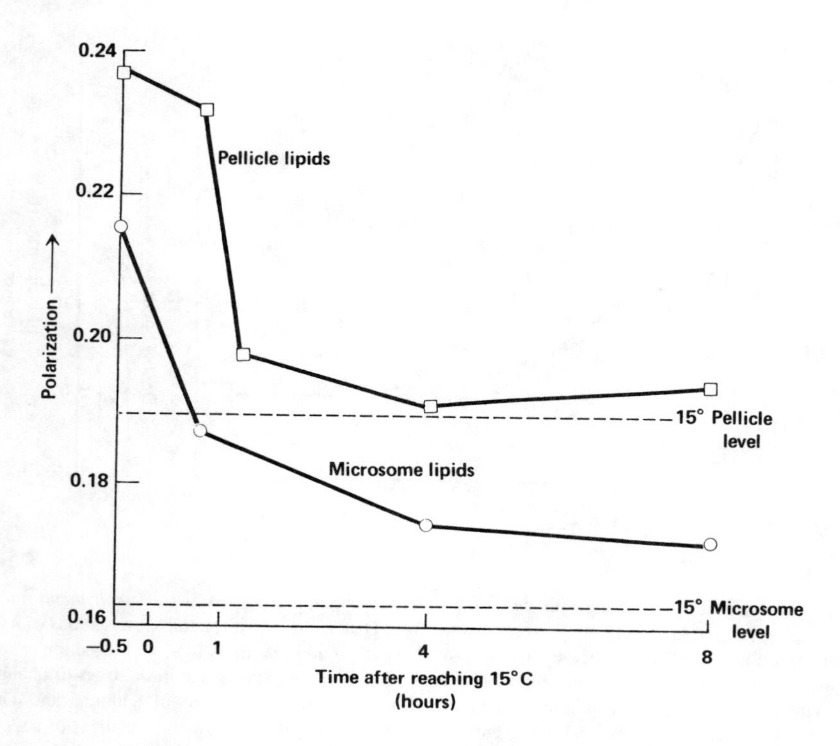

FIGURE 4. Time course of changes in 1,6-diphenyl-1,3,5-hexatriene (DPH) polarization in membrane lipids of 39.5°C-acclimated *Tetrahymena* cells following a shift to 15°C. Cells were grown at 39.5°C, shifted to 15°C over a 30-min period, and then harvested and fractionated at the indicated times for membrane lipid isolation and DPH polarization measurements at 15°C. Dashed lines show the polarization values in cells fully acclimated to 15°C. (From Martin, C. E. and Thompson, G. A., Jr., *Biochemistry*, 17, 3581, 1978. With permission.)

Increases in the level of fatty acid unsaturation, although readily detectable in less than 1 h, are not sufficiently pronounced to account for the almost immediate change in the physical state of *Tetrahymena* microsomal lipids in chilled cells.[28] The physical alterations noted within 15 min by fluorescence polarization measurements are more likely to result from the rapid retailoring of phospholipid molecular species (see Chapter 4) induced by the chilling process.[29,30] Evidence suggests that the rate-limiting step of molecular species retailoring in *Tetrahymena* cilia is the activity of phospholipase A in producing lysophospholipids for use in formulating products containing novel fatty acid pairings.[29] Cells may rely on this recombination of existing fatty acids into new, more flexible phospholipids to restore function to the membrane until increased fatty acid desaturation can allow for a more permanent fluidization.

A key fact to remember from all these studies is that analysis of whole cell lipid alterations during rapid responses to temperature change may provide a rather blurred picture of the degree and rapidity of response in specific portions of the cell. It seems logical that in animals the initial reactions of lipid metabolic enzymes to low temperatures should be detectable in the microsomal membranes, where fatty acid desaturases are characteristically located. Changes initiated by a rise in temperature should also appear first in the microsomes, since it is in these membranes that all new phospholipids arise. Under such circumstances, reduced lipid unsaturation can only occur through dilution of preexisting lipids via *de novo* synthesis of more saturated phospholipids. In the case of *Tetrahymena,* the first impact of both elevated and reduced temperature on membrane lipid composition is on phospholipid fatty acid

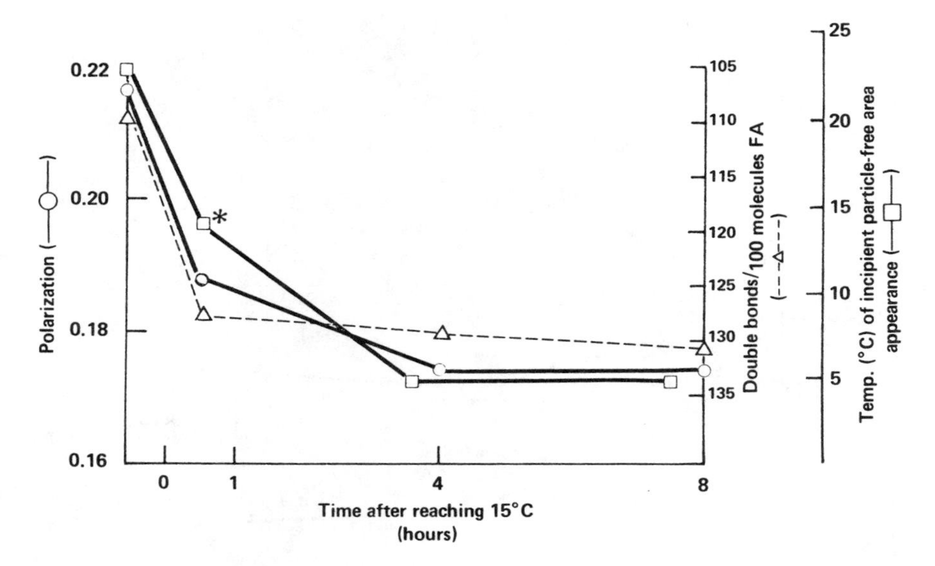

FIGURE 5. Comparison of the changes of three independently measured properties of *Tetrahymena* microsomal membranes during temperature acclimation from 39.5 to 15°C. Data from freeze-fracture observations of membrane particle redistribution (□), fluorescence polarization of DPH in membrane lipids (○), and the number of double bonds in phospholipid fatty acids (△) have been plotted to show the correspondence in the rates of change of the three parameters. Freeze-fracture data were recalculated from previous experiments of Kitajima and Thompson (1977); the point marked with an asterisk represents a new observation of freeze-fracture replicas of cells fixed at 15°C only. Since no particle-free areas were present, this can be taken as an upper limit. The true temperature of incipient particle-free area appearance is probably 1 to 2° lower. (From Martin, C. E. and Thompson, G. A. Jr., *Biochemistry,* 17, 3581–3591, 1978. With permission.)

composition. Modification of the phospholipid polar head groups (Table 4) cannot be detected during the initial period of fatty acid change, and the level of the cholesterol-like triterpenoid, tetrahymanol, remains constant over a wide temperature range.

Photosynthetic eukaryotes respond to temperature change in a manner rather similar to that described above for *Tetrahymena.* Thus the unicellular green alga *Dunaliella salina* experiences a retailoring of phospholipid molecular species and an increase in fatty acid unsaturation in both chloroplast and microsomal lipids as its first lipid alterations during the initial hours after chilling.[31]

3. Temperature Responses in Higher Plants

Many experiments have been conducted to examine the possible role of membrane lipids in the low temperature acclimation or hardening of plants. The adaptation might conveniently be subdivided into two related phenomena; the process of gaining resistance to chilling injury by temperatures above freezing and the process of becoming resistant to subfreezing temperatures involving potential physical damage to the cell. Freeze-thaw injury is now known to affect primarily the plasma membrane as the formation of extensive extracellular ice crystals causes a sometimes irreversible cell dehydration and shrinkage of sensitive cells.

Recent studies have focused attention on changes in the plasma membrane lipid composition as an exposure to low but above freezing temperatures hardens them to withstand subsequent freezing.[32] It has only recently been discovered that the plant plasma membrane lipids contain a predominance of nonphospholipid components. In rye plasma membranes free sterols, steryl glucosides and acylated steryl glucosides together account for >50 mol% of the structural lipids.[33] Another 16 mol% of the lipids are made up of glucocerebrosides, particularly those containing C_{22}-C_{24} hydroxy fatty acids linked to trihydroxy long chain bases.[34] A number of plasma membrane lipid compositional changes accompany the development of cold acclimation.[33] Noteworthy is the decline in glucocerebroside content from 16 to 7%. This decrease, along with an increase in phospholipids with which the glucocerebrosides are poorly miscible, may be in part responsible for the increased tolerance to osmotic damage observed to develop in the plasma membrane.

Many laboratories have concentrated their attention on the considerable damage that occurs when certain plants are exposed to temperatures in the 0 to 15°C range.[35] Increases in fatty acid unsaturation are generally found during acclimation to chilling temperatures but progress in understanding the molecular factors regulating these changes has been slow. New insights have recently emerged from studies using lipid mutants of *Arabidopsis thaliana*. For example, Kunst et al.[36] studied an *Arabidopsis* mutant deficient in a chloroplast desaturase that normally introduces a double bond at the omega 9 position of palmitate linked to monogalactosyldiacylglycerol (MGDG). Although the mutation caused a large increase in MGDG palmitate and a reduced level of unsaturated C_{16} fatty acids, possible deleterious consequences of this defect on thylakoid physical properties were reduced by an enhanced importation of MGDG made in the endoplasmic reticulum by a pathway utilizing C_{18} polyunsaturates.

Interestingly, this same mutant, which exhibited an overall reduction in the unsaturation of chloroplast lipids, grew more rapidly than wild type *Arabidopsis* at temperatures above 28°C and contained chloroplasts that were unusually resistant to high temperature inactivation of photosynthetic electron transport.[37]

Special attention has been directed to the occurrence of high proportions of two phosphatidylglycerol molecular species, *sn* 1,2-dipalmitoyl phosphatidylglycerol and *sn* 1-palmitoyl, 2-(*trans* $\Delta 3$)hexadecenoyl phosphatidylglycerol, which have thermotropic phase transitions at 41 and 32°C, respectively. These two molecular species together accounted for 26 to 65% of chloroplast phosphatidylglycerol in 9 species of chilling sensitive plants and 19% or less in 12 chilling resistant plants.[38] It would seem logical that damage to the chloroplast might result from a lipid phase separation of these high melting and abundant phospholipids. This interpretation is clouded by the recent discovery of some exceptions to the generalization that the high melting phosphatidylglycerols are abundant only in chilling sensitive plants.[39] Also, calorimetric studies[40] did not show a correlation between chilling sensitivity of photosynthesis and the detection of a phase separation of bulk membrane lipids.

The relationship between chilling sensitivity and phosphatidylglycerol composition is probably more complex than originally envisioned. Lipids such as certain high melting molecular species of phosphatidylglycerol could exert a more telling effect on chilling sensitivity through their interaction with membrane proteins than by their perturbation of membrane fluidity per se.[41]

A possible case in point involves the apparent stabilization of the light harvesting complex of proteins associated with the photosystem II of chloroplasts by high levels of phosphatidylglycerol molecular species containing *trans* -Δ^3-hexadecenoic acid. In wild type and mutant *Chlamydomonas*[42] and in winter rye (*Secale cereale* L.) grown at different temperatures,[43] the proportion of the photosystem II light harvesting complex that is found in thylakoid extracts as the oligomeric (functional) form is positively correlated with the content of *trans*-Δ^3-hexadecenoate. The presence of this unusual fatty acid, which occurs in no other phospholipid besides phosphatidylglycerol, is not essential for photosynthesis, as evidenced by the normal growth of a *trans*-Δ^3-hexadecenoate-deficient *Arabidopsis* mutant under standard conditions, but the stabilizing effect it provides even in *Arabidopsis* appears to be of considerable value under some conditions such as high temperature.[37]

B. CELLULAR ACCLIMATION TO CHANGES IN PRESSURE

While pressure changes experienced by most terrestrial organisms are insignificant, creatures inhabiting the deep sea may sustain large pressure variations during migration or through involuntary vertical transport. Pressures of 500 atm are common and have effects on membranes which, when added to those caused by the uniformly low (0 to 4°C) marine temperatures, produce the equivalent of –6°C for a surface organism.[44]

Relatively little is known regarding the effects of pressure change on membrane lipids. A *Vibrio* type deep sea bacterium contained more unsaturated fatty acids when grown at increasing pressures.[45] The facultative anaerobe was thought to achieve this regulation by modulating 3-oxoacyl-[acyl-carrier-protein] synthase II, the same enzyme shown in *E. coli* to determine the ratio of saturated to unsaturated fatty acids (see Chapter 2).

Marine animals also respond to increasing pressure as they would to decreasing temperature. In a detailed study of liver mitochondrial phospholipids from 13 species of teleost from habitats 200 to 4000 m deep, Cossins and MacDonald[46] found a correlation between increasing pressure and increasing proportions of unsaturated fatty acids in phosphatidylcholine, phosphatidylethanolamine, and phosphatidylserine. The changes, largely a replacement of saturates by monoenes, with polyenoic fatty acid levels remaining relatively constant, were greater than would be expected from the slight temperature differences at the depths of collection. Other examples of pressure effects on membrane lipid composition have been reviewed by Hazel.[21]

C. RESPONSES TO CHANGES IN INORGANIC ION CONCENTRATIONS

I have postulated that the response of cells to membrane-perturbing environmental effects is to restore optimal physical properties by altering membrane lipid composition. If this is a general phenomenon, then detectable changes in lipid metabolism might be expected to accompany fluctuations in cation concentrations in much the same way as they accompany temperature variations.

The fluidity of artificial lipid bilayers may be strongly affected by differing concentrations of cations.[47] Although any brief explanation of the interactions is necessarily oversimplified, especially when applied to a natural membrane, divalent cations, such as calcium or magnesium, can be considered to interact with single negative charges on the head group of two adjacent phospholipid molecules, linking them together somewhat more rigidly than would be the case if each phospholipid were independently associated with a monovalent cation instead. Monovalent and divalent cations, when present together, compete with each other for the charged polar head groups of membrane phospholipids, particularly negatively charged molecules such a phosphatidylserine or phosphatidylglycerol.

The efficient regulatory systems of multicellular organisms ensure that the ionic environment both inside and outside most cells remains almost constant. The few observations regarding the effects of salts on membrane lipid metabolism have generally involved unicellular organisms. Thus *Staphlococcus aureus* grown in medium containing 10% (by weight) of NaCl contained a dramatically altered phospholipid composition in which cardiolipin accounted for 50% of the total lipid phosphorus rather than the 10% found in normal (0.05% NaCl) medium.[48] The increased cardiolopin was offset by decreased amounts of phosphatidylglycerol and lysophosphatidylglycerol. Although the overall fatty acid pattern for total phospholipids showed little change, compositional alterations within classes were large, e.g., branched-chain fatty acids in cardiolipin increased from 45% in normal medium to 77% in high-salt medium.

The lipids of an even more halotolerant bacterial species, *Staphlococcus epidermidis,* have been analyzed following cell growth in varying levels of NaCl.[49] Little change was noted in the phospholipid composition until the NaCl level reached 15%, but by 25% NaCl, cardiolipin increased to 11 mol% of the total polar lipids (vs. 0.5% at low-salt concentrations). Less extensive differences were noted in the fatty acid content.

It would be interesting to compare the salt responses of the above organisms with those of extreme halophiles, such as *Halobacterium halobium* (see p. 145). However, the latter bacteria, which form only ether-linked lipids in their very saline medium, have not yet been successfully grown in low-salt concentrations.

The membrane lipid composition of the eukaryote *Tetrahymena* is altered by growth in either high NaCl or high $CaCl_2$.[50,51] The effect of growth in 0.3 *M* NaCl was much more pronounced in the surface membranes of the cell, presumably because the cytoplasmic concentration of Na^+ can be maintained nearly constant even in a medium of high salinity.[52] The content of 2-aminoethylphosphonolipid increased in the surface membranes from 18 mol% of the total lipid phosphorus in normal medium to 27 mol% in 0.3 *M* NaCl medium. The change was offset principally by a fall in the phosphatidylcholine concentration. There was also a shift in the phospholipid fatty acid pattern towards somewhat lower unsaturation in the salt-grown cells.

Dunaliella salina is a eukaryotic alga capable of withstanding extremes of salinity ranging from 0.86 *M* up to 4.3 *M* NaCl.[53] Because it offsets the osmotic effects of high external NaCl by maintaining an appropriately high internal glycerol concentration, only its plasma membrane is in direct contact with the potentially damaging excesses of Na^+. Analysis of plasma membranes isolated from *D. salina* grown in 0.85 *M*, 1.7 *M*, and 3.4 *M* NaCl medium revealed only modest changes in polar lipid classes.[54] The largest difference was a 10% increase in the proportion of diacylglyceroltrimethylhomoserine in the plasma membranes of 3.4 *M* NaCl-grown cells. This particular lipid, although not widespread in nature, is also found in nonhalotolerant algae, such as *Chlamydomonas reinhardtii* [55] and is presumably not specifically associated with membrane adaptation to high salinity. Small decreases in fatty acid unsaturation were also noted in several lipid classes. Considering the lack of any major alteration of plasma membrane lipid composition over such a wide range of salinity, it would seem unlikely that changes in lipid distribution play a crucial role in salinity acclimation.

In its response to sudden osmotic stress, *D. salina* does employ lipid changes of another sort. Within seconds following a dilution from 1.7 *M* to 0.85 *M* NaCl, 30% of the cells' phosphatidylinositol 4,5-bisphosphate is hydrolyzed to yield inositol 1,4,5-trisphosphate.[56] A transient rise in plasma membrane diacylglycerol accompanies the hydrolysis.[57] The transient operation of this phosphatidylinositol-specific phospholipase C-mediated signal transduction pathway (see Chapter 4 for details) may be expected to activate one or more protein kinases, but the physiological consequences of the activation are not yet known. Interestingly, exposure of *D. salina* to hyperosmotic stress, by suddenly raising the NaCl level from 1.7 *M* to 3.4 *M*, triggers a short term rise in the concentration of polyphosphoinositides.[58]

Considering the practical benefits to be gained, it is surprising how little basic information has been gathered on the response of higher plant membranes to salt stress. Some evidence is available to suggest that the roots of salt-tolerant plants contain more sterols and sterol derivatives than salt-sensitive species. This trend was observed in a comparison of the salt-sensitive bean (*Phaseolus vulgaris* L. cv. Saxa), the less salt-sensitive barley (*Hordeum vulgaris* L. cv. Wisa), and the salt-tolerant sugar beet (*Beta vulgaris* L. cv. Kawemono).[59] Differences were noted in other lipid constituents, such as fatty acids, but their significance is difficult to assess in view of species differences that might be superimposed upon any true salt effect.

Perhaps more meaningful was an analysis of root lipids from five grape varieties that differed in the extent to which their leaves accumulated chloride.[60] The most salt-sensitive variety, which accumulated the most chloride, contained an extremely low amount of sterols (9% of total lipids vs. approximately 20% for other varieties), in keeping with the finding quoted above.[59] Phosphatidylcholine was present in lowest amounts in the more salt-sensitive varieties, while monogalactosyldiacylglycerol showed the opposite trend. Particularly interesting was the high level of very long-chain fatty acids, especially lignoceric acid ($C_{24:0}$) in phospholipids of the salt-tolerant plants.

The kind of sterol present in roots may be important in determining its tolerance to salinity. In *Plantago maritima*, a halophytic species, the sitosterol-cholesterol ratio in roots decreased with increasing salinity, mainly due to a decrease in the sitosterol level.[61] This suggested to the authors that cholesterol, thought to be more effective than sitosterol in limiting permeability, may become more influential in regulating ion fluxes. High sitosterol/stigmasterol ratios have been correlated with efficient Cl^- exclusion of three *Citrus* rootstock varieties.[62]

Information concerning the responses of cellular lipid metabolism to inorganic ions is still too fragmentary to warrant the construction of a general hypothesis concerning regulatory mechanisms. Many of the available reports have not concerned themselves with possibly very important interactions between the specific ions being tested and other ions also present in the system. These undetermined relationships doubtless contribute to the present confused state of our understanding.

D. EFFECTS OF CERTAIN UNNATURAL CHEMICALS ON LIPID METABOLISM

In-depth studies of drug effects on lipid composition and metabolism are few and far between. The "take-home lesson" can be disclosed here at the outset: unnatural chemicals can induce profound alterations in membrane lipid metabolism, but the molecular mechanisms underlying the changes are virtually unknown.

For this reason, it does not seem worthwhile to discuss in detail the various studies that have been done. I shall simply present a representative sample of experimental findings so as to indicate the extent of lipid changes observed.

1. Chlorinated Hydrocarbons

Abnormal lipid metabolism has been reported in rats fed dieldrin (1,2,3,4,10,10-hexachloro-6,7-epoxy-1,4,4a,5,6,7,8,8a-octahydro-1,4-endo-exo-5,8-dimethanonaphthalene) and other chlorinated aromatic hydrocarbons.[63] The pesticide *o,p′*-DDT [1-(*O*-chlorophenyl)-1-(*p*-chlorophenyl)-2,2,2-trichloroethane] has been shown to disrupt normal lipid metabolism in humans [64] and rats.[65] In the latter study, DDT was found to be distributed throughout the membranous organelles of liver cells. Rats fed diets containing 5 ppm DDT exhibited a markedly suppressed desaturation of essential fatty acids (Table 5). The remarkably low levels of linoleic acid (18:2) in the phospholipids despite its presence at a sufficient level in the diet suggested a defect in its incorporation into phospholipids.

2. Ethanol

Because the physiological effect of ethanol is of such great medical interest, a number of studies have been conducted on the effects of intoxicating doses on membranes. Large effects on lipid composition have been found in unicellular organisms grown in the presence of alcohols. Ingram[65] grew *E. coli* in various concentrations of a homologous series of aliphatic alcohols, including ethanol. The phospholipids of cells grown in the presence of alcohols having 5 to 10 carbon atoms contained a much higher than normal content of saturated fatty acids. However, lower alcohols (C_1 to C_4) induced a different response altogether. The major *E. coli* unsaturated fatty acid, vaccenic acid (18:1), underwent a major increase at the expense

TABLE 5
Effects of 5 ppm *O,p´*-DDT in the Diets of Rats upon Pattern of Fatty Acids in Phospholipids

Fatty acid	Lab chow	Lab chow + DDT
n	5	5
16:1	1.5[a]	5.0
18:1	9.3	24.4
20:3ω9	1.5	18.0
18:2ω6	17.7	3.4
20:2ω6	2.0	nd[b]
20:3ω6	1.6	tr[c]
20:4ω6	22.5	10.2
Total ω6 metabolites	26.0[d]	10.6
20:3ω9λ20:4ω6	0.06	1.8

[a] Percentage of total fatty acids from phospholipids.
[b] Not detectable.
[c] Trace, less than 0.01%.
[d] Total ω6 acids minus 18:2ω6.

of saturated acids. The effects may tentatively be explained by the following factors: (1) a direct effect of the short-chain alcohols on β-hydroxydecanoyl thioesterase dehydrase, the key enzyme regulating the proportion of *E. coli* fatty acid precursors directed into unsaturated vs. saturated fatty acids (see p. 50) and (2) a production of more saturated fatty acids as a kind of homeoviscous adaptation[67] to the fluidizing effects of longer-chain alcohols.

Nandini-Kishore et al.[68] analyzed the changes in lipid composition induced by the growth of *T. pyriformis* in 1.6% ethanol. Phospholipid-bound hexadecenoic acids (16:1 and 16:2) were decreased from 23 to 5%, and linoleic acid rose from 14 to 25%. Phosphatidylethanolamine rose from 39% of the total phospholipids to 46% in ethanol-grown cells, while 2-aminoethylphosphonolipid dropped from 16 to 6%. A comparison of two membrane fluidity-related properties, namely, aggregation patterns of intramembranous particles as seen in freeze-fracture electron micrographs and fluorescence polarization of the probe 1,6-diphenylhexatriene in membrane-derived liposomes, indicated unexpectedly that ethanol-induced lipid changes rendered the membrane lipids more fluid than those of normal cells. It is not clear at the present time whether the responses of *Tetrahymena* and *E. coli* to ethanol should be considered as an acclimation, returning certain properties of the affected membranes to a more nearly normal physical state, or as a physiologically purposeless manifestation of ethanol toxicity.

Mammals chronically exposed to intoxicating levels of ethanol have not been found to sustain any sizable modification in their membrane lipids. However, Taraschi, Rubin and associates have reported titilating evidence that relatively minor and, in fact, still undetermined changes in anionic lipid classes are responsible for the acquisition of "membrane tolerance" in rats. The membranes of animals fed ethanol for 28 to 35 days developed a resistance to the disordering effect (as inferred using electron spin resonance probes) that membranes from control rats displayed when treated *in vitro* with 50 to 100 m*M* ethanol. The same effect could be seen using phospholipid vesicles purified from the tissues. In a recent report[69] it was demonstrated that by reconstituting the phospholipid mixture from liver

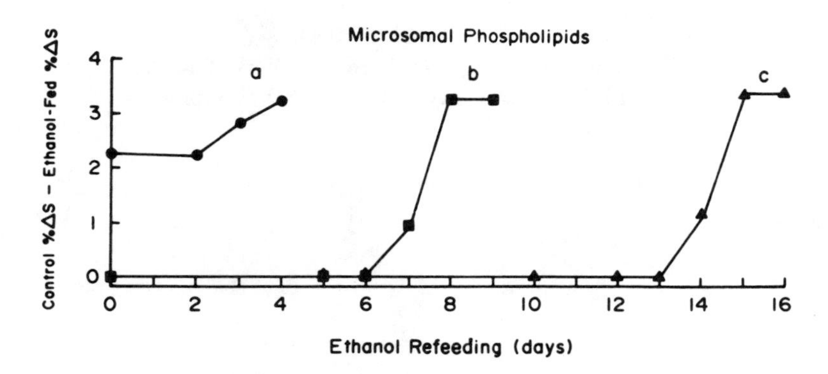

FIGURE 6. Typical time-course for the reappearance of membrane tolerance in liposomes composed of liver microsomal phospholipids from rats withdrawn for 2 (●—●), 3 (■—■) or 4 (▲—▲) days and re-fed ethanol for up to 20 days. In the ordinate Control %ΔS represents the percent difference in the ESR order parameters, S, obtained for the spin probe in microsomal phospholipids from control rats between 0 and 100 mM ethanol. Ethanol-fed %ΔS represents the percent difference in the order parameters obtained between 0 and 100 m*M* ethanol for rats refed the ethanol diet after withdrawal. (From Taraschi, T. F., Ellingson, J. S., Wu-Sun, A., and Rubin, E., *Biochim. Biophys. Acta,* 1021, 51–55, 1990. With permission.)

mitochondria of control rats, but substituting all or part of its cardiolipin with cardiolipin from ethanol-fed mitochondria, tolerance to ethanol perturbation could be achieved. Likewise, tolerance could be conferred on liver microsomal phospholipids by substituting in phosphatidylinositol from the equivalent preparation of ethanol-fed rats.

The as yet unknown changes that gradually accrue during a month of chronic ethanol ingestion appear to be reversed within a very few days following alcohol withdrawal. But if the animals are refed ethanol shortly thereafter tolerance returns with a relatively short delay, depending strictly upon the elapsed withdrawal time (Figure 6)[70] Hopefully, detailed analysis of the phosphatidylinositol and cardiolipin fractions found to confer ethanol tolerance in microsomal and mitochondrial membranes, respectively, will reveal the subtle features responsible for the effects.

Apart from its function as an intoxicant, ethanol also acts as an anesthetic, albeit a very poor one. For comparison, information is available concerning the effects of methoxyflurane, a much more efficient general anesthetic, on lipid metabolism in *Tetrahymena*.[71] In the presence of anesthetizing concentrations of methoxyflurane, fatty acid desaturation was decreased, resulting in an accumulation of phospholipid-bound palmitate and a large decrease in the desaturation of linoleic acid to γ-linolenic acid, as determined by the labeling patterns following administration of [^{14}C]-acetate. This change was postulated to be part of an acclimation to the superoptimal membrane fluidity induced by the anesthetic.

A number of other chemicals have been shown to trigger changes in membrane lipid composition and metabolism. For example, exposure of *E. coli* to a nonbacteriostatic concentration of phenethyl alcohol led to a reduced incorporation of fatty acids, especially saturated fatty acids, into phospholipids.[72] Thus, the fatty acyl composition was altered appreciably. A similar trend was observed in *Tetrahymena*.[73] In the protozoan cell, phenethyl alcohol (8 m*M*) also induced a rise in the proportion of phosphatidylcholine at the expense of phosphatidylethanolamine and the phosphonolipid.

Ingram[74] has demonstrated marked changes in the fatty acid composition of *E. coli* membranes following growth of the cells in medium containing one of 23 different organic solvents and food additives of assorted types. The changes in fatty acid and phospholipid head group composition were in some cases considerable and suggested an acclimation to the perturbing effect of the added compound.

III. CONCLUSIONS

Drawing any sort of logical conclusion from the diverse findings described in this chapter is obviously premature. There is a thread of continuity tying some of the observations together. For example, many effects of temperature on lipid composition agree with the concept that a cell can, through altered lipid metabolism, restore optimal or nearly optimal fluidity to a membrane lipid bilayer whose physical state has been adversely affected by high or low temperature. Efforts to establish the same molecular mechanism as an explanation for salt- and drug-induced changes have thus far met with mixed success. It may be counterproductive to extend such studies much further without taking more into account the direct effects of the perturbants upon membrane structural proteins.

For the immediate future, we might well concentrate on determining whether an environmentally produced change in lipid metabolism really contributes to an acclimation, in the sense that it helps restore normal function. Once such responses are separated from functionally useless lipid alterations arising from some toxic effect on enzyme action, the pertinent mechanisms can be examined with more confidence.

REFERENCES

1. **Hazel, J. R. and Williams, E. E.,** The role of alterations in membrane composition in enabling physiological adaptation of organisms to their physical environment, *Prog. Lipid Res.,* 29, 167–227, 1990.

1a. **Platt-Aloia, K.,** Freeze-fracture evidence of stress-induced phase separations in plant cell membranes, in *Physiological Regulation of Membrane Fluidity,* Aloia, R. C., Curtain, C. C., and Gordon, L. M., Eds., Alan R. Liss, New York, 1988, 259–292.

2. **Quinn, P. J.,** Principles of membrane stability and phase behavior under extreme conditions, *J. Bioenergetics Biomembranes,* 21, 3–19, 1989.
3. **Shinitsky, M.,** Membrane fluidity and cellular functions, in *Physiology of Membrane Fluidity,* Vol. 1, Shinitsky, M., Ed., CRC Press, Boca Raton, 1984, 1–51.
4. **McElhaney, R. N.,** The relationship between membrane lipid fluidity and phase state and the ability of bacteria and mycoplasmas to grow and divide at various temperatures, in *Membrane Fluidity,* Kates, M. and Manson, L. A., Eds., Plenum Press, New York, 1984, 249–278.
5. **Okuyama, H., Yamada, K., Kameyama, Y., Ikezawa, H., Akamatsu, Y., and Nojima, S.,** Regulation of membrane lipid synthesis in *Escherichia coli* after shifts in temperature, *Biochemistry,* 16, 2668–2673, 1977.
6. **Cronan, J. E., Jr.,** Molecular biology of bacterial membrane lipids, *Annu. Rev. Biochem.,* 47, 163–189, 1978.
7. **Fulco, A. J.,** The biosynthesis of unsaturated fatty acids by bacilli. IV. Temperature-mediated control mechanisms, *J. Biol. Chem.,* 247, 3511–3519, 1972.
8. **Murata, N.,** Low-temperature effects on cyanobacterial membranes, *J. Bioenerg. Biomemb.,* 21, 61–75, 1989.
9. **Wada, H., Gombos, Z., and Murata, N.,** Enhancement of chilling tolerance of a cyanobacterium by genetic manipulation of fatty acid desaturation, *Nature,* 347, 200–203, 1990.
10. **Russell, N. J.,** The regulation of membrane fluidity in bacteria by acyl chain length changes, in *Membrane Fluidity,* Kates, M. and Manson, L. A., Eds., Plenum Press, New York, 1984, 329–347.
11. **Irving, L.,** Animal adaptation to cold, in *Cold Injury, Transactions of the 5th Conference on Cold Injury,* Ferrer, M. I., Ed., Madison Printing, Madison, New Jersey, 1958, 11–60.
12. **Henriques, V. and Hansen, C.,** Vergleichende Untersuchungen Über die Chemische Zusammensetzung des tierischen Fettes, *Skand. Arch. Physiol.,* 11, 151–165, 1901.
13. **Eybel, C. E. and Simon, R. G.,** Fatty acid composition of the neutral lipids and individual phospholipids of muscle of cold-stressed arctic mice, *Lipids,* 5, 590–596, 1970.
14. **Aloia, R. C.,** Lipid, fluidity, and functional studies of the membranes of hibernating animals, in *Physiological Regulation of Membrane Fluidity,* Alan R. Liss, New York, 1988, 1–39.
15. **Swan, H. and Schätte, C.,** Phospholipid fatty acids of cerebral cortex, heart, and liver in hibernating and active ground squirrel, *Citellus tridecemlineatus, Am. Nat.,* 111, 802–806, 1977.

16. **Ferguson, K. A., Glaser, M., Bayer, W. H., and Vagelos, P. R.,** Alteration of fatty acid composition of LM cells by lipid supplementation and temperature, *Biochemistry,* 14, 146–151, 1975.
17. **Williams, R. E., Rittenhouse, H. G., Iwata, K. K., and Fox, C. F.,** Effects of low temperature and lipid modification on the proliferation of cultured mammalian cells, *Exp. Cell Res.,* 107, 95–104, 1977.
18. **Johnston, P. V. and Roots, B. I.,** Brain lipid fatty acids and temperature acclimation, *Comp. Biochem. Physiol.,* 11, 303–309, 1964.
19. **Roots, B. I.,** Phospholipids of goldfish (*Carassius auratus* L.) brain: the influence of environmental temperature, *Comp. Biochem. Physiol.,* 25, 457–466, 1968.
20. **Roots, B. I. and Johnston, P. V.,** Plasmalogens of the nervous system and environmental temperature, *Comp. Biochem. Physiol.,* 26, 553–560, 1968.
21. **Hazel, J. R.,** Homeoviscous adaptation in animal cell membranes, in *Physiological Regulation of Membrane Fluidity,* Alan R. Liss, New York, 1988, 149–188.
22. **Thompson, G. A., Jr. and Nozawa, Y.,** The regulation of membrane fluidity in *Tetrahymena,* in *Membrane Fluidity,* Kates, M. and Manson, L. A., Eds., Plenum Press, New York, 1984, 397–432.
23. **Lozina-Lozinskii, L. K.,** *Studies in Cryobiology,* John Wiley & Sons, New York, 1974, 55.
24. **Fukushima, H., Martin, C. E., Iida, H., Kitajima, Y., Thompson, G. A., Jr., and Nozawa, Y.,** Changes in membrane lipid composition during temperature adaptation by a thermotolerant strain of *Tetrahymena pyriformis, Biochim, Biophys. Acta,* 431, 165–179, 1976.
25. **Martin, C. E. and Thompson, G. A., Jr.,** Use of fluorescence polarization to monitor intracellular membrane changes during temperature acclimation. Correlation with lipid compositional and ultrastructural changes, *Biochemistry,* 17, 3581–3591, 1978.
26. **Nozawa, Y. and Umeki, S.,** Regulation of membrane fluidity in unicellular organisms, in *Physiological Regulation of Membrane Fluidity,* Aloia, R. C., Curtain, C. C., and Gordon, L. M., Eds., Alan R. Liss, New York, 1988, 239–357.
27. **Umeki, S., Fukushima, H., Watanabe, T., and Nozawa, Y.,** Temperature acclimation mechanisms in *Tetrahymena pyriformis:* effects of decreased temperature on microsomal electron transport, *Biochem. Int.,* 4, 101–107, 1982.
28. **Dickens, B. F. and Thompson, G. A., Jr.,** Rapid membrane response during low temperature acclimation. Correlation of early changes in the physical properties and lipid composition of *Tetrahymena* microsomal membranes, *Biochim. Biophys. Acta,* 644, 211–218, 1981.
29. **Ramesha, C. S. and Thompson, G. A., Jr.,** The mechanism of membrane response to chilling. Effect of temperature on phospholipid deacylation and reacylation reactions in the cell surface membrane, *J. Biol. Chem.,* 259, 8706–8712, 1984.
30. **Kameyama, Y., Yoshioka, S., and Nozawa, Y.,** Mechanism for adaptive modification during cold acclimation of phospholipid acyl chain composition in *Tetrahymena, Biochim. Biophys. Acta,* 793, 28–33, 1984.
31. **Lynch, D. V. and Thompson, G. A., Jr.,** Chloroplast phospholipid molecular species alteration during low temperature acclimation in *Dunaliella, Plant Physiol.,* 74, 198–203, 1984.
32. **Steponkus, P. L. and Lynch, D. V.,** Freeze/thaw-induced destabilization of the plasma membrane and the effects of cold acclimation, *J. Bioenerg. Biomemb.,* 21, 21–41, 1989.
33. **Lynch, D. V. and Steponkus, P. L.,** Plasma membrane lipid alterations associated with cold acclimation of winter rye seedlings (*Secale cereale* L. cv Puma), *Plant Physiol.,* 83, 761–767, 1987.
34. **Cahoon, E. B. and Lynch, D. V.,** Analysis of glucocerebrosides of rye (*Secale cereale* L. cv Puma) leaf and plasma membrane, *Plant Physiol.,* 95, 58–68, 1991.
35. **Lynch, D. V.,** Chilling injury in plants: the relevance of membrane lipids, in *Environmental Injury to Plants,* Katterman, F., Ed., Academic Press, New York, 1990, 17–34.
36. **Kunst, L., Browse, J., and Somerville, C.,** A mutant of *Arabidopsis* deficient in desaturation of palmitic acid in leaf lipids, *Plant Physiol.,* 90, 943–947, 1984.
37. **Kunst, L., Browse, J., and Somerville, C.,** Enhanced thermal tolerance in a mutant of *Arabidopsis* deficient in palmitic acid unsaturation, *Plant Physiol.,* 91, 401–408, 1989.
38. **Murata, N.,** Molecular species composition of phosphatidylglycerols from chilling-sensitive and chilling-resistant plants, *Plant Cell Physiol.,* 24, 81–86, 1983.
39. **Roughan, P. G.,** Phosphatidylglycerol and chilling sensitivity in plants, *Plant Physiol.,* 77, 740–746, 1985.
40. **Low, P. S., Ort, D. R., Cramer, W. A., Whitmarsh, J., and Martin, B.,** Search for an endotherm in chloroplast lamellar membranes associated with chilling-inhibition of photosynthesis, *Arch. Biochem. Biophys.,* 231, 336–344, 1984.
41. **Li, G., Knowles, P. F., Murphy, D. J., and Marsh, D.,** Lipid-protein interactions in thylakoid membranes of chilling-resistant and -sensitive plants studied by spin label electron spin resonance spectroscopy, *J. Biol. Chem.,* 265, 16867–16872, 1990.
42. **Maroc, J., Trémolières, A., Garnier, J., and Guyon, D.,** Oligomeric form of the light-harvesting chlorophyll a + b-protein complex CPII, phosphatidyldiacylglycerol, Δ3-*trans*-hexadecenoic acid, and energy transfer in *Chlamydomonas reinhardtii* wild type and mutants., *Biochim. Biophys. Acta,* 893, 91–99, 1987.

43. **Huner, N. P. A., Krol, M., Williams, J. P., Maissan, E., Low, P. S., Roberts, D., and Thompson, J. E.,** Low temperature development induces a specific decrease in *trans*-Δ^3-hexadecenoic acid content which influences LHC II organization, *Plant Physiol.,* 84, 12–18, 1987.
44. **MacDonald, A. G. and Cossins, A. R.,** The theory of homeoviscous adaptation of membranes applied to deep sea animals, in *Physiological Adaptation of Marine Animals,* Laverack, M. S., Ed., *Sym. Soc. Exp. Biol. XXXIX,* 301–322, 1985.
45. **DeLong, E. F. and Yayanos, A. A.,** Adaptation of the membrane lipids of a deep-sea bacterium to changes in hydrostatic pressure, *Science,* 228, 1101–1103, 1985.
46. **Cossins, A. R. and MacDonald, A. G.,** Homeoviscous adaptation under pressure. III. The fatty acid composition of liver mitochondrial phospholipids of deep-sea fish, *Biochim. Biophys. Acta,* 860, 325–335, 1986.
47. **Traüble, H. and Eibl, H.,** Molecular interactions in lipid bilayers, in *Functional Linkage in Biomolecular Systems,* Schmitt, F. O., Schneider, D. M., and Crothers, D. M., Eds., Raven Press, New York, 1975, 59–90.
48. **Kanemasa, Y., Yoshioka, T., and Hayashi, H.,** Alteration of the phospholipid composition of *Staphlococcus aureus* cultured in medium containing NaCl, *Biochim. Biophys. Acta,* 280, 444–450, 1972.
49. **Komaratat, P. and Kates, M.,** The lipid composition of a halotolerant species of *Staphlococcus epidermidis, Biochim. Biophys. Acta,* 398, 464–484, 1975.
50. **Mattox, S. M. and Thompson, G. A., Jr.,** The effect of inorganic cations on the membranes of *Tetrahymena pyriformis, J. Cell Biol.,* 75, 215a, 1977.
51. **Mattox, S. M. and Thompson, G. A., Jr.,** The effects of high concentrations of sodium or calcium ions on the lipid composition and properties of *Tetrahymena* membranes, *Biochim. Biophys. Acta,* 599, 24–31, 1980.
52. **Dunham, P. B. and Kropp, D. L.,** Regulation of solutes and water in *Tetrahymena,* in *Biology of Tetrahymena,* Elliott, A. M., Ed., Dowden, Hutchinson, and Ross, Stroudsburg, Pennsylvania, 1973, 165–198.
53. **Ginsburg, M.,** *Dunaliella:* a green alga adapted to salt, *Adv. Botanical Res.,* 14, 93–183, 1987.
54. **Peeler, T. C., Stephenson, M. B., Einspahr, K. J., and Thompson, G. A., Jr.,** Lipid characterization of an enriched plasma membrane fraction of *Dunaliella salina* grown in media of varying salinity, *Plant Physiol.,* 89, 970–976, 1989.
55. **Eichenberger, W. and Boschetti, A.,** Occurrence of 1(3),2-diacylglyceryl-(3)-0-4′-(N,N,N-trimethyl)-homoserine in *Chlamydomonas reinhardi, FEBS Lett.,* 88, 201–204, 1978,
56. **Einspahr, K. J., Peeler, T. C., and Thompson, G. A., Jr.,** Rapid changes in polyphosphoinositide metabolism associated with the response of *Dunaliella salina* to hypoosmotic shock, *J. Biol. Chem.,* 263, 5775–5779, 1988.
57. **Ha, K. S. and Thompson, G. A., Jr.,** Diacylglycerol metabolism in the green alga *Dunaliella salina.* A possible role of diacylglycerols in phospholipase C-mediated signal transduction, *Plant Physiol.,* in press.
58. **Einspahr, K. J., Maeda, M., and Thompson, G. A., Jr.,** Concurrent changes in *Dunaliella salina* ultrastructure and membrane phospholipid metabolism after hyperosmotic shock, *J. Cell Biol.,* 107, 529–538, 1988.
59. **Stuiver, C. E. E., Kuiper, P. J. C., and Marschner, H.,** Lipids from bean, barley, and sugar beet in relation to salt resistance, *Physiol. Plant.,* 42, 124–128, 1978.
60. **Kuiper, P. J. C.,** Lipids in grape roots in relation to chloride transport, *Plant Physiol.,* 43, 1367–1371, 1968.
61. **Erdei, L., Stuiver, B., (C.E.E.), and Kuiper, P. J. C.,** The effect of salinity on lipid composition and on activity of Ca^{2+}-and Mg^{2+}-stimulated ATPases in salt-sensitive and salt-tolerant Plantago species, *Physiol. Plant.,* 49, 315–319, 1980.
62. **Douglas, T. J. and Walker, R. R.,** 4-Desmethylsterol composition of citrus rootstocks of different salt exclusion capacity, *Physiol Plant.,* 58, 69–74, 1983.
63. **Bhatia, S. C. and Venkitasubramanian, J.,** Mechanism of dieldren-induced fat accumulation in rat liver, *J. Agric. Food Chem.,* 20, 993–996, 1972.
64. **Geyer, G.,** Erfolgreiche behandlung eines falles von Cushing-syndrom mit o,p-DDD, *Acta Endochrinol.,* 40, 332–348, 1962.
65. **Darsie, J., Gosha, S. K., and Holman, R. T.,** Induction of abnormal fatty acid metabolism and essential fatty acid deficiency in rats by dietary DDT, *Arch. Biochem. Biophys.,* 175, 262–269, 1976.
66. **Ingram, L. O.,** Adaptation of membrane lipids to alcohols, *J. Bacteriol.,* 125, 670–678, 1976.
67. **Sinensky, M.,** Temperature control of phospholipid biosynthesis in *Escherichia coli, J. Bacteriol.,* 106, 449–455, 1971.
68. **Nandini-Kishore, S. G., Mattox, S. M., Martin, C. E., and Thompson, G. A., Jr.,** Membrane changes during growth of *Tetrahymena* in the presence of ethanol, *Biochim. Biophys. Acta,* 551, 315–327, 1979.
69. **Ellingson, J. S., Taraschi, T. F., Wu, A., Zimmerman, R., and Rubin, E.,** Cardiolipin from ethanol-fed rats confers tolerance to ethanol in liver mitochondrial membranes, *Proc. Natl. Acad. Sci. U.S.A.,* 85, 3353–3357, 1988.
70. **Taraschi, T. F., Ellingson, J. S., Wu-Sun, A., and Rubin, E.,** Rats withdrawn from ethanol rapidly re-acquire membrane tolerance after resumption of ethanol feeding, *Biochim. Biophys. Acta,* 1021, 51–55, 1990.

71. **Nandini-Kishore, S. G., Kitajima, Y., and Thompson, G. A., Jr.,** Membrane fluidizing effects of the general anesthetic methoxyflurane elicit an acclimation response in *Tetrahymena, Biochim. Biophys. Acta,* 471, 157–161, 1977.
72. **Nunn, W. D.,** Fatty acid synthesis in *Escherichia coli* is indirectly inhibited by phenethyl alcohol, *Biochemistry,* 16, 1077–1081, 1977.
73. **Nozawa, Y., Kasai, R., and Sekiya, T.,** Modification of membrane lipids: phenethyl alcohol-induced alteration of lipid composition in *Tetrahymena* membranes, *Biochim. Biophys. Acta*, 552, 38–52, 1979.
74. **Ingram, L. O.,** Changes in lipid composition of *Escherichia coli* resulting from growth with organic solvents and with food additives, *Appl. Environ. Microbiol.,* 33, 1233–1236, 1977.

INDEX

A

B

C

F

K

L

M

N

O

P

U

V

X

Y

Z